AUTOMOTIVE ENGINE REPAIR AND REBUILDING

FRANK J. THIESSEN

PRENTICE HALL
ENGLEWOOD CLIFFS, NEW JERSEY, 07632

Library of Congress Cataloging-in-Publication Data

Thiessen, F. J.
    Automotive engine repair and rebuilding / Frank J. Thiessen.
        p.   cm.
    Includes index
    ISBN 0-13-051012-2
    1. Automobiles—Motors—Maintenance and repair. I. Title.
TL210.T46   1992
629.25′04—dc20   91-30442
                                                         CIP

Editorial/production supervision and
    interior design: Eileen M. O'Sullivan
Cover design: Ben Santora
Manufacturing buyer: Ed O'Dougherty
Prepress buyer: Ilene Levy
Page layout: Anne Ricigliano
Acquisitions Editor: Robert Koehler

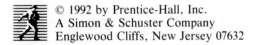
© 1992 by Prentice-Hall, Inc.
A Simon & Schuster Company
Englewood Cliffs, New Jersey 07632

Printed in the United States of America
10 9 8 7 6 5 4 3 2 1

ISBN   0-13-051012-2

PRENTICE-HALL INTERNATIONAL (UK) LIMITED, *London*
PRENTICE-HALL OF AUSTRALIA PTY. LIMITED, *Sydney*
PRENTICE-HALL CANADA INC., *Toronto*
PRENTICE-HALL HISPANOAMERICANA, S.A., *Mexico*
PRENTICE-HALL OF INDIA PRIVATE LIMITED, *New Delhi*
PRENTICE-HALL OF JAPAN, INC., *Tokyo*
SIMON & SCHUSTER ASIA PTE. LTD., *Singapore*
EDITORA PRENTICE-HALL DO BRASIL, LTDA., *Rio de Janeiro*

Contents

PREFACE   xi

ACKNOWLEDGMENTS   xiii

1 ENGINE IDENTIFICATION AND INFORMATION   1

Objectives, 1
Vehicle Identification Number, 1
Engine Identification Plate, 3
Casting Numbers, 5
Service Manuals, 6
Tool and Equipment Manufacturers' Publications, 6
Parts Supplier Publications, 6
Engine Rebuilders' Organizations, 7
Remanufactured Engines, 7
Engine Kits, 8
Ordering Parts, 9
Parts Prices, 10
Core Credits, 10
Heat Tabs, 10
Review Questions, 11

2 CLEANING METHODS AND EQUIPMENT   12

Objectives, 12
Introduction, 12
Hand Cleaning Tools, 12
Power Cleaning Tools, 13
Cleaning Valve Guides, 14
Cleaning Valves, 14
Cleaning Valve Springs, Retainers, and Rocker Assemblies, 16

Cleaning Pistons, 16
Cleaning Cylinder Blocks and Heads, 16
Cleaning Crankshafts and Camshafts, 16
Cleaning Oil Galleries, 16
Steam Cleaning, 17
Chemicals for Cleaning, 18
Cold-Soak Cleaning, 20

Hot-Tank Cleaning, 20
Descaling Engine Blocks and Heads, 20
Glass Bead Cleaning, 20
Airless Shot Blasting, 21
Thermal Cleaning, 23
High-Pressure Spray Cleaning, 23
Review Questions, 23

3 MEASURING ENGINE COMPONENTS 24

Objectives, 24
Engine and Engine Component
    Measurements, 24
English and Metric Dimensions, 24
Cylinder Wear, 25
Cylinder Wall Thickness, 25
Measuring Engine Cylinders, 26
Measuring Main Bearing Bores (Saddle
    Bores), 27
Block Deck Height and Clearance, 28
Measuring Crankshafts, 30
Crankshaft Grinding Limits, 32

Measuring Connecting Rods, 32
Measuring Piston Diameter, 34
Measuring Ring Groove Wear, 35
Measuring Camshafts, 35
Hydraulic Lifter Testing, 37
Valve Spring Measuring and Testing, 37
Measuring Valve Guide Wear, 38
Measuring Cylinder Head and Block Deck
    Surface, 40
Measuring the Intake and Exhaust Manifolds,
    41
Review Questions, 41

4 BLUEPRINTING AN ENGINE 43

Objectives, 43
Introduction, 43
Preparing the Block, 47
Preparing the Heads, 48
Calculating a Specific Compression Ratio, 49
Measuring Combustion Chamber Volumes, 50
Balancing Combustion Chamber Volumes, 50
High Performance Camshafts, 50
Degreeing in a Camshaft, 51

Correcting Camshaft Timing, 53
Cam Centerline Method of Determining
    Correct Cam Installation Position, 54
Checking Piston-to-Valve Clearance, 55
Balancing Valve Spring Pressures, 55
Balancing the Engine, 56
Assembling the Engine, 56
Review Questions, 57

5 PARTS FAILURE ANALYSIS 58

Objectives, 58
Inspecting Parts for Causes of Excessive Oil
    Consumption, 58
Flywheel Problems, 62
Vibration Damper Problems, 63
Crankshaft Failure, 64
Bearing Failure, 65
Piston Failure, 69
Piston Ring Failure, 72
Connecting Rod Failure, 73

Piston Pin Failure, 74
Valve Failure, 75
Hydraulic Lifter Failure, 79
Camshaft Drive Mechanism Failure, 80
Valve Spring Failure, 82
Checking Pushrods, 83
Checking Rocker Arms, 83
Checking Rocker Arm Studs, 84
Checking Rocker Arm Shafts, 84
Review Questions, 85

6 CRACK AND THREAD REPAIR 86

Objectives, 86

Introduction, 86

Causes of Cracks, 86

Crack Detection, 87

Crack Repair, 88

Ceramic Sealing, 93

Repairing Threads with a Thread Insert, 93

Repairing a Damaged Spark Plug Hole Thread, 94

Cutting Tool Lubricants, 95

Removing Broken Bolts, 95

Review Questions, 96

7 CYLINDER BLOCK RECONDITIONING 98

Objectives, 98

Introduction, 98

Cylinder Wall Thickness, 98

Decking the Block, 99

Block Deck Height and Clearance, 99

Main Bearing Bore Alignment, 102

Main Bearing Bore Stretch, 102

Bearing Spin, 102

Measuring Main Bearing Bores (Saddle Bores), 103

Correcting Bore Alignment, 104

Cylinder Reconditioning, 106

Cylinder Wall Finish, 106

Piston Sizes, 108

Cylinder Boring, 108

Sleeving a Cylinder, 110

Rigid Honing of Cylinders, 111

Chamfering Cylinder Bores, 114

Deglazing Cylinders, 114

Honing the Lifter Bores, 115

Threaded Holes, 115

Deburring the Block, 116

Cleaning the Block, 116

Core Plugs, 117

Heat Tabs, 118

Painting the Block, 118

Review Questions, 119

8 CRANKSHAFTS, FLYWHEELS, AND BALANCERS 120

Objectives, 120

Crankshaft Materials, 120

Crankshaft Design, 120

Flywheel, 124

Harmonic Balancer, 124

Balance Shafts or Gears, 125

Crankshaft Inspection, 126

Measuring the Crankshaft, 127

Crankshaft Straightening, 129

Factory-Undersized Crankshafts, 130

Nitrided Crankshaft Check, 130

Rebuilding Journals, 130

Shot Peening, 132

Cross Drilling, 132

Crankshaft Grinding, 132

Vibration Damper Inspection, 138

Flywheel Inspection, 138

Ring Gear Replacement, 140

Pilot Bushing Replacement, 140

Review Questions, 141

9 CONNECTING RODS AND PISTON PINS 143

Objectives, 143

Connecting Rod Function, 143

Connecting Rod Materials and Design, 143

Rod Balancing, 144

Connecting Rod Offset, 144

Oil Holes, 145

Connecting Rod Failure Analysis, 145

Checking and Measuring Connecting Rods, 145

Reconditioning the Connecting Rod, 147
Piston Pin Function, 151
Piston Pin Materials and Design, 151
Piston Pin Mounting Methods, 152
Piston Pin Offset, 153

Pin Fitting, 154
Assembling the Piston and the Rod, 154
Full-Floating Pin Installation, 155
Press-Fit Pin Installation, 155
Review Questions, 158

10 PISTONS AND PISTON RINGS 159

Objectives, 159
Piston Function, 159
Piston Terminology, 159
Piston Material, 163
Piston Skirt Finish, 163
Piston Temperatures, 163
Piston Sizes, 164
Piston Selection, 165
Using Used Pistons, 165
Piston Failure Analysis, 166

Piston Ring Function, 166
Piston Ring Types, 167
Piston Ring Action and Design, 167
Piston Ring Materials, 168
Piston Ring Coatings, 168
Piston Ring Gaps, 169
Piston Ring Clearance, 170
Piston Ring Selection, 171
Piston Ring Installation, 172
Review Questions, 174

11 CAMSHAFTS, LIFTERS, AND CAMSHAFT DRIVES 177

Objectives, 177
Camshaft Function, 177
Camshaft Location, 177
Camshaft Materials, 178
Camshaft Design, 179
Camshaft End Thrust, 179
Cam Lobe Design, 180
Cam Action, 181
Valve Timing, 182
Valve Lifter Function, 183
Hydraulic Lifters, 183
Hydraulic Lifter Failure Analysis, 187
Camshaft Bearing Wear, 187

Camshaft Inspection and Measuring, 187
Camshaft Regrinding and Straightening, 189
Installing the Camshaft in the Block or Head, 190
Camshaft Drive Function, 193
Camshift Drive Methods, 193
Camshaft Drive Failure Analysis, 196
Timing Gear Installation, 196
Timing Chain and Sprocket Installation, 197
Timing Belt and Sprocket Installation, 199
Degreeing in a Cam, 200
Review Questions, 200

12 ENGINE BEARINGS 202

Objectives, 202
Engine Bearing Function, 202
Bearing Operation, 202
Crankshaft Bearings, 202
Bearing Design, 204
Bearing Materials, 205

Bearing Characteristics, 205
Camshaft Bearings, 206
Bearing Replacement Guidelines, 206
Bearing Installation Procedure, 208
Review Questions, 210

13 GASKETS, SEALANTS, AND SEALS 211

Objectives, 211
Introduction, 211
Gasket Properties, 212
Gasket Materials, 212
Gasket Packages, 214
Handling Gaskets, 214
Sealants, 214

Cylinder Head Gaskets, 217
Head Gasket Installation Guidelines, 218
Valve Cover Gaskets, 218
Oil Pan Gaskets, 219
Manifold Gaskets, 221
Other Engine Gaskets, 223
Oil Seals, 223
Review Questions, 225

14 OIL PUMPS AND OIL FANS 227

Objectives, 227
Lubrication System Function, 227
Lubrication System Components, 227
Types of Lubrication Systems, 228
Oil Pump Design and Operation, 230
Oil Filters, 233

Oil Coolers, 233
Oil Pump Problems, 233
Checking Oil Pump Clearances, 235
Installing the Oil Pump, 236
Engine Oil Pans, 237
Review Questions, 238

15 CYLINDER HEAD RECONDITIONING 239

Objectives, 239
Resurfacing Cylinder Heads and Manifolds, 239
Surfacing Methods, 240
Port Alignment on V Engines, 245
Maximum Stock Removal, 247
Straightening Aluminum Heads, 248
Valve Guide Service, 248
Replacing Removable Valve Guides, 248
Replacing Bronze Guides, 250
Reaming Worn Guides, 251
Knurling Valve Guides, 251

Repairing Integral Guides with Thin-Walled Inserts, 253
Bronze Guide Inserts, 255
Head Shops, 256
Rocker Arm Stud Service, 256
Valve Stem Seal Function and Design, 258
Machining Guides for Positive Type Seals, 260
Valve Seat Insert Service, 261
Repairing Spark Plug Hole Threads, 264
Review Questions, 264

16 VALVE TRAIN COMPONENTS 265

Objectives, 265
Introduction, 265
Valve Train Requirements, 265
Pushrod Function and Design, 266
Pushrod Inspection, 267
Rocker Arm Function and Design, 267
Rocker Arm Materials, 267
Rocker Arm Mounting Methods, 268
Rocker Arm Lubrication, 271

Rocker Arm Valve Train Adjustment Methods, 272
Rocker Arm Service, 272
Valve Spring Function and Materials, 273
Valve Spring Vibration Control, 273
Valve Spring Inspection and Testing, 274
Split Locks or Keepers, 276
Valve Rotators, 276
Review Questions, 277

17 VALVE AND SEAT SERVICE 279

Objectives, 279
Intake and Exhaust Valves, 279
Number of Valves per Cylinder, 280
Valve and Valve Stem Design, 281
Valve Temperatures, 283
Valve Seats, 284
Valve, Seat, and Guide Wear, 285
Cracked, Burned, and Broken Valves, 286

Valve Failure Analysis, 286
Inspecting and Measuring Valves, 286
Valve Grinding, 286
Valve Stem Height, 293
Valve Seat Reconditioning, 294
Valve Seat-to-Face Contact and
    Concentricity, 301
Review Questions, 303

18 CYLINDER HEAD ASSEMBLY AND INSTALLATION 304

Objectives, 304
Introduction, 304
Installing the Core Plugs, 304
Installed Spring Height, 305
Balancing Valve Spring Pressures, 306

Installing the Valves, 307
Installing the Cylinder Head, 309
Installing the Overhead Camshaft, 311
Installing the Pushrods and Rocker Arms, 314
Review Questions, 314

19 FINAL ASSEMBLY, INSTALLATION, AND BREAK-IN 316

Objectives, 316
Final Assembly, 316
New Parts versus Used Parts, 316
Service and Replacement Guidelines, 317
Test Stand Run-In, 318
Adjusting the Valves, 320
Adjusting Solid Lifter Valves (Tappet
    Clearance), 320
Nonadjustable Hydraulic Lifters, 321

Adjusting Hydraulic Lifters (Adjustable
    Type), 321
Adjusting the Overhead Cam Valves, 322
Installing the Valve Cover, 323
Engine Installation, 323
Engine Startup, 324
Engine Break-in, 324
Review Questions, 324

APPENDIX 326

Decimal Equivalents and Tap Drill Sizes, 327
Pipe Thread Sizes, 327
Drill Sizes, 328

Torque Conversion, 329
English–Metric Equivalents, 330
English–Metric Conversion, 331

INDEX 333

Preface

This book is designed to provide the basis for building a career in automotive engine rebuilding. It is designed to guide the student through a wide range of learning experiences in failure diagnosis and rebuilding of automotive engines. It describes how to perform the different service procedures as well as how to avoid errors.

The subject matter is organized into 19 chapters. Each chapter deals with a number of closely related topics and procedures detailed in a set of objectives. Review questions at the end of each chapter can be used to check progress or as a guide to chapter review. The illustrations are keyed to the text to help visualize the many concepts and procedures.

The highly technical nature of rebuilding and remanufacturing automotive engines requires a dedication to precision and accuracy found only in those who are serious about their work. The success of the engine rebuilding industry depends on that dedication.

A great deal of assistance, cooperation, and advice from a number of individuals, equipment manufacturers, engine manufacturers, engine rebuilders, and colleagues has gone into the development of this book, which would not have been possible without their help. I am most grateful to all of these people and organizations for their generosity. I am deeply indebted to my wife, Margret, for her diligence and long hours at the computer, typing and proofreading the text. My sincere appreciation to her and all those who have helped in so many ways. Thanks.

IMPORTANT SAFETY NOTICE

Appropriate service methods and proper repair procedures are essential for the safe, reliable operation of all equipment as well as the personal safety of the individual doing the work.

There are numerous variations in procedures, techniques, tools, parts, and equipment as well as in the skill of the individual doing the work. This book cannot possibly anticipate all such variations and provide advice or cautions as to each.

The variety and complexity of the procedures and equipment described in this book are such that each individual attempting to perform them must first be fully acquainted with them to ensure the safety of the individual as well as that of the equipment. Some general precautions follow.

- Always wear safety glasses for eye protection.
- Use safety stands whenever a procedure requires you to be under a vehicle.
- Be sure that the vehicle ignition switch is always in the OFF position, unless otherwise required by the procedure.
- Set the parking brake when working on the vehicle. With an automatic transmission, set it in PARK. With a manual transmission, it should be in FIRST. Place blocks of a 4″ × 4″ size to the front and rear surfaces of the tires to prevent inadvertent vehicle movement.
- Operate the vehicle's engine only in a well-ventilated area to avoid the danger of carbon monoxide. Use exhaust collection equipment.
- Keep yourself and your clothing away from moving parts.
- Keep all shields and protective devices in place on all equipment.
- To prevent serious burns, avoid contact with hot metal parts, hot fluids, caustic or acidic solutions.
- Do not smoke in areas where restricted.
- To avoid injury, do not wear rings, watches, loose hanging jewelry, and loose clothing that can get caught in moving or rotating equipment.

Acknowledgments

This book reflects a major contribution of time, effort, and assistance from a number of individuals and organizations in the automotive engine rebuilding industry. Rebuilding plants, equipment companies, industry organizations, and colleges have made their facilities and personnel available to the author for countless hours to discuss operations, take photographs, collect information, answer questions, and observe and discuss procedures. It is their contribution as much as that of the author that has made this book possible. Without their assistance a project of this kind is simply not possible. I am deeply grateful to the following individuals for their generous assistance and support.

Brian McAughey, President, Western Engine Ltd.

John Kolochuk, Plant Foreman, Western Engine Ltd.

Doug Bawel, President, Jasper Engine and Transmission Exchange.

Mike Pfau, Advertising Manager, Jasper Engine and Transmission Exchange.

B. James Hartford, Manager, Public Relations, Ford of Canada.

J. S. McIntire, Manager, Warranty Administration, Chrysler Canada Ltd.

Janine M. Sine, Manager, Planning/Technical Services Department, TRW, Inc.

Dean Barber, Sealed Power Corporation.

Patrick D. Crowley, Technical Communications Supervisor, Federal-Mogul Corporation.

Gene T. Hailey, Manager, Service Engineering, Muskegon Piston Ring Company.

Lyle H. Haley, Sales Manager, Peterson Machine Tool, Inc.

Ed Liebler, National Sales Manager, Storm Vulcan Co.

Remi L. Wrona, Vice President, Sunnen Products Company.

Norman Pugh, Sales/Production Engineer, Hastings Manufacturing Company.

H. W. Watt, Director of Communications, OTC Division, Sealed Power Corporation.

Jeff Stearns, Sales Manager, Kwik-Way Manufacturing Company.

David E. Colburn, Executive Vice President, Krizeman, Inc., McQuay Norris, Inc.

Ronald C. Roeschlaub, President, Irontite Products Co., Inc.

W. H. Penheit, Advertising Manager, Sioux Tools, Inc.

Roland R. Gagnon, Advertising/Merchandising Manager, The L.S. Starrett Company.

Myron R. May, Vice President, Marketing, Ammco Tools, Inc.

Laurie Kraus, Advertising/Production Coordinator, Loctite Corporation.

F. B. Gelinas, Marketing Manager, Ingersoll-Rand Canada, Inc.

J. Haimish, Advertising And Promotions Manager, Easco/KD Tools, Inc.

Kirk MacKenzie, Marketing Manager, Snap-On Tools of Canada Ltd.

R. W. Ashley, Advertising Manager, Mac Tools, Inc.

D. N. Dales, Department Head Auto/Diesel Dept., Red River Community College.

Dirk VanTongeren, Instructor, Red River Community College.

Roger Locken, Instructor, Red River Community College.

Dick Chitty, Technical Training Manager, Toyota Motor Sales USA, Inc.

Norm Thurston, Technical Training Manager, Toyota Canada, Inc.

In addition, the following companies and organizations provided valuable assistance and support.

American Hammered Automotive Replacement Division, Sealed Power Corporation, Muskegon, Michigan.

Ammco Tools, Inc. North Chicago, Illinois.

Automotive Engine Rebuilders Association, Glenview, Illinois.

Bear Service Equipment, Applied Power, Inc., Milwaukee, Wisconsin.

Brush Research Manufacturing Company, Inc., Los Angeles, California.

Chrysler Canada Ltd., Windsor, Ontario.

Easco/KD Tools, Inc., Lancaster, Pennsylvania.

Federal-Mogul Corporation, Detroit, Michigan.

Fel-Pro Incorporated, Skokie, Illinois.

Ford of Canada, Oakville, Ontario.

Ford Parts and Service Division, Oakville, Ontario.

Guspro, Inc., Chatham, Ontario.

Hastings Manufacturing Company, Hastings, Michigan.

Hayes-Dana Parts Co. Ltd., Beamsville, Ontario.

Automedia Enterprises, Steinbach, Manitoba.

Helicoil Division, Emhart Fastening Systems Group, Danbury, Connecticut.

Ingersoll-Rand Canada, Inc., Don Mills, Ontario.

Irontite Products Co., Inc., El Monte, California.

Jasper Engine and Transmission Exchange, Jasper, Indiana.

K.O. Lee Company, Aberdeen, South Dakota.

Krizeman, Inc., Mishawaka, Indiana.

Kwik-Way Manufacturing Company, Marion, Iowa.

Loctite Corporation, Cleveland, Ohio.

Mac Tools, Inc., Washington Court House, Ohio.

Magnaflux Corporation, Chicago, Illinois.

McQuay Norris, Inc., St. Louis, Missouri.

Muskegon Piston Ring Company, Muskegon, Michigan.

OTC Division, Sealed Power Corporation, Owatonna, Minnesota.

Peterson Machine Tool, Inc., Shawnee Mission, Kansas.

Prentice-Hall, Inc., Englewood Cliffs, New Jersey.

Production Engine Remanufacturers Association, Glendale, California.

Red River Community College, Winnipeg, Manitoba.

Sealed Power Corporation, Muskegon, Michigan.

Silver Seal Products Company, Inc., Trenton, Michigan.

Sioux Tools, Inc., Sioux City, Iowa.

Snap-On Tools of Canada Ltd., Concord, Ontario.

Society of Automotive Engineers, Warrendale, Pennsylvania.

South Winnipeg Technical Institute, Winnipeg, Manitoba.

Stanley Canada, Inc., Burlington, Ontario.

The L.S. Starrett Company, Athol, Mass.

Storm Vulcan Co., Dallas, Texas.

Sunnen Products Company, St. Louis, Missouri.

TRW Inc., Automotive Aftermarket Division, Cleveland, Ohio.

Toyota Canada Inc., Scarborough, Ontario.

Toyota Motor Sales USA Inc., Torrance, California.

Western Engine Ltd., Winnipeg, Manitoba.

Winona Van Norman Machine Company, Winona, Minnesota.

TRW Inc. Replacement Parts Division, Cleveland, Ohio, for the following illustrations:

Figures 5-4, 5-44, 5-53, 5-56, 5-59, 5-62, 5-64, 5-65, 5-70, 5-71, 5-74, 5-76, 5-83, 5-87, 5-88, 5-89, 5-91, 5-92, 5-93, 7-15, 7-34, 7-37, 7-51, 7-52, 11-31, 11-56, 11-63, 11-65, 14-12, 14-13, 14-14, 14-15, 14-16, 14-17, 14-23, 14-24, 14-25, 14-26, 16-10, 16-15, 17-9, 17-63

taken from the publication ENGINE DIAGNOSIS AND INSTALLATION GUIDE, Copyright 1978 and

Federal-Mogul Corporation, Detroit, Michigan for the following illustrations:
Figures 3-9, 3-11, 3-12, 3-20, 3-21, 3-31, 5-27, 5-28, 5-29, 5-30, 5-32, 5-33, 5-34, 5-35, 5-36, 5-37, 5-40, 5-41, 5-42, 5-43, 5-49, 7-11, 8-22 (top), 8-25, 8-26, 10-19, 10-46, 10-48, 12-7, 12-14.

Over two dozen of these companies also provided illustrations, for which they are individually credited in the captions below the figures. All such figures are reproduced with the permission of the copyright holder, with all rights reserved.

CHAPTER 1

Engine Identification and Information

OBJECTIVES

1. To develop the ability to identify the various automotive engines.
2. To develop the ability to locate and use the appropriate reference materials and specifications for any particular engine.
3. Utilize information and specifications, to order correct engine parts.

INTRODUCTION

To make all necessary measurements, repairs, and adjustments to an engine requires the ability to identify accurately the vehicle and engine being repaired. Incorrect identification can result in the wrong parts being ordered or the wrong specifications and adjustments being used. The consequences are delays and possible engine damage. To avoid these problems, learn how to identify vehicles, their engines, and their equipment and how to find information and specifications for any kind of engine.

VEHICLE IDENTIFICATION NUMBER (VIN)

Automobile manufacturers identify each vehicle they produce with a serial number. This number is usually stamped on a plate mounted on top of the dash near the bottom left side of the car windshield (Figure 1-1). The number is read from the outside of the vehicle through the windshield glass. On other vehicles the plate may be attached to the left front door pillar or the firewall in the engine compartment.

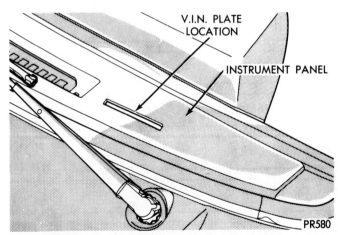

FIGURE 1-1 Location of VIN (Vehicle Identification Number) plate. (Courtesy of Chrysler Canada Ltd.)

The vehicle identification number (VIN) contains coded information identifying the type of engine and equipment used in the vehicle. The applicable factory service manual provides complete VIN decoding information. (See Figures 1-2 to 1-5 for examples.) The

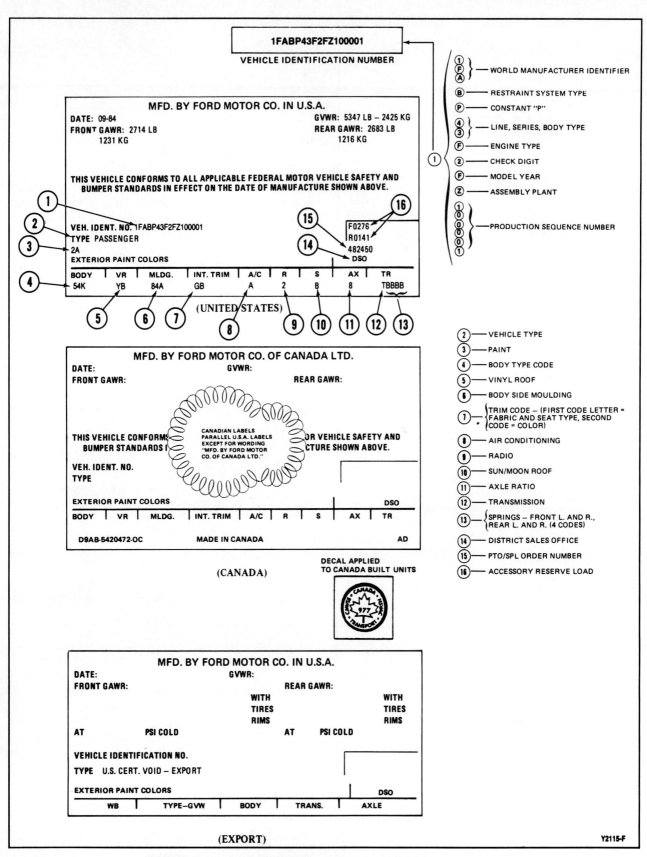

FIGURE 1-2 VIN coding information and certification labels typical for Ford Motor Co. (Courtesy of Ford of Canada.)

1988 V.I.N. CODE CHART
(FRONT-WHEEL-DRIVE VEHICLES)

Position	Code Options			Interpretation
1	1 = United States 2 = Canada	3 = Mexico J = Japan		Country of Origin
2	B = Dodge C = Chrysler	P = Plymouth		Make
3	3 = Passenger Car	7 = Truck		Type of Vehicle
4	B = Manual Seat Belts	D = 1-3,000 Lbs. GVW		Passenger Safety System
5	D = Dynasty C = New Yorker Landau J = Caravelle E = 600 T = New Yorker Turbo	V = Daytona X = Lancer H = LeBaron GTS J = LeBaron P = Reliant	D = Aries C = LeBaron S = Sundance S = Shadow	Line
6	1 = Economy 2 = Low	3 = Medium 4 = High	5 = Premium 6 = Special	Series
7	1 = 2 Dr. Sedan 3 = 2 Dr. Hardtop 4 = 2 Dr. Hatchback 5 = 2 Dr. Convertible	6 = 4 Dr. Sedan 8 = 4 Dr. Hatchback 9 = 4 Dr. Wagon		Body Style
8	C = 2.2L D = 2.2L E.F.I.	E = 2.2L Turbo K = 2.5L	3 = 3.0L	Engine
9*	(1 thru 9, 0 or X)			Check Digit
10	J = 1988			Model Year
11	A = Outer Drive C = Jefferson D = Belvidere E = Modena	F = Newark G = St. Louis 1 N = Sterling R = Windsor	T = Toluca W = Kenosha X = St. Louis 2	Engine
12 thru 17	(6 Digits)			Sequence Number

*Digit in position 9 is used for V.I.N. verification.

FIGURE 1-3 VIN coding information typical for Chrysler vehicles. (Courtesy of Chrysler Canada Ltd.)

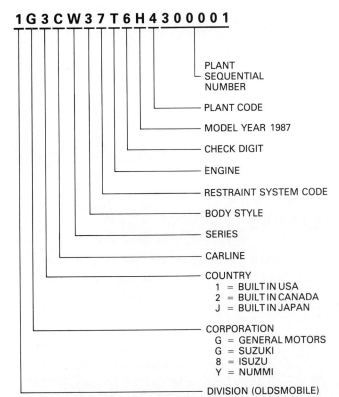

FIGURE 1-4 VIN coding information typical for General Motors vehicles.

A	= 1980	M	= 1991	2	= 2002
B	= 1981	N	= 1992	3	= 2003
C	= 1982	P	= 1993	4	= 2004
D	= 1983	R	= 1994	5	= 2005
E	= 1984	S	= 1995	6	= 2006
F	= 1985	T	= 1996	7	= 2007
G	= 1986	V	= 1997	8	= 2008
H	= 1987	W	= 1998	9	= 2009
J	= 1988	X	= 1999	A	= 2010
K	= 1989	Y	= 2000	B	= 2011
L	= 1990	1	= 2001	C	= 2012

FIGURE 1-5 The tenth character in the VIN is always the model year. Letters and numbers for the years 1980 to 2012 are shown here.

original equipment engine type can be identified from the VIN plate. If a different replacement engine has been installed in the vehicle, the engine code on the VIN plate no longer applies and engine identification must be made by other methods.

ENGINE IDENTIFICATION PLATE

On some engines the engine alpha code letter used on the VIN plate also appears on the engine block. An identification tag attached to the engine is often used;

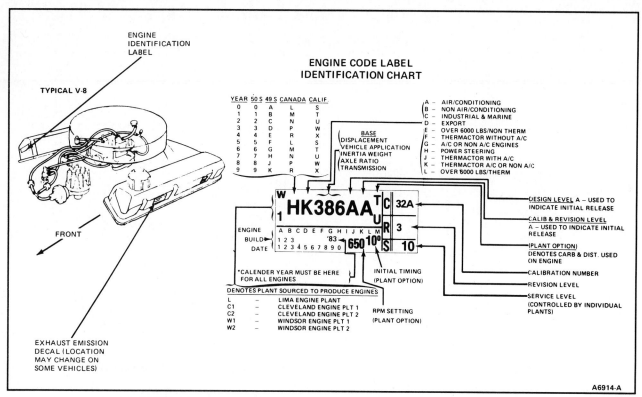

FIGURE 1-6 Location of engine ID label and decoding information typical for Ford. (Courtesy of Ford of Canada.)

FIGURE 1-7 Engine ID is located on left front of cylinder block on this engine. (Courtesy of Chrysler Canada Ltd.)

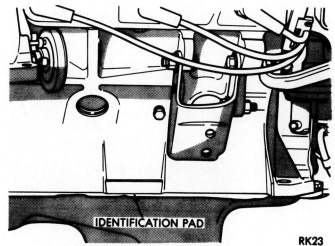

FIGURE 1-8 Engine ID on this engine is located on pad on lower left center of block. (Courtesy of Chrysler Canada Ltd.)

however, these tags can become lost or the part to which the tag is attached can be replaced, thus invalidating the existing tag. For example, the tag may be attached to an intake manifold bolt, coil mounting bolt, or to a valve cover (Figure 1-6). This type of identification is obviously not permanent. Other engines have the ID number stamped on a pad on the engine block, which is more permanent (Figures 1-7 and 1-8). Figure 1-9 shows examples of engine ID number decoding information.

VIN Code	Displacement		Number of Cylinders	Producer
	liters	in³		
AMC				
B	2.5	151	4	Pontiac
C	4.2	258	I-6	AMC
Chrysler				
A	1.7	105	4	Chrysler
B	2.2	135	4	Chrysler
D	2.6	156	4	Chrysler
E	3.7	225	I-6	Chrysler
J	5.2	318	EFI V-8	Chrysler
K	5.2	318	2-bbl V-8	Chrysler
M	5.2	318	4-bbl V-8	Chrysler
Ford				
2	1.6	98	4	Ford
A	2.3	140	4	Ford
T	2.3	140	4 Turbo	Ford
B	3.3	200	I-6	Ford
D	4.2	255	V-8	Ford
F	5.0	302	V-8	Ford
G	5.8	351W	V-8	Ford
General Motors				
A	3.8	231	V-6	Buick
H	5.0	305	V-8	Chevrolet
J	4.4	267	V-8	Chevrolet
N	5.7	350	Diesel	Oldsmobile
X	2.8	173	V-6	Chevrolet
Y	5.0	307	V-8	Oldsmobile
3	3.8	231	V-6 Turbo	Buick
4	4.1	252	V-6	Buick
5	2.5	151	4	Pontiac

FIGURE 1-9 Typical engine ID number decoding information.

CASTING NUMBERS

When engine identification cannot be established by the VIN plate and engine ID plate, the block casting number must be used for identification. The casting number may be located at the top front, top rear, bottom side, or top side of the block. Locate the casting number and compare it to a listing of casting numbers available from many engine rebuilders. It should also

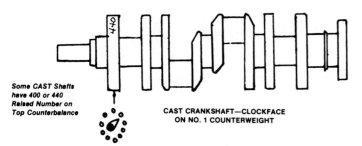

FIGURE 1-10 Typical crankshaft markings on V-8 Chrysler engine crankshafts. (Courtesy of Jasper Engine and Transmission Exchange.)

be noted that positive identification of cylinder heads, crankshafts, and camshafts is also established by their casting numbers or certain identifying design features (Figures 1-10 to 1-12).

318 & 360 HEAD IDENTIFICATION
Intake Manifold Side

Air Heated Intake Manifold

Water Heated Intake Manifold - 318 only

Exhaust Manifold Side
1974-67, 318-360 Dodge

1984-75, 318-360 Dodge W/O A.I.R.

1984-75, 318-360 Dodge W/A.I.R.

FIGURE 1-11 Cylinder head identification of Chrysler 318 and 360 engines. (Courtesy of Jasper Engine and Transmission Exchange.)

GM 6-cyl.
DIPSTICK LOCATIONS

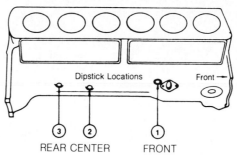

GM 6-cyl.
CLUTCH PEDAL LINKAGE MOUNTINGS

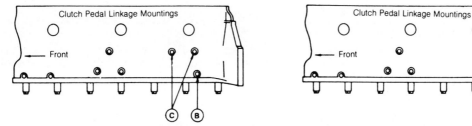

FIGURE 1-12 Identifying features of various General Motors in-line six-cylinder engine blocks. (Courtesy of Jasper Engine and Transmission Exchange.)

SERVICE MANUALS

Service manuals provide information on problem diagnosis, repair procedures, and specifications for engines. Often, two or three manuals may be required to cover all aspects of service for a single model and year of vehicle. For example, engine performance diagnosis and repair may be in one manual while engine mechanical repair procedures may be in another.

Vehicle manufacturers annually produce service manuals for each type of vehicle they manufacture. Manufacturers' service manuals can be purchased from the manufacturers or their subsidiaries. Order forms for service manuals are usually available from authorized automobile dealers. Several independent publishers produce their own general repair manuals, which include repair procedures and specifications. Three of the most common are the following:

- *Chilton's Auto Repair Manual*
- *Mitchell Manuals—National Service Data*
- *Motor's Auto Repair Manual*

Manuals are available for both domestic and imported automobiles.

TOOL AND EQUIPMENT MANUFACTURERS' PUBLICATIONS

A great variety of publications, training materials, video tapes, slides, and audio tapes, and the like are available from different tool and equipment manufac-

turers. Since it is impossible to list them all in this book, the best way to obtain information and prices is to talk to your local tool and equipment supplier representatives. They will be able to provide the necessary information or will be able to get it to you. You may want to check the classified pages in the telephone directory for the supplier nearest to you. Among the more prominent tool and equipment companies are the following:

- Sioux Tools, Inc.
- Sunnen Products Company
- Kwik-Way Manufacturing
- Black and Decker Manufacturing
- K.O. Lee Company
- Van Norman Machine Company
- Rottler Manufacturing Company
- Tobin Arp Manufacturing Company
- Storm-Vulcan, Inc.
- Irontite Products Company, Inc.
- Peterson Machine Tool, Inc.

PARTS SUPPLIER PUBLICATIONS

Companies that manufacture and distribute engine parts have a variety of publications, specifications, installation procedure booklets, parts catalogs, manuals, and updating bulletins available. Contact your nearest parts supplier for information and prices of their publications. Among the best known of these suppliers are the following.

- Dana Corporation
- Federal-Mogul Corporation
- Gould, Inc.
- Hastings Manufacturing Company
- McQuay-Norris Manufacturing Company
- Muskegon Piston Ring Company
- NAPA Parts
- TRW, Inc., Replacement Parts Division
- Sealed Power Corporation
- Fel-Pro Incorporated
- Imperial Clevite
- Silver Seal Products Company, Inc.

Excellent information is available to members of TRW's Tech Team organization. Members receive the *TechTopics* magazine, which publishes technical information on automotive engines and other subjects. Problem-solving shop techniques and hints are also provided. A toll-free ''Hot Line'' provides answers to technical questions as well as cataloging information to TRW customers. For information on how to join the TRW TechTeam, write to

TRW
Dept. TTH
P.O. Box 429
Cleveland, OH 44107-0429

ENGINE REBUILDERS' ORGANIZATIONS

The Automotive Engine Rebuilders Association (AERA) provides publications listing engine block, camshaft, and cylinder head casting numbers, as well as technical bulletins and shop tips. The association holds a national convention annually at which a great deal of valuable information regarding the engine rebuilding industry is presented for members.

The Production Engine Remanufacturers Association (PERA) began as the Western Engine Rebuilders Association in 1946. Membership in all categories is now over 250. The goal of the association is to ''provide its members with the opportunity to exchange the ideas, methods, and procedures necessary to produce remanufactured products which are equal or superior to their original in quality and performance.'' This association provides a *Common Numbering Parts Catalog for Passenger Car and Truck Engines* containing a listing of engine block, crankshaft, and cylinder head casting numbers and their application. This publication also includes helpful tips on identification of these components.

The addresses of these two organizations are as follows.

Automotive Engine Rebuilders Association
234 Waukegan Road
Glenview, IL 60025
Production Engine Remanufacturers Association
Suite 311
512 East Wilson Avenue
Glendale, CA 91206

REMANUFACTURED ENGINES

Remanufactured or rebuilt engines are assembled in engine rebuilding plants as complete engines, short block assemblies, and short-short block assemblies. The extent to which new parts are used varies somewhat between rebuilders and remanufacturers. Figure 1-13 shows a typical remanufactured engine.

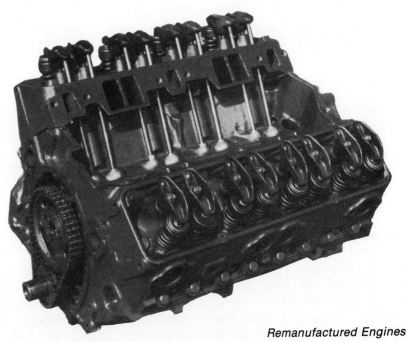

FIGURE 1-13 Remanufactured engine ready for installation. (Courtesy of Jasper Engine and Transmission Exchange.)

Short-Short Block Assembly

The short-short block assembly includes a rebored and machined cylinder block with new oversize pistons and rings and new piston pins. New bearings are used throughout the engine. A new or reground crankshaft and camshaft are installed. New timing gears or a new timing chain or belt and sprockets, reconditioned connecting rods, a new fuel pump eccentric, and new core plugs are also installed. All oil gallery plugs are in place.

Short Block Assembly

The short block assembly includes all the items in the short-short block, plus the following: a new oil pump, pickup tube and screen, a new oil pump drive if applicable, oil pan and gasket, timing cover, gasket and seal, and all the gaskets and seals needed to complete the engine assembly.

Complete Engine Assembly

The complete engine assembly includes all the parts in the short block assembly, plus the following: remanufactured cylinder head assemblies, complete with rocker arms and pivots (ball studs, pedestals, or shafts as applicable); cylinder head covers; and cylinder head cover gasket.

ENGINE KITS

Several kinds of engine repair kits are available from engine rebuilders and parts suppliers. Included are the following.

Rebuilt Cylinder Head

The rebuilt cylinder head consists of a core that has been checked for cracks, pressure tested, and resurfaced. Core plugs have been replaced and the valve guides reconditioned or replaced. Where applicable, damaged rocker arm studs are replaced, and the seats are ground and narrowed to specifications. The valves are either reconditioned or replaced. Valve springs may be tested and used again if acceptable, or they may be replaced, while new valve stem seals are used during assembly. A core charge may be included for which credit is issued upon receipt of a rebuildable core.

Crankshaft and Bearing Kit

A crankshaft and bearing kit includes a reconditioned and reground crankshaft with main and rod bearing journals reground and polished to 0.010, 0.020, or 0.030

in. undersized. The kit includes main and connecting rod bearings to fit the undersized journals.

Camshaft and Lifter Kit

This kit includes a reground camshaft with the same lift and duration as the original camshaft. The camshaft is surface treated by acid etching or moly spray coating to aid in retaining lubrication during the lobe/lifter break-in period. A new set of lifters comes with the kit, as well as a special lubricant to be used during assembly (Figure 1-14).

FIGURE 1-14 Camshaft and lifter kit, including assembly lube. (Courtesy of Sealed Power Corporation.)

Engine Bearing Kit

A complete engine bearing kit comes with all main bearings, connecting rod bearings, and camshaft bearings in the sizes specified by the purchaser: standard or undersize (Figure 1-15).

FIGURE 1-15 Engine crankshaft bearing kit. (Courtesy of Sealed Power Corporation.)

Engine Rebuild Gasket and Seal Kit

A complete engine overhaul gasket and seal kit includes the cylinder head gaskets, rocker arm cover gaskets, timing cover gaskets, manifold gaskets, fuel

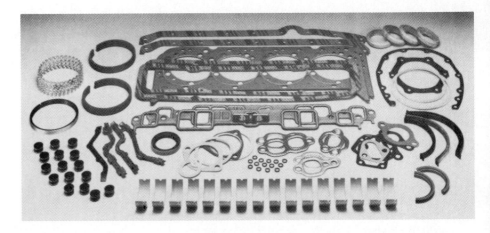

FIGURE 1-16 Typical engine rebuild gasket kit. (Courtesy of Sealed Power Corporation.)

pump gasket, oil pan gasket, oil drain plug gasket, and water manifold gaskets, if applicable. Also included are the rear main bearing oil seal, timing cover seal, distributor housing seal—if applicable, and the valve stem seals. Valve grind gasket kits are also available which include the cylinder head gaskets, valve cover gaskets, valve stem seals, and thermostat housing gasket (Figure 1-16).

ORDERING PARTS

Ordering engine parts requires providing positive vehicle and engine identification information, types and location of certain engine equipment, a knowledge of pricing practices, and core credit policy.

Vehicle and Engine Identification Information

Information on the following items is often required in order to obtain the correct engine or engine parts.

- Make of vehicle
- Model year and type (passenger car or truck)
- Engine code letter from VIN
- Block casting number
- Cylinder head casting number
- Number of cylinders
- Engine displacement (CID or L)
- Cylinder bore size and spacing
- Type of transmission/transaxle (manual/automatic)
- Fuel system, carburetor type (single-, two-, or four-barrel), fuel injection
- With or without air conditioning
- With or without air pump
- Oil pan (sump—front, center, rear, double)
- Oil drain plug (front, rear, or side)
- Rocker cover bolt holes (number and pattern)
- Bell housing bolt holes (number and pattern)
- Oil dipstick location (on oil pan, on block, right or left, front or rear)
- Timing cover type (stamped steel, cast iron, or aluminum)
- Clutch linkage mounting (location and type)
- Flywheel mounting (number and size of bolts)
- Motor mounts (number, location, number, and pattern of bolt holes)
- Core plugs (number and location)
- Oil filter mounting (type and location)
- Accessory mounting bolt holes on front of cylinder heads (number and pattern of bolt holes)
- Starter mounting bolt holes (number and pattern)
- Crankshaft snout design and forging number
- Camshaft forging number
- Hydraulic or solid lifters

It is important to have the necessary information on hand before ordering the parts. The parts supplier may ask for any of the above information to ensure correct identification of the engine so that the proper parts can be supplied.

Dimensional Abnormalities

As a cost cutting measure engine manufacturers are reducing the number of parts being scrapped during the manufacturing process. Parts that have been accidentally overmachined are often salvaged by machining to the next undersize, as in the case of crankshaft bearing journals, or to the next oversize, as in the case of engine cylinders or lifter bores. The following are the more common examples of this practice.

- one or more lifter bores machined to oversize and fitted with oversize lifters.
- a crankshaft with one or more main or rod bearing journals machined to undersize and fitted with undersize bearings.

- one or more cylinder bores slightly oversize and fitted with high limit pistons.
- all main bearing saddle bores machined to oversize and fitted with special thick walled bearings.
- a cylinder deck height that is lower than standard and fitted with special head gaskets that are thicker than standard to restore the compression ratio to normal.

Production Changes

Engines may undergo several modifications during the production year. Vehicle manufacturers notify their dealers of any such changes to keep them informed of the latest technology. Items such as cylinder head bolts, rear main bearing oil seals, valve springs, valve stem oil seals and timing cover seals are often affected. It is important to use the latest updated parts when repairing or rebuilding engines to ensure the most trouble free service life. Parts suppliers are usually well informed about such changes and will normally advise their customers of any changes or will substitute with the latest version of the affected parts.

PARTS PRICES

Vehicle manufacturers' parts departments generally use three different price levels for replacement parts:

1. Dealer net price (lowest)
2. Trade price (intermediate)
3. List price (highest)

Authorized dealers buy parts at the dealer net (or wholesale) price. Other automotive businesses generally pay the trade price, while the general public pays the list price.

Companies in the replacement parts business also have three price levels.

1. Jobber price (lowest)
2. Net price (intermediate)
3. List price (highest)

The price level at which a particular business can purchase parts is determined by the type of business and the volume of parts purchased. Every automotive business negotiates for the lowest possible price obtainable with the parts supplier.

CORE CREDITS

Core credits can be given for used parts such as cylinder heads, cylinder blocks, crankshafts, camshafts, connecting rods, flywheels, and the like, and they are generally available from most engine remanufacturing companies. There are two common policies regarding the acceptability of cores for credit.

1. Cores must be rebuildable in order to receive credit. Cores with cracks, breaks, overbored blocks, over-machined crank journals, and the like are not eligible for credit. The engine rebuilder reserves the right to determine whether core credit will be granted. Generally, a core charge is included in the price of a rebuilt engine and credit is issued to the purchaser only after an acceptible core has been received by the engine rebuilder.
2. Core credit is given by some engine rebuilders regardless of the internal condition of the engine if the engine has not been disassembled and has no major exterior damage.

Since not all cores received are rebuildable, engine remanufacturing companies must often purchase additional cores for rebuilding. Auto wreckers are one source of supply for cores. Some companies are in the business of supplying cores to the rebuilders.

HEAT TABS

Some engine rebuilders use heat tabs on rebuilt engines and cylinder heads to provide visual evidence that an engine has been overheated. Heat tabs have a special heat-sensitive material in the center of the tab, which when an engine reaches a temperature of 250°F, melts and falls out. Overheating of an engine usually voids any warranty. The heat tab is the best method of establishing evidence of overheating (Figure 1-17). A special structural adhesive is used to attach the heat tabs to the engine or cylinder head. This adhesive is not affected by temperature up to 300°F and transfers heat readily.

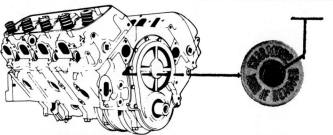

HEAT SENSITIVE MATERIAL

FIGURE 1-17 Heat tab installed on rebuilt block or cylinder head can establish whether engine failure has occurred due to overheating. (Courtesy of Silver Seal Products Company, Inc.)

REVIEW QUESTIONS

1. Why is it important to be able to identify engines and engine components accurately?

2. Where is the VIN most likely to be located?

3. What type of information is obtained from service repair manuals?

4. What positive method of identification is used for major engine components such as crankshafts, cylinder heads, and camshafts?

5. What parts are included in a short block assembly?

6. Name the three levels of prices for replacement parts from the highest to the lowest.

7. What are core credits?

8. What is the purpose of using heat tabs on engine blocks and cylinder heads?

CHAPTER 2

Cleaning Methods and Equipment

OBJECTIVES

To be able to clean engine parts by using:

1. Hand cleaning tools
2. Power cleaning tools
3. Steam cleaning equipment
4. Cold-soak cleaning equipment
5. High-pressure spray cleaning equipment
6. Hot-tank immersion cleaning equipment
7. Descaling equipment
8. Glass bead cleaning equipment
9. Airless shot blasting equipment
10. Thermal cleaning equipment

INTRODUCTION

Many different cleaning operations are needed in the engine rebuilding process. This includes cleaning by hand and by using special equipment and cleaning materials. After the engine is disassembled, the various parts are thoroughly cleaned. Different methods are used for different parts. It is important that the right methods, materials, and equipment be used in each case since using improper methods or materials can be harmful to engine parts. Rust and corrosion result from improper methods and materials used for cleaning. The particular methods used in any shop reflect the volume and speed required to clean parts. A production shop must use methods and equipment that are less labor intensive and provide greater productivity.

HAND CLEANING TOOLS

Scrapers, brushes, aerosol sprays, and solvents are used to clean carbon, varnish, sludge, gasket sealer, gasket material, and paint from engine parts. People with sensitive skin should avoid prolonged exposure to cleaning solvents or wear rubber gloves to protect the hands. Always wear eye or face protection when cleaning parts.

Hand-held scrapers and wire brushes can be used

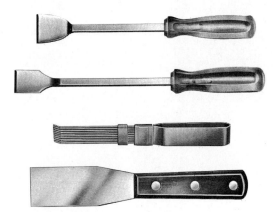

FIGURE 2-1 Scrapers are used to remove carbon and gasket material from parts. A and B are rigid scrapers. C consists of a number of round steel scrapers attached to a handle. A sliding bracket controls scraper flexibility. D is the common putty knife type of scraper. (Courtesy of Ingersoll-Rand Canada, Inc.)

to clean carbon and varnish from parts (Figure 2-1). Special care must be taken when cleaning aluminum heads and other aluminum parts not to damage machined surfaces with scrapers. Oil and sludge can be cleaned in a solvent tank with a recirculating pump and nozzle using a nylon bristle brush (Figures 2-2 and 2-3). The solvent tank should be equipped with a lid and a fusible link. The fusible link melts if the solvent catches fire and causes the lid to close and smother the flames. The cleaning solvent in the tank must be replaced when it is dirty and ineffective. Disposal of waste solvent and sludge must be done in a manner that meets all local, state, and federal regulations re-

FIGURE 2-4 Using a spray-on gasket remover. (Courtesy of Loctite Corporation.)

garding hazardous materials disposal. Check with them to ensure compliance and avoid legal charges. A spray-on gasket remover is shown in Figure 2-4.

POWER CLEANING TOOLS

A variety of power-driven tools such as wire cleaning brushes and scrapers are used to clean engine parts. Carbon can be removed from combustion chambers with an electric or air-driven wire brush (Figures 2-5 to 2-7). Valve guides are cleaned with power-driven scrapers first to remove carbon and varnish, followed by a final cleaning with a narrow hole-type brush. Power-driven abrasive discs are used to clean carbon and varnish from machined surfaces; however, care must be exercised not to abrade any metal during the procedure. Discs specially designed for cleaning aluminum surfaces are available. These discs are less abrasive and therefore less likely to damage relatively soft aluminum parts such as cylinder heads, manifolds, and thermostat housings.

FIGURE 2-2 Parts washing tank with pump that recirculates cleaning solvent. (Courtesy of Winona Van Norman Machine Company.)

FIGURE 2-3 Polypropylene bristle parts cleaning brush for use in parts washing tank (top). Wire bristle brush for cleaning carbon, etc. (bottom). (Courtesy of Easco/KD Tools, Inc.)

FIGURE 2-6 Cleaning carbon from combustion chamber with power-driven heavy-duty brush. This should be done before removing valves to avoid seat damage. (Courtesy of Ford of Canada.)

FIGURE 2-5 Rotary wire cleaning brushes for use with an electric drill motor. Brush at top has stiff twisted bristles for heavy-duty cleaning. Bottom two brushes have softer bristles. (Courtesy of Easco/KD Tools, Inc.)

FIGURE 2-7 Using a soft-bristled wire brush to polish engine block after degreasing. (Courtesy of Western Engine Ltd.)

CLEANING VALVE GUIDES

After cylinder head disassembly and before measuring valve guide wear, the carbon and varnish must be removed from the valve guides. This is best accomplished with a scraper-type cleaner driven by a portable electric drill motor followed by a final cleaning with a nylon brush (Figures 2-8 to 2-10). Solvent or carburetor cleaner may be used to aid the cleaning process. Make sure that all the varnish is removed. Inspect each guide with a light at one end while looking into the guide at the other. The inside of the guide should appear polished if clean.

CLEANING VALVES

Valves often have heavy deposits of carbon and varnish that must be removed if the valves are to be used again. The head and fillet areas can be cleaned with the wire wheel of a bench grinder, but care must be

FIGURE 2-8 Cleaning valve guide with a power-driven guide cleaner. (Courtesy of Sioux Tools, Inc.)

FIGURE 2-9 Expandable wire type of valve guide cleaner. (Courtesy of Easco/KD Tools, Inc.)

FIGURE 2-10 Expandable blade type of guide cleaner (top) and brush (bottom). (Courtesy of Mac Tools, Inc.)

taken not to damage the valve face by excessive abrasion. The valve stem can be cleaned in the same way; however, only light pressure should be used to prevent damage to the stem. Crocus cloth can be used to finish cleaning and polishing the stem.

Valves can also be cleaned by the cold soak method using carburetor cleaning fluid or other cold soak cleaning fluids. Valves are placed in a strainer basket and immersed in the fluid for several hours, after which they are rinsed and blown dry (Figures 2-11 and 2-12). Another method is to use glass bead blasting equipment, described later in the chapter.

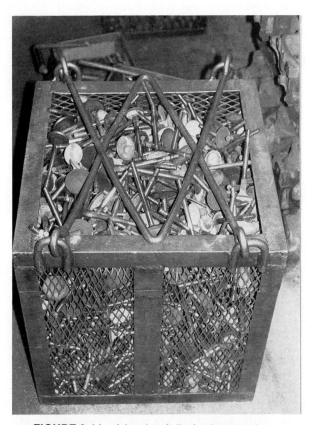

FIGURE 2-11 A bucket full of valves ready to go into the cleaning tank. (Courtesy of Western Engine Ltd.)

FIGURE 2-12 Typical parts cleaning chemical can be used to clean a variety of engine parts. (Courtesy of Red River Community College.)

CLEANING VALVE SPRINGS, RETAINERS, AND ROCKER ARM ASSEMBLIES

Valve springs, retainers, and rocker arms and shafts can be cleaned by hand using a nylon bristle brush and cleaning solvent. Springs that are painted or treated to prevent acid etching should not be cleaned in carburetor cleaner. Carburetor cleaner will remove any acid-preventive paint. Unpainted springs, spring retainers, valve keepers, and rocker arm assemblies are quickly cleaned by immersing in carburetor cleaner for 30 minutes or so. An alternate method is to clean them in a hot-tank solution; however, some hot-tank chemicals will destroy aluminum parts. Make sure that the chemical used is compatible with aluminum, or avoid cleaning aluminum parts this way.

CLEANING PISTONS

Pistons that are to be reused can be cleaned by hand brushing with a soft-bristle brush and brushing in a solvent tank. The ring grooves require special attention when hand cleaning methods are used. The hard carbon deposits in the bottom of the ring grooves must be removed with a ring groove scraper. Care must be exercised not to remove any metal from either the bottom of the grooves or from the sides. Scraping metal from the bottom of the ring groove weakens the piston, while scraping metal from the sides of the grooves prevents proper sealing of the piston ring against the ring lands. The oil-drain-back holes or slots must also be cleaned to ensure proper oil drain back. Make sure that drain-back holes are not enlarged by cleaning (Figures 2-13 to 2-15). A drill bit of the appropriate size may be used. Pistons can also be cleaned by immersing in a cold soak cleaner or by glass bead blasting. Glass shot blasting is described later in the chapter.

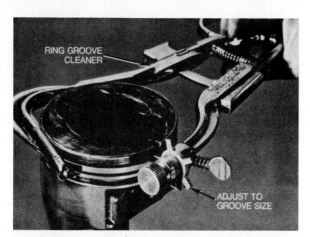

FIGURE 2-13 Using a ring groove cleaner to remove carbon from the bottom of piston ring grooves. (Courtesy of Ford of Canada.)

FIGURE 2-14 Cleaning out the oil drain back holes. (Courtesy of Ford of Canada.)

CLEANING CYLINDER BLOCKS AND HEADS

After complete disassembly, including the removal of oil gallery plugs and freeze plugs, cylinder blocks and heads are normally cleaned by high pressure spraying, thermal methods or hot tank immersion (Figure 2-16). Most hot tanks are equipped with a form of agitation to enhance the cleaning process (Figures 2-17 to 2-21). After cleaning, the block or head is rinsed thoroughly to remove all cleaning solution, then blown dry and immediately sprayed with a corrosion inhibitor to prevent rust. Aluminum cylinder heads require cleaning fluids compatible with aluminum.

CLEANING CRANKSHAFTS AND CAMSHAFTS

Crankshafts and camshafts can be washed by hand in the solvent tank with a nylon brush and a rifle brush for the oil passages. After cleaning they should be blown dry with compressed air. Crankshafts, camshafts, and connecting rods may also be cleaned in the hot tank, after which they should be rinsed with water and sprayed with corrosion inhibitor to prevent rust (Figures 2-22 to 2-24).

CLEANING OIL GALLERIES

Oil passages in crankshafts, connecting rods, engine blocks, rocker shafts, camshafts, pushrods, rocker arms, and the like must be cleaned to ensure proper lubrication. A rifle-type brush such as those used to clean valve guides can be used. Solvent can be used to aid the process. After removing all carbon and varnish, use compressed air to clear the passages.

FIGURE 2-15 Typical hot tank with rinse tank next to it. (Courtesy of Kwik-Way Manufacturing Company.)

Tank shown with optional rinse booth
↓

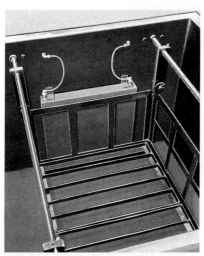

FIGURE 2-16 Interior of hot tank, showing agitator and hot water coil. (Courtesy of Kwik-Way Manufacturing Company.)

FIGURE 2-17 Hot tank in action. (Courtesy of Western Engine Ltd.)

Small parts basket

Head and crank rack

FIGURE 2-18 Large items are usually placed directly in main basket shown in Figure 2-16; smaller items are placed in small parts baskets and racks shown here. (Courtesy of Kwik-Way Manufacturing Company.)

STEAM CLEANING

Steam cleaning is essentially a high-pressure vapor system utilizing water, heat, soap, and high pressure. The machine consists of a motor-driven high-pressure pump, a water coil, coil heater, detergent tank and

FIGURE 2-19 Engine block being removed from hot tank. (Courtesy of Western Engine Ltd.)

dispenser, and a discharge hose and nozzle. Tap water is connected to the machine and pressurized to about 100 psi by the pump. The heater heats the water in the coil to about 300°F. The nozzle has a restriction that causes back pressure to increase. A high-pressure stream of vapor is directed at the area to be cleaned. The detergent concentration can be varied to any desired level. Steam cleaning is an effective method of cleaning the outside of engines and engine compartments. To operate the steam cleaner, follow the directions provided with the machine. Use protective clothing to prevent injury due to hot high pressure back spray.

CHEMICALS FOR CLEANING

There are three basic types of chemical cleaners for automotive parts cleaning: (1) alkaline chemicals, (2) emulsion chemicals, and (3) acid-based chemicals.

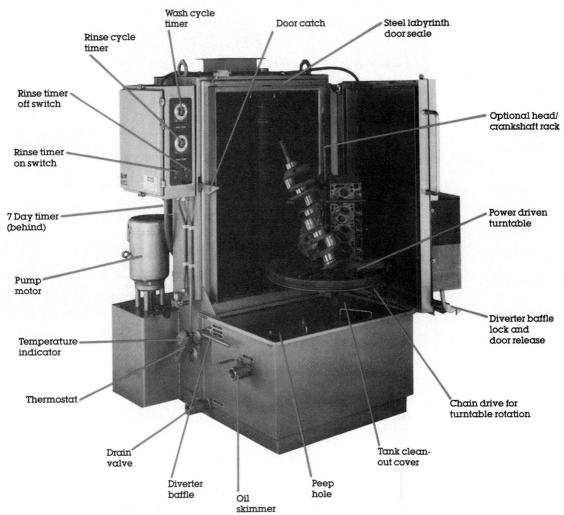

FIGURE 2-20 Hot-spray cleaning equipment is used for engine block cleaning as well as for other engine parts. (Courtesy of Kwik-Way Manufacturing Company.)

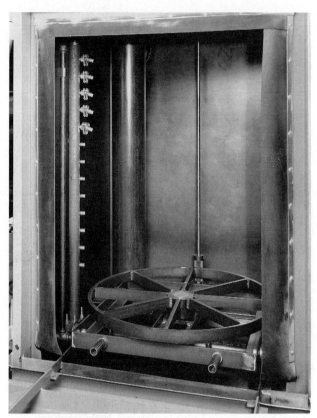

FIGURE 2-21 Interior of jet spray parts cleaner, showing spray nozzles. (Courtesy of Storm Vulcan Co.)

FIGURE 2-22 Camshafts and crankshafts in parts cleaning baskets. (Courtesy of Western Engine Ltd.)

Alkaline Chemicals

The alkaline chemical most people are familiar with is ordinary lye. Combined with heat the alkaline and water solution is used to remove organic deposits and light rust. Alkaline chemicals have been widely used for automotive cleaning. The strength of alkaline cleaners is rated according to their pH level, a chemical designation of hydrogen-ion activity. The pH scale

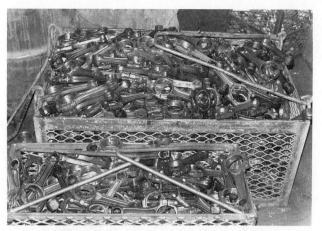

FIGURE 2-23 Connecting rods in parts cleaning baskets. (Courtesy of Western Engine Ltd.)

FIGURE 2-24 Cleaning and flushing the lubrication passages in the crankshaft. (Courtesy of Jasper Engine and Transmission Exchange.)

ranges from 0 to 14, with 7 being the neutral point. The pH of pure water, for example, is 7. Anything above 7 is alkaline and anything below 7 represents increased acidity. The strongest acid is represented as 1 and the strongest alkaline is represented by 14. Commercial alkaline cleaners range from 10 to 12 on the pH scale. These alkaline cleaners work best at temperatures over 140°F.

Emulsion Chemicals

Emulsion-type chemical cleaners are often called "degreasers." These emulsifiable solvents do not require heat and perform well at room temperature (65 to 85°F). Wetting agents in emulsion solvents break down the petroleum binders in oil-based deposits. This allows the rinse water to convert the deposits to soap, which is water soluble. Emulsion chemicals are normally mixed with water for cleaning engine parts.

Acid-Based Chemicals

Descaling and rust removal requires the use of an acid-based cleaner. Acid cleaners are used only where scale and rust must be removed from cylinder blocks and heads removed from engines that use only water as a coolant. A muriatic acid and water solution is used after degreasing has been done. Acid cleaners do not remove grease and oil-based deposits. Acid cleaners have a pH level of 1.5 to 2.

COLD-SOAK CLEANING

Carburetor cleaning fluid and other special chemical solutions are used for cleaning parts. Parts are placed in a dip basket and then immersed in the cleaning solution. For good cleaning the solution should be mixed by raising and lowering the parts and the dip basket several times in the dip tank. This also ensures the release of any trapped air. The parts are left to soak until clean, then rinsed with water and blown dry with compressed air. Cold-soak cleaning fluids irritate the skin and can cause eye damage. Avoid skin contact and wear eye protection. Use water immediately to wash off any solution.

HOT-TANK CLEANING

Hot-tank cleaning is a common procedure in engine rebuilding shops. A variety of equipment types is available. The tank is filled to the specified level with the appropriate solution for the type of metals to be cleaned. Caustic solutions destroy aluminum and magnesium parts. Cleaning materials suppliers are able to recommend the proper cleaner for each application. Hot tanks are normally equipped with some means of fluid agitation. These include the following.

1. A load tray on which the parts to be cleaned are placed. After loading the tray, the agitator is switched on. An electric motor and oscillator cause the loaded tray to move up and down in the solution to create a washing effect.
2. An electrically driven pump, which circulates the solution in the tank to provide the washing action.
3. A controlled volume of compressed air, sent through the solution to impart turbulence.

Most shops use a prepared alkaline cleaning agent composed of about 50% each of sodium hydroxide and sodium, which is dangerous to eyes and skin. The solution acts as a detergent in that it makes oil water soluble. The ratio of caustic cleaner to water is about 1 lb to 1 gallon of water; however, follow the directions on the product container. With use the solution loses its cleaning power and more caustic cleaner must be added.

Always handle caustic solutions and materials with extreme care. Wear rubber gloves and face protection to prevent injury. Caustic solutions are also harmful to such metals as aluminum and magnesium. These parts must be cleaned using noncaustic cleaning agents. Consult your supplier.

DESCALING ENGINE BLOCKS AND HEADS

Lime and calcium deposits in engine water jackets cannot be effectively removed by caustic hot-tank immersion cleaning methods. However, scale and rust must be removed to ensure proper cooling system operation. Scale is a good insulator so if not removed, heat is retained to a much higher degree in engine cylinders. Scale and rust usually form in engines that use only water as a coolant.

Descaling requires the use of an acid solution. An immersion tank is used to immerse the head and block in the acid solution after degreasing in the hot tank. Muriatic acid is generally used for descaling. Follow the supplier's directions in handling the acid and in preparation of the descaling solution. Muriatic acid is extremely corrosive and must not be allowed to contact the skin or eyes. Acid fumes must not be inhaled. After the descaling process the acid must be neutralized either in the hot tank or in a neutralizing solution.

GLASS BEAD CLEANING

Carbon deposits are not always easily removed by hot-tank immersion, especially the very dry hard carbon. Glass bead blasting is frequently used to remove carbon from engine parts such as combustion chambers, pistons, valves, and the like. Glass bead blasting, however, should not be used on parts where these abrasives are not easily removed. Residual cleaning abrasives can seriously damage an otherwise properly rebuilt engine. Glass bead blasting is often used to remove carbon, rust, and gasket material instead of scraping and brushing by hand, as the process is faster and does a better job. Glass bead cleaning is usually done after preliminary cleaning by hand or hot-tank immersion.

The glass bead cleaner consists of a cabinet, a storage tank for glass beads, a blasting nozzle, two sleeved and gloved access holes for handling the nozzle and parts, and a window to observe the operation. Some units are equipped with a bead reclaiming device (Figures 2-25 to 2-27). The glass beads are kept under

FIGURE 2-25 Typical glass bead cleaning equipment. (Courtesy of Kwik-Way Manufacturing Company.)

FIGURE 2-26 Benchtop glass bead cleaner with the cabinet open showing gloves. (Courtesy of Peterson Machine Tool, Inc.)

constant compressed air pressure in the storage tank. When the nozzle valve is triggered, a metered amount of glass bead material is forced out of the nozzle by compressed air. The operator directs the nozzle at the area to be cleaned. The cabinet retains all the glass beads and dust.

All abrasives must be cleaned from parts exposed to glass bead cleaning, to prevent them from entering the engine. The problem of possible residual abrasives is the major drawback of the glass bead cleaning method. Glass bead cleaning should only be done on parts that are easily hand cleaned, to ensure removal of all glass bead residue.

AIRLESS SHOT BLASTING

Airless shot blasting is a method of blasting steel or zinc shot or pellets without air pressure. Shot is propelled by a motor-driven impeller at high speed against the parts being cleaned. The impeller turns at about 2500 rpm. The parts are placed in a revolving drum basket in one model or on a rotating tray in another. This exposes all sides of the parts to the blasting process. This method is used to clean cylinder heads, manifolds, valves, camshafts, and the like. The operating cycle is approximately 15 minutes but can be increased or decreased depending on the types of parts being cleaned and their sensitivity to damage from the shot. The shot is approximately 0.030 in. in diameter. All parts to be cleaned must first be cleaned in the hot-tank degreaser, just as with the glass bead cleaner. After cleaning a tumble cycle is used to help remove all shot from the parts.

FIGURE 2-27 Photomicrograph of machined metal surface before glass shot treatment (left) and after (right). (Courtesy of Peterson Machine Tool, Inc.)

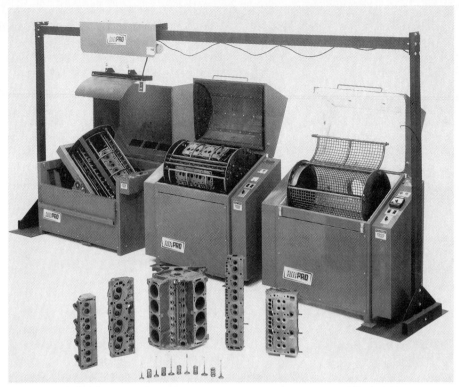

FIGURE 2-28 Complete thermal parts cleaning equipment setup consisting of a thermal oven (left), an airless shot blaster (center), and a shaker/cool-down unit (right). (Courtesy of Storm Vulcan Co.)

FIGURE 2-29 High-pressure spray cleaning equipment. Parts to be cleaned are placed on turntable, then positioned in cabinet and the door closed. Cleaning takes only 10 minutes or less. (Courtesy of Storm Vulcan Co.)

THERMAL CLEANING

The thermal cleaning oven consists of a specially designed heating compartment and an afterburner. Parts are placed in the heating compartment, where they are initially heated to near 400°F, at which point the afterburner cuts in to oxidize the exhaust vapors at about 1300°F. The temperature in the heating compartment continues to rise to about 600 to 700°F to continue the cleaning process. After the cleaning cycle the parts are left to cool for several hours. Sludge and grease deposits are reduced to a dry ash that is easily removed with compressed air. One of the advantages of this cleaning method is that no hazardous caustic solution or sludge waste material is left to be disposed of by the shop. A complete thermal cleaning system consists of three units: a thermal oven, an airless shot blaster, and a shaker/cool-down unit (Figure 2-28).

HIGH-PRESSURE SPRAY CLEANING

High-pressure spray cleaners use hot or cold cleaning solutions and a tap water rinse cycle. High-pressure jet streams (around 1000 psi) of hot detergent are directed at the parts from all angles. Parts placed on the turntable rotate during the cleaning process. Cleaning cylinder heads and crankshafts can take as little as 10 minutes (Figure 2-29).

REVIEW QUESTIONS

1. Name three types of scrapers used to clean parts.
2. Dirty solvent and sludge should be disposed of in a manner that meets _____, _____, and _____ regulations.
3. Describe how to clean valve guides.
4. Name two methods that can be used to clean valves.
5. How should valve springs and rocker arms be cleaned?
6. When cleaning pistons by hand use a:
   (a) ring groove scraper to clean the ring grooves
   (b) soft bristle brush to clean the skirt
   (c) drill bit to clean the drain back holes
   (d) all of the above
7. What method is normally used to clean cylinder blocks?
8. Describe how to clean the oil passages in a crankshaft.
9. What is a steam cleaner often used for?
10. A steam cleaner utilizes:
    (a) water, heat, soap, and high pressure
    (b) water, heat, soap, and low pressure
    (c) solvent, heat, detergent, and high pressure
    (d) solvent, heat, detergent, and low pressure
11. Name three types of cleaning chemicals used in engine rebuilding shops.
12. Technician A says that acid-based chemicals have a low pH level. Technician B says that acid chemicals and alkaline chemicals have similar pH levels. Who is right?
    (a) technician A
    (b) technician B
    (c) both are right
    (d) both are wrong
13. What is the effect of caustic hot-tank cleaning solutions on aluminum parts?
14. What three methods of fluid agitation are used on the various types of hot-tank parts cleaners?
15. What is glass bead blasting used for?
16. What is the major drawback of glass bead cleaning?
17. True or false: Airless shot blasting equipment uses a motor-driven impeller to propel shot at high speeds.
18. What is one of the major advantages of thermal cleaning of engine parts?
19. The three steps used in thermal cleaning of parts are:
    (a) heating, shot blasting, shaker/cool-down
    (b) heating, cold chemical soak, rinse
    (c) heating, rinsing, drying
    (d) heating, rinsing, cool-down
20. True or false: High-pressure spray cleaning directs hot or cold spray against parts placed on a turntable.

CHAPTER 3

Measuring Engine Components

OBJECTIVE

To develop the ability to accurately measure all engine components in accordance with the manufacturer's recommended procedures to determine their serviceability and to ensure accuracy in establishing tolerances and clearances during engine assembly.

ENGINE AND ENGINE COMPONENT MEASUREMENTS

Successful engine rebuilding requires great accuracy when measuring and machining engine components. A variety of precision measuring tools are used. Several conditions must be met when measuring to ensure that measurements will be accurate. They are:

1. Parts to be measured must be free of foreign matter such as carbon and varnish. Make sure that only metal is being measured, not dirt.
2. To maintain consistency, all parts should be at room temperature (about 70°F) while measuring. Measuring a cold cylinder and a warm piston while fitting pistons could result in a poor fit. When machining or honing parts, be sure before measuring to cool down parts that are heated by these procedures.
3. Be sure to develop your skill in measuring so that you are able to take a measurement several times over and come up with the same results each time. This requires that the technician develop a certain sense of "feel" when taking measurements that is consistent each time.

ENGLISH AND METRIC DIMENSIONS

Technicians need to be familiar with both measurement systems. Due to the current internationalization of the automotive industry, an increasing number of automotive components dimensioned in metric measurements have appeared. Although it is important to be able to use each system of measurement independently, a comparison between the two systems may be helpful.

1 inch (in.) = 25.40 millimeters (mm)

1 millimeter = 0.0394 inch

1 cubic centimeter (cm³ or cc) = 0.06102 cubic inch

1 cubic inch = 16.39 cubic centimeters

1 centimeter = 10 millimeters

$$1 \text{ liter} = 1000 \text{ cubic centimeters}$$
$$1 \text{ liter} = 61.02 \text{ cubic inches}$$
$$1 \text{ cubic inch} = 0.01639 \text{ liter}$$

Refer to the Appendix for more information on measurement systems.

CYLINDER WEAR

In operation, the cylinder block is subject to great changes in temperature, pressure of combustion, stress from expansion and contraction, cylinder wear from piston thrust, ring pressure, abrasives (possible scoring), and distortion. Major thrust forces of the piston against the cylinder wall occur as a result of combustion pressures against the piston and the angle of the connecting rod during the power stroke. Piston thrust is considerably less during the other strokes. Piston thrust contributes to cylinder wear.

Piston rings push and slide against cylinder walls as the piston moves up and down. Small particles of carbon and other abrasives that may enter the lubricating oil can cause wear. Heat and pressure are most severe when the piston is near the top on the power stroke. Lubrication at this point and under these conditions is also least effective. Consequently, most cylinder wear resulting in cylinder taper takes place at the very top of ring travel in the cylinder. This wear results in a cylinder ridge developing at the top of the cylinder (Figure 3-1).

ton ring can be transferred to the cylinder wall and cause piston, cylinder, and ring scoring.

A loose piston pin, broken rings, dirt, and carbon can also damage cylinder walls. An excessively rich fuel mixture can cause lubrication to be washed away and increase cylinder wear. Incorrectly tightened cylinder head bolts or thermal shock (sudden and extreme temperature change) can cause cylinder distortion. Coolant seepage into the combustion chamber can cause corrosion, as can combustion by product acids, especially if the engine is not serviced frequently enough.

CYLINDER WALL THICKNESS

The thickness of the cylinder walls must be established prior to cylinder resizing. Cylinder walls must maintain minimum thickness limits after reboring. A cylinder wall that is too thin after reboring is subject to distortion and failure. Previously rebored cylinders may already be at the maximum oversize possible. In some instances the cylinder wall may be thinner on one side than it is on the other due to core shift during production boring of the block.

Several methods are used to check cylinder wall thickness—a dial-type thickness gauge is operated like a spring-loaded clamp. It has one movable jaw which registers its position on the dial gauge (Figure 3-2). The ultrasonic tester is a meter equipped with a probing device. The probe is placed on the cylinder wall and held in place with heavy grease. Readings taken are accurate within 0.010 in.

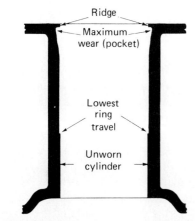

FIGURE 3-1 Cylinder wear is greatest at the top of ring travel area. (Courtesy of Sunnen Products Company.)

Hot spots can develop as a result of rust and scale buildup in the water jacket. This prevents good heat transfer to the engine coolant and causes this area of the cylinder to overheat. Distortion and a wavy cylinder surface can result. If severe, metal from the pis-

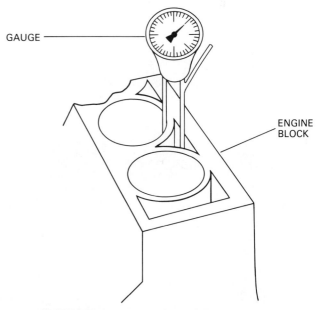

FIGURE 3-2 Measuring cylinder wall thickness to determine if cylinders can be rebored. (Courtesy of FT Enterprises.)

MEASURING ENGINE CYLINDERS

Engine cylinders are measured to determine their service-ability. This may require one or more measurements as follows.

1. *Bore size*. Have the cylinders been rebored or are they still standard? If they were rebored, how much oversize are they, and can they be rebored to a larger oversize?
2. *Taper*. The difference between the minimum and maximum cylinder diameters.
3. *Out-of-round*. The difference between the minimum and maximum cylinder diameter taken at the same height in the cylinder. Maximum out-of-round usually occurs just below the cylinder ridge.
4. *Waviness*. Irregularity in the vertical surface plane of the cylinder wall at any point.

An outside micrometer is set to the engine's standard bore diameter (as specified in the shop manual). The cylinder gauge is then set to the same specification and

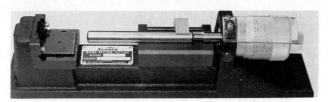

FIGURE 3-4 Cylinder bore gauge setting fixture with micrometer adjustment. (Courtesy of Red River Community College.)

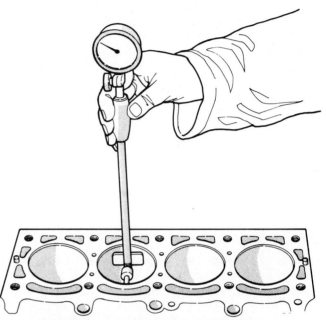

FIGURE 3-5 Measuring a cylinder bore. (Courtesy of Chrysler Canada Ltd.)

the dial gauge is set at zero while positioned in the micrometer (Figures 3-3 and 3-4). The cylinder gauge is then inserted into the cylinder and pushed down to the area below ring travel (Figure 3-5). Since this area is not subject to wear, it will still be standard bore size if the engine has not been rebored. The cylinder gauge, when positioned exactly across the cylinder, should register zero ($+$ or $-$ 0.002 in. or 0.0508 mm for production tolerance) in this case. The gauge should be placed across the cylinder (90° to the crankshaft).

If the engine has been rebored to 0.030 in. (0.762 mm) oversize, the cylinder gauge will read 0.030 in. (0.762 mm) on the plus ($+$) side of the scale. This indicates the piston and the ring sizes required if the cylinders are not to be rebored. This information also indicates whether cylinders that have been rebored previously can be rebored again. Some engines can be rebored to a larger oversize than others. Cylinder measurements must be compared to the manufacturer's maximum allowable rebore size to determine whether reboring is feasible. Most automotive engines cannot be rebored beyond 0.030 in. oversize.

FIGURE 3-3 Cylinder bore gauge and set of contact points. (Courtesy of Sunnen Products Company.)

Cylinder Taper

The difference in the bore diameter at the top of ring travel compared to the bore diameter below ring travel is the amount of taper or wear. If excessive, cylinders should be rebored and fitted to new pistons. Most manufacturers allow a maximum of 0.002 to 0.005 in. (0.05 to 0.13 mm) before reboring is required. Some diesel engines may allow a maximum of only 0.001 in. (0.025 mm) of taper (Figures 3-6 and 3-7).

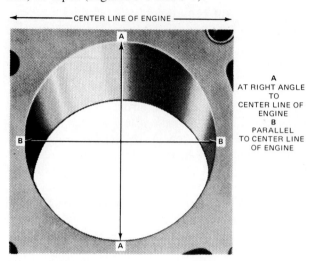

← CENTER LINE OF ENGINE →

A AT RIGHT ANGLE TO CENTER LINE OF ENGINE
B PARALLEL TO CENTER LINE OF ENGINE

1. OUT-OF-ROUND = DIFFERENCE BETWEEN **A** AND **B**
2. TAPER = DIFFERENCE BETWEEN THE **A** MEASUREMENT AT TOP OF CYLINDER BORE AND THE **A** MEASUREMENT AT BOTTOM OF CYLINDER BORE

FIGURE 3-6 Out-of-round and taper measurements. (Courtesy of Ford of Canada.)

3EN060

FIGURE 3-7 Cylinder measurements should be made at three levels. (Courtesy of Chrysler Canada Ltd.)

Cylinder Out-of-Round

Cylinders should be measured parallel to the piston pin and at right angles to the piston pin to determine cylinder out-of-round. This should be done at the top of

ring travel, in the center of the cylinder, and at the bottom of ring travel. If out-of-round exceeds the manufacturer's wear limits (approximately 0.001 to 0.002 in. or 0.025 to 0.05 mm), cylinders should be rebored and fitted to oversized pistons.

Cylinder Waviness

Cylinder waviness can be detected by carefully moving the cylinder gauge the full length of ring travel in the cylinder and observing the dial. If waviness is excessive, cylinders should be rebored and fitted to oversized pistons.

MEASURING MAIN BEARING BORES (SADDLE BORES)

Main bearing bores (saddle bores) must be measured to determine their roundness and alignment (Figures 3-8 and 3-9). To do this the main bearing caps are torqued in place in their proper positions with the bearing inserts removed. Saddle bore out-of-round can be checked with a dial bore gauge, inside micrometer, or

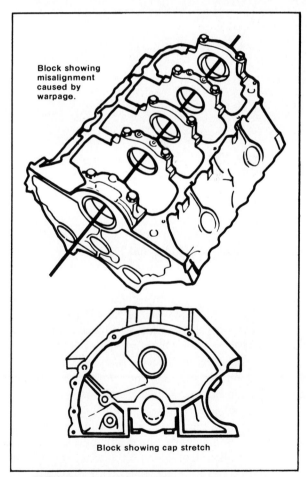

Block showing misalignment caused by warpage.

Block showing cap stretch

FIGURE 3-8 Main bearing saddle bore misalignment and out of round. (Courtesy of Sunnen Products Company.)

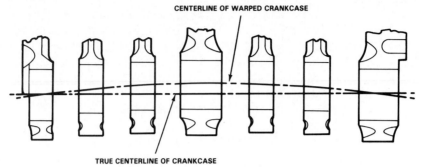

CENTERLINE OF WARPED CRANKCASE

TRUE CENTERLINE OF CRANKCASE

FIGURE 3-9 Main bearing saddle bore misalignment. (Courtesy of Federal-Mogul Corporation.)

FIGURE 3-10 Measuring main bearing bore out-of-round with a dial bore gauge. (Courtesy of Sunnen Products Company.)

telescoping gauge and outside micrometer. Measurements are taken vertically and horizontally. The difference between these two measurements is the amount of out-of-round. Each saddle bore must be measured separately. A maximum of 0.001 in. of out-of-round is normally allowed (Figure 3-10).

To measure saddle bore alignment, either of two methods may be used. One is to use a round precision arbor 0.001 in. less in diameter than the standard saddle bore diameter and long enough to extend through all the saddle bores. With the main bearing caps removed, place the arbor in the saddle bores and torque the bearing caps into place. If you can turn the arbor with a 12-in.-long handle, the saddle bores are considered to be in alignment limits. If the arbor will not turn, the bores are out of alignment (Figure 3-11).

Another method is to use a precision straightedge and a 0.0015-in. feeler gauge (Figure 3-12). While hold-

ing the straightedge firmly against the sides of the bores, try to insert the feeler gauge at each bore. This checks the lateral alignment. Do the same while holding the straightedge against the top of the bores. This checks the vertical alignment of the saddle bores. If the feeler gauge cannot be inserted at any saddle bore in both straightedge positions, the saddle bores are in acceptable alignment limits. If they are not, the caps must be shaved and the bores align bored or honed as described in Chapter 7.

BLOCK DECK HEIGHT AND CLEARANCE

The deck height of a cylinder block is the distance from the main bearing saddle bore centerline to the deck or top of the block (Figure 3-13). This height is an important factor in that it affects the engine's compression ratio. The deck surface must be parallel to the main bearing centerline to ensure that the compression ratio will be equal in all cylinders. On V engines the deck height dimension must be equal for both cylinder banks. If they are unequal, cylinder balance will be affected.

Deck clearance is the difference in height of the deck and the top of the piston when it is at the TDC position. Deck clearance can be positive—the pistons at TDC being higher than the deck, or negative—the pistons at TDC being lower than the deck (Figures 3-14 and 3-15). Factors that affect deck clearance are:

- Deck height
- Connecting rod length
- Piston height: pin centerline to top of piston
- Crank pin stroke
- Vertical positioning of crankshaft in block

See Chapter 7 for more about deck height, deck clearance, and machining the block deck.

FIGURE 3-11 Measuring saddle bore alignment with a precision arbor. (Courtesy of Federal-Mogul Corporation.)

FIGURE 3-12 Measuring saddle bore alignment with a straightedge and feeler gauge. (Courtesy of Federal-Mogul Corporation.)

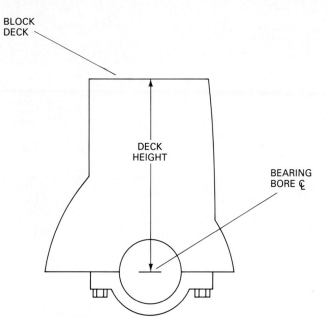

FIGURE 3-13 Cylinder block deck height. (Courtesy of FT Enterprises.)

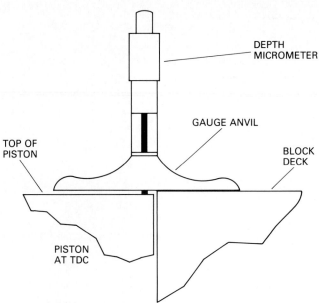

FIGURE 3-15 Measuring deck clearance with a depth micrometer. (Courtesy of FT Enterprises.)

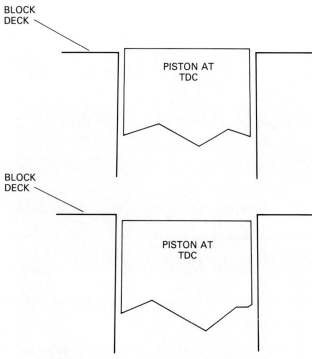

FIGURE 3-14 Positive (top) and negative (bottom) deck clearance. (Courtesy of FT Enterprises.)

MEASURING CRANKSHAFTS

Crankshaft main and rod bearing journals must be measured with an outside micrometer and a dial indicator to determine the following:

Journal Size and Whether Standard or Undersized. If the crankshaft has been reground, the journals will be undersized. Some engine crankshafts may have one or more main or rod journals that are several thousandths inch undersize from the factory.

Journal Out-of-Round. The difference between vertical and horizontal diameters of each journal taken at the front and rear determines their roundness. A maximum out-of-round limit of 0.001 in. is usually acceptable.

Journal Taper. The difference between the vertical or horizontal measurements at the front and rear

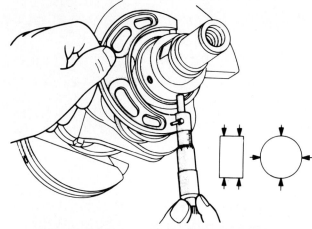

FIGURE 3-16 Measuring a crankshaft main bearing journal for size, taper, and out-of-round. (Courtesy of Chrysler Canada Ltd.)

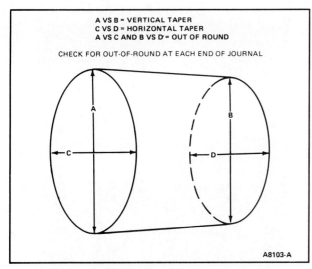

FIGURE 3-17 Crankshaft journal measuring points to determine taper and out-of-round. (Courtesy of Ford of Canada.)

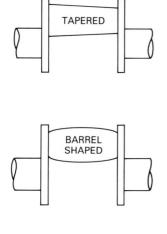

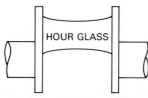

FIGURE 3-18 Crankshaft main and rod journals may wear in different ways, resulting in tapered, barrel-shaped, or hourglass-shaped journals.

of each journal is the amount of taper present. To be acceptable, this should not exceed 0.001 in. Out-of-round and taper in excess of 0.001 in. indicates that the crankshaft must be reground (Figures 3-16 to 3-18).

Barrel-Shaped Wear. This is checked vertically and horizontally at the front, center, and rear of each journal. A maximum of 0.001 in. is acceptable.

Ridging. Ridging is the result of journal wear occurring on both sides of a grooved bearing. If ridging exceeds 0.0003 in. the crankshaft should be reground (Figure 3-19).

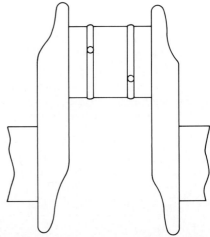

FIGURE 3-19 Crankshaft journal ridging results from grooved bearings. Journals do not wear in bearing groove area. (Courtesy of FT Enterprises.)

FIGURE 3-20 Measuring crankshaft runout (bend) with a dial indicator and V blocks. (Courtesy of Federal-Mogul Corporation.)

Crankshaft Runout. This must be checked with a dial indicator and a set of V blocks (Figure 3-20). The crankshaft is supported by V blocks at the two end main journals. (The V blocks must be smooth and well oiled to prevent the journals from being scored.) A dial indicator is mounted with the plunger travel perpendicular to the journal surface being measured. All unsupported main journals and the crank snout should be checked for runout as the crankshaft is turned. Total indicator reading for one revolution is the amount of runout for each position measured. Maximum runout should not exceed 0.001 or 0.002 in. Check against crankshaft specifications in the service manual. If runout is excessive, the crankshaft must be straightened in a press fixture designed for the purpose.

Caution: Do not bend crankshafts that are nitrided since cracking may be induced. See chapter 8 "Nitrided Crankshaft Check" to determine whether the crankshaft has been nitrided.

Thrust Surface Wear. This can be measured with an inside micrometer or a telescoping gauge and outside micrometer. Measure between the thrust surfaces parallel to the journal (Figure 3-21). Compare measurements with specifications. Excessive thrust surface wear requires crankshaft replacement or reconditioning by welding to replace lost metal and machining or the use of thicker thrust bearings.

FIGURE 3-21 Measuring crankshaft thrust surface wear with an inside micrometer. (Courtesy of Federal-Mogul Corporation.)

CRANKSHAFT GRINDING LIMITS

Most automotive crankshafts are not reground beyond 0.030 in. undersized. Removing too much metal weakens the crankshaft and can result in breakage. Larger industrial and transport engine crankshafts can be reground to 0.060 in. or more undersized, depending on the make, model, and size of the engine. A minimum of about 0.008 in. of metal is removed during the grinding procedure. This means that this must be taken into consideration when measuring crankshaft journals to determine whether regrinding is feasible. See chapter 8 for crankshaft grinding procedure.

MEASURING CONNECTING RODS

Connecting rods must be checked for alignment to see if they are bent or twisted. This is done on a special alignment fixture (Figures 3-22 and 3-23). If the big-end bore and small-end bore are parallel, the rod is not twisted or bent. Maximum bend or twist is 0.001 in. per 6 in. of rod length.

To measure the big-end bore for out-of-round,

FIGURE 3-22 Measuring connecting rod alignment with piston attached. (Courtesy of Red River Community College.)

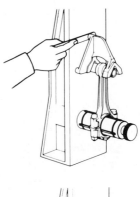

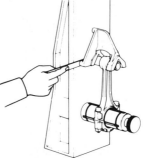

FIGURE 3-23 Measuring connecting rod alignment with piston removed. Checking for bend (top) and for twist (bottom). (Courtesy of Ford of Canada.)

FIGURE 3-24 Connecting rod clamped in rod holding fixture prevents rod twist when tightening connecting rod bolts. (Courtesy of Red River Community College.)

FIGURE 3-25 Measuring connecting rod big-end bore for stretch or out-of-round with a dial bore gauge. (Courtesy of Federal-Mogul Corporation.)

torque the rod caps into place first. Out-of-round or stretch normally does not occur across the cap-to-rod parting lines, but takes place vertically or to about 30° off vertical. Measurements should therefore be made vertically and horizontally just off the parting edge. A maximum of 0.001 in. out-of-round is allowed. Figures 3-24 to 3-27.

Measure the small-end bore of the rod with an

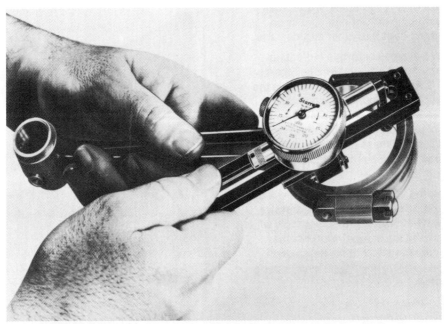

FIGURE 3-26 Measuring big-end bore out-of-round with shallow dial bore gauge. (Courtesy of The L.S. Starrett Company.)

Measuring Connecting Rods **33**

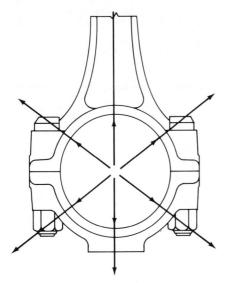

FIGURE 3-27 Measuring points for checking connecting rod big-end bore. (Courtesy of Sunnen Products Company.)

FIGURE 3-28 Precision bore gauge measures both big-end and small-end connecting rod bores to an accuracy of 0.0001 in. (Courtesy of Sunnen Products Company.)

inside micrometer or precision gauge that measures accurately to 0.0001-in. increments (Figure 3-28). Compare measurements to specifications. Small-end bores are honed to fit oversized piston pins available in 0.0015-, 0.003-, and 0.005-in. oversized. For full-floating pins the small-end bushing is honed to provide from 0.0003 to 0.0005 in. of clearance, while the piston pin boss is honed to provide from 0.0001 to 0.0003 in. of clearance. If the pin is a press fit in the rod, the rod eye is honed to 0.0008 to 0.0012 in. smaller than pin diameter to provide the interference fit needed to keep the pin tight in the rod.

MEASURING PISTON DIAMETER

Piston measurements are made to determine piston size and skirt collapse. The piston size or diameter is normally measured across the thrust surfaces (90° to the pin bore) with an outside micrometer (Figures 3-29 to 3-30). The point on the piston where measurement is made varies with engine and piston design and manufacturer. Some specifications call for measurement to be made at the centerline of the piston pin bore. Others require measurement to be made just below the oil ring groove or at the bottom of the skirt. Be sure to measure at the recommended height and compare to specifications.

Pistons that are not collapsed will generally measure from 0.0005 to 0.003 in. less than the cylinder bore diameter. A greater difference usually indicates a collapsed piston. Collapsed pistons result in excessive piston-to-cylinder clearance, allowing the piston to rock back and forth in the bore. This results in piston slap, excessive blowby, increased oil consumption and plug fouling, and eventual cracking of the piston.

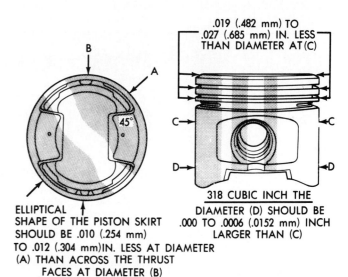

ELLIPTICAL SHAPE OF THE PISTON SKIRT SHOULD BE .010 (.254 mm) TO .012 (.304 mm) IN. LESS AT DIAMETER (A) THAN ACROSS THE THRUST FACES AT DIAMETER (B)

.019 (.482 mm) TO .027 (.685 mm) IN. LESS THAN DIAMETER AT (C)

318 CUBIC INCH THE DIAMETER (D) SHOULD BE .000 TO .0006 (.0152 mm) INCH LARGER THAN (C)

FIGURE 3-29 Typical piston measurements. (Courtesy of Chrysler Canada Ltd.)

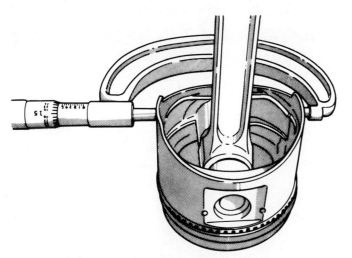

FIGURE 3-30 Measuring a piston across thrust faces at bottom of skirt. (Courtesy of Chrysler Canada Ltd.)

FIGURE 3-32 Measuring ring groove wear with a new piston ring and feeler gauge. (Courtesy of Chrysler Canada Ltd.)

MEASURING RING GROOVE WEAR

Excessive ring groove wear results in increased ring-to-land clearance, increased blowby and oil consumption, plug fouling, ring wear, and ring and land breakage. Most ring groove wear occurs in the top ring groove because it gets the least lubrication, runs the hottest, and absorbs the most dynamic pressure.

All upper ring grooves should be measured for wear with a ring groove wear gauge (Figure 3-31). A new piston ring and feeler gauge may also be used to make this check (Figure 3-32). If a 0.006-in. feeler gauge can be inserted between the new ring and ring land, the piston must be replaced or the ring groove machined and a spacer installed. See Chapter 10 for this procedure.

MEASURING CAMSHAFTS

If the cam lobes and bearing journals are not chipped or otherwise damaged and the integral distributor drive gear and fuel pump eccentric are not damaged, the camshaft should be carefully measured for wear and runout. An outside micrometer is used to measure cam lobe and bearing journal wear. To measure lobe lift, measure across the heel and nose of the lobe and again across the lobe at 90° to the first measurement. The difference in these two measurements is the lobe lift (Figures 3-33 and 3-34). There should be no more than 0.006-in. variation between lobes and lift should be within service manual specifications. Measure each bearing journal to determine wear and out-of-round conditions (Figures 3-35 and 3-36). Maximum allow-

FIGURE 3-31 Measuring piston ring groove wear with groove wear gauge. (Courtesy of Federal-Mogul Corporation.)

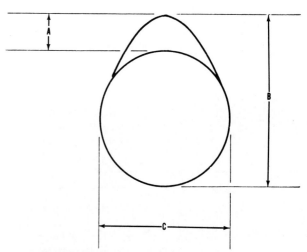

FIGURE 3-33 Cam lobe wear is determined by measuring lobe lift. Measurement at B minus measurement at C equals lobe lift A. (Courtesy of Ford of Canada.)

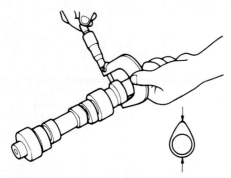

FIGURE 3-34 Measuring lobe wear with a micrometer. (Courtesy of Chrysler Canada Ltd.)

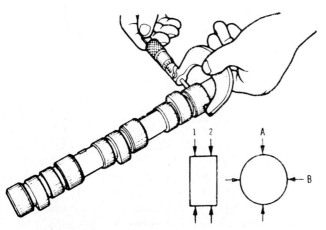

FIGURE 3-35 Measuring cam bearing journal wear. (Courtesy of Chrysler Canada Ltd.)

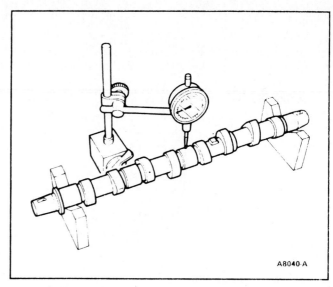

FIGURE 3-37 Measuring camshaft runout (bend). (Courtesy of Ford of Canada.)

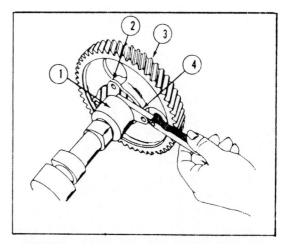

FIGURE 3-38 Measuring gear driven camshaft end play with feeler gauge between thrust surfaces. (Courtesy of Ford of Canada.)

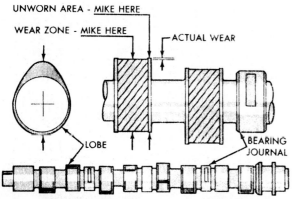

FIGURE 3-36 On some engines cam lobe wear does not extend to the edges, due to cam follower design. Be sure to measure in worn section only. (Courtesy of Chrysler Canada Ltd.)

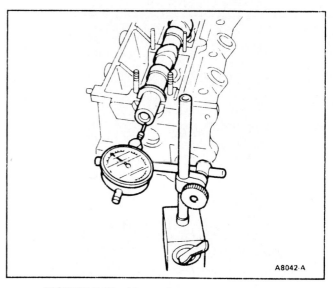

FIGURE 3-39 Measuring camshaft end play with a dial indicator. (Courtesy of Ford of Canada.)

able journal wear or out of round is 0.001 in. Use V blocks and a dial indicator to measure camshaft runout (Figure 3-37). Support the camshaft in the V blocks on the two end journals. Mount the dial indicator and measure the runout at each unsupported bearing journal. Maximum allowable runout is 0.002 in. Camshaft end play is measured between the thrust plate and the thrust surface of the camshaft with a feeler gauge or by moving the camshaft endwise while measuring with a dial indicator (Figures 3-38 and 3-39).

HYDRAULIC LIFTER TESTING

Hydraulic valve lifters can be tested for their leakdown rate. Lifter leakdown is the leakage of oil from the oil chamber in the lifter past the lifter plunger. The leakdown rate of hydraulic lifters is designed to be quite precise. If the leakdown rate is too slow, lifters may pump up and hold the valve open when it should be closed. A leakdown rate that is too fast reduces valve opening and causes valve train noise.

To use the leakdown tester, fill the cup with tester fluid to a level high enough to cover the oil inlet port. Place a steel ball on the lifter pushrod seat to prevent damage to the seat. Pump the lifter full of test fluid by

raising the weight and lowering it to depress the lifter plunger several times (Figure 3-40). Adjust the indicator to indicate plunger travel. With the lifter full of test fluid, time the plunger travel with the weight applied. Allow the weight to depress the plunger. Compare the results with specifications. Generally, plunger travel of 0.100 in. should not take less than 10 seconds or more than 60 seconds.

VALVE SPRING MEASURING AND TESTING

Valve springs should be measured for free length and squareness (Figure 3-41). Springs should all be within $\frac{1}{16}$ in. of each other and not more than $\frac{1}{16}$ in. out of length specifications. A spring that is shorter than specified has collapsed and should be replaced. Each spring should also be checked for squareness. An out-of-square spring will pull sideways on a valve, causing poor valve seating and excessive guide and stem wear. Springs between 2 and $2\frac{1}{2}$ in. in height should not be out of square more than $\frac{1}{16}$ in. (Figure 3-42).

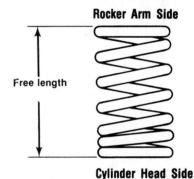

FIGURE 3-41 Valve spring free length is spring length when not compressed. (Courtesy of Chrysler Canada Ltd.)

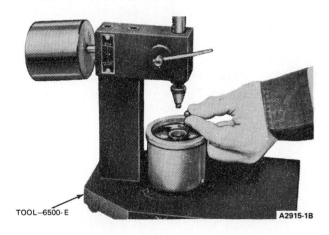

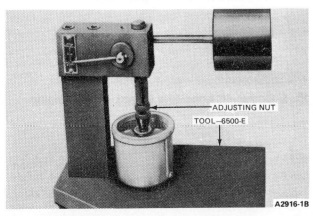

FIGURE 3-40 Hydraulic lifter testing. (Courtesy of Ford of Canada.)

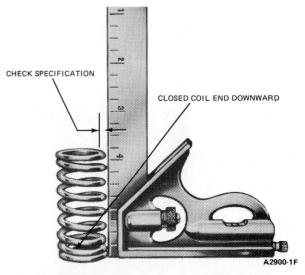

FIGURE 3-42 Checking valve spring squareness. (Courtesy of Ford of Canada.)

Valve springs must have the proper spring tension to ensure proper functioning of the valves. Excessive spring tension increases cam lobe wear, valve train wear, and valve face and seat wear. Insufficient valve spring tension causes poor valve action and hydraulic lifter pump-up. Valve spring pressures are tested at two spring heights: the valve-closed (seated) spring height and the valve-open spring height (Figure 3-43). Obviously, the compressed height of a valve spring is less with the valve open compared to its height when the valve is closed.

Before testing the springs set the tester gauge to read zero. To check the valve-seated spring pressure, proceed as follows.

1. Place the spring on the tester base plate.
2. Turn the handle to lower the pressure plate to the specified test height, then set the stop clamp securely. (*CAUTION:* Do not set the

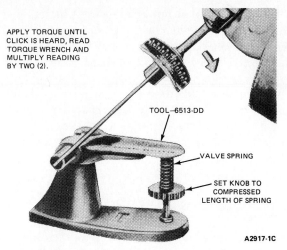

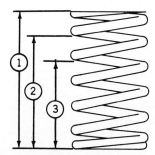

FIGURE 3-43 1, Valve spring free height; 2, valve closed spring height; 3, valve open spring height. (Courtesy of Ford of Canada.)

FIGURE 3-44 Testing valve spring pressure. (Courtesy of Ford of Canada.)

APPLY TORQUE UNTIL CLICK IS HEARD, READ TORQUE WRENCH AND MULTIPLY READING BY TWO (2).

TOOL—6513-DD

VALVE SPRING

SET KNOB TO COMPRESSED LENGTH OF SPRING

A2917-1C

FIGURE 3-45 Torque wrench type of valve spring tester. (Courtesy of Ford of Canada.)

tester without the spring on the base plate since this results in an inaccurate setting.)

3. Turn the handle to compress the spring until the stop is reached and read the scale (Figure 3-44).
4. Test all the springs at the two heights specified: (a) the valve-closed spring height and (b) the valve-open spring height.

A different type of spring tester is shown in Figure 3-45. A torque wrench is needed to use this tester. Adjustment for spring height is provided by turning the platform wheel. Pull the torque wrench down carefully until a "ping" is heard from the tester. At that precise point take the torque wrench reading and multiply it by 2. A 60-lb reading on the torque wrench, for example, would be a spring tension of 120 lb.

MEASURING VALVE GUIDE WEAR

As valve guides wear the valve stem to guide clearances increases. Excessive clearance allows valves to cock in the guide. A cocked valve seats on only one side, resulting in leakage and burning of the valve and seat. Valve guides do not wear evenly. Due to rocker arm action, bell mouthing of the guide occurs. Bell mouthing allows increased cocking of the valve in the guide.

Specifications for valve stem-to-guide clearance provided in manufacturers' service manuals vary considerably depending on engine design and on the method used to measure clearance. It should be noted that the specifications given in the service manual are valid only for the measuring method employed in the manual. Before measuring, the guides must be thoroughly cleaned with a scraper and brush to remove all carbon and varnish. Methods used to measure clearance include the following.

1. Using a dial indicator, measure the side-to-side movement of the valve head with the valve off its seat and compare to specifications (Figure 3-46).

2. Using a dial indicator, measure the side-to-side movement of the valve stem on the spring side, close to the guide, with the valve off its seat and compare to specifications.

3. Using a dial indicator and a special clearance

checking tool installed on the valve stem, measure the side-to-side movement of the valve and compare to specifications (Figure 3-47).

4. Use an inside micrometer to measure the valve bore guide diameter and compare to specifications.

5. Use a split ball gauge and outside micrometer to measure the guide bore diameter and compare to specifications (Figures 3-48 to 3-50).

6. Use a special dial indicator bore gauge to measure the guide bore diameter and compare to specifications (Figures 3-51 and 3-52).

FIGURE 3-46 Using a dial indicator to measure valve head side-to-side movement. (Courtesy of Chrysler Canada Ltd.)

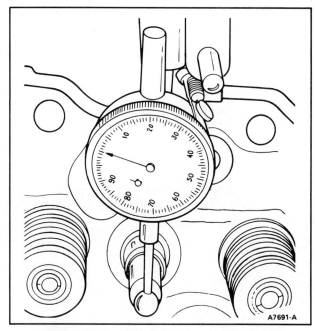

FIGURE 3-47 Using a special device on the valve stem and a dial indicator to measure valve stem side-to-side movement. (Courtesy of Ford of Canada.)

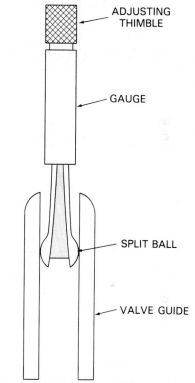

FIGURE 3-48 Split ball gauge method of measuring guide wear.

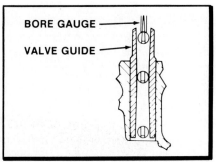

FIGURE 3-49 Measuring valve guide at three positions with split ball gauge. (Courtesy of Chrysler Canada Ltd.)

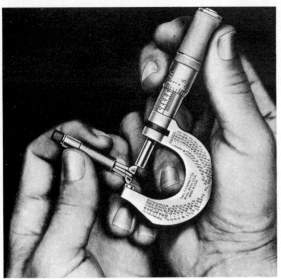

FIGURE 3-50 Measuring split ball gauge setting with outside micrometer. (Courtesy of The L.S. Starrett Co.)

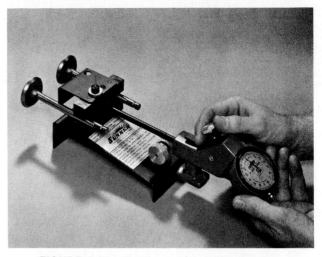

FIGURE 3-52 Setting valve guide gauge in setting fixture. Gauge can also be set with an outside micrometer. (Courtesy of Sunnen Products Company.)

MEASURING CYLINDER HEAD AND BLOCK DECK SURFACE

The cylinder head and block deck surfaces must be inspected for etching, erosion, or scoring. If the etching, erosion, or scoring are plainly visible and can be felt by running a fingernail across the surface, the head or block deck should be resurfaced. The same check must be made on the manifold mounting surfaces of the cylinder head.

To check the cylinder head or block deck for warpage, place a straightedge across the surface in several positions, as shown in Figure 3-53 and check for clearance under the straightedge with a feeler gauge. Warpage can occur in any direction. The surfaces should be checked in each position shown. The maximum warpage allowed is 0.002 in. over any 6 in. of length and 0.006 in. overall.

FIGURE 3-51 Checking valve guide wear with a dial gauge. (Courtesy of Sunnen Products Company.)

To determine the stem-to-guide clearance when using the guide bore measurement, subtract the valve stem diameter from the guide bore diameter.

Caution: Some engines are equipped with exhaust valves with tapered valve stems. The stem diameter is usually 0.001 in. smaller at the head end than it is at the retainer end. Tapered valve stems reduce scuffing, scoring, and sticking while maintaining relatively low stem-to-guide clearance. Since the valve runs hotter at the head end, expansion of the stem at that end is greater. The smaller stem diameter in this area allows for this difference.

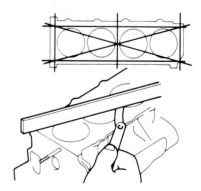

FIGURE 3-53 Checking cylinder block deck warpage with a straightedge and feeler gauge. Cylinder head must be checked in a similar manner. (Courtesy of Chrysler Canada Ltd.)

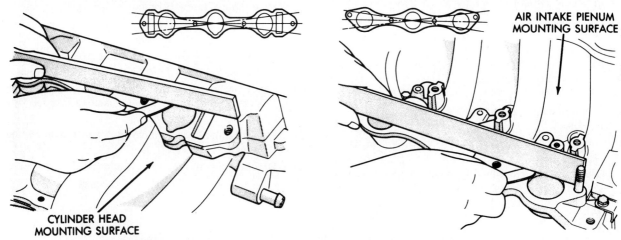

Cylinder Head Mounting Surface
Standard: .10 mm (.003 inch)
Maximum: .2 mm (.005 inch)

Intake Plenum Mounting Surface
Standard: .15 mm (.004 inch)
Maximum: .3 mm (.008 inch)

AIR INTAKE PLENUM
MOUNTING SURFACE

CYLINDER HEAD
MOUNTING SURFACE

FIGURE 3-54 Measuring intake plenum and cylinder head mounting surfaces for warpage. (Courtesy of Chrysler Canada Ltd.)

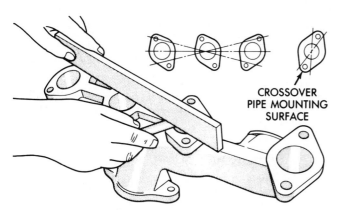

CROSSOVER
PIPE MOUNTING
SURFACE

FIGURE 3-55 Measuring exhaust manifold warpage. (Courtesy of Chrysler Canada Ltd.)

MEASURING THE INTAKE AND EXHAUST MANIFOLDS

The mounting surfaces of intake and exhaust manifolds are measured with a straightedge and feeler gauge (Figures 3-54 and 3-55). Surfaces must be straight and flat within 0.010 in. overall. There should be no major surface irregularities, pitting, or erosion. Mounting surfaces can be refinished by grinding or milling, if required.

REVIEW QUESTIONS

1. Engine parts to be measured should be clean and at around _____ °F.

2. Engine cylinder wear is defined as:
   (a) taper, out-of-round, and waviness
   (b) taper, waviness, and ridging
   (c) ridging, out-of-round, and taper
   (d) out-of-round, waviness, and ridging

3. Cylinder wear is caused by:
   (a) friction
   (b) abrasives
   (c) piston thrust
   (d) all of the above

4. Why is cylinder wall thickness a factor in cylinder block reconditioning?

5. Engine cylinders are measured for:
   (a) size, taper, and ridging
   (b) out-of-round, size, and taper
   (c) ridging, out-of-round, and size
   (d) taper, angularity, and size

6. To measure the main bearing saddle bore alignment, use a _____ and a _____.

7. Why should saddle bore alignment be checked at the top and sides?

8. Define "block deck height."

9. Technician A says that deck clearance can be positive or negative. Technician B says that deck clearance is the distance from the deck to the crankshaft centerline. Who is right?
   (a) technician A
   (b) technician B
   (c) both are right
   (d) both are wrong

10. Crankshaft bearing journal wear can cause the journals to become:

(a) tapered or out-of-round
(b) barrel-shaped or ridged
(c) tapered or ridged
(d) any of the above

11. The crankshaft bearing journals should be measured with:
(a) a dial indicator
(b) an outside micrometer
(c) a bearing bore gauge
(d) a divider

12. Why should the big-end bore of connecting rods be measured?

13. Connecting rod misalignment is defined as _____ and _____ .

14. Piston size is determined by measurement across the _____ surfaces.

15. Name two methods used to measure piston ring groove wear.

16. Define "cam lobe lift."

17. How should camshaft runout be measured?

18. The term "leakdown rate" refers to _____ testing.

19. Valve springs should be tested to determine what four conditions?

20. Excessive valve-to-guide clearance allows the valve to _____ in the guide, resulting in poor valve _____ .

21. List three methods of measuring valve stem-to-guide clearance.

22. Describe how to measure cylinder head warpage.

CHAPTER 4

Blueprinting an Engine

OBJECTIVE

To develop the knowledge and expertise required to rebuild an engine accurately to blueprinting standards.

INTRODUCTION

Engine blueprinting is a specialized art requiring a very high degree of attention to every detail. Blueprinting is both expensive and time consuming. Simply stated, it is the process of rebuilding the engine in a manner that will restore all its components critical dimensions and clearances within exact factory tolerances (Figure 4-1).

Blueprinting requires fitting components precisely, often using maximum specified clearances and minimum specified volumes within the range of specifications provided by the engine manufacturer (Figure 4-2). Engine blueprinters, however, are not always in agreement as to the extent to which the procedure should be applied, nor do they always agree on whether to stay strictly within factory tolerances. Many competition engine rebuilders include minor variations that they feel provide that extra winning edge. Based on wide experience with a particular engine type, they are able to fine tune their tolerances and specifications to suit their particular needs. Some of these variations include piston, ring gap, and bearing clearances. Before attempting any alterations to an engine, be sure to check with all the local and federal authorities to make sure that what you have in mind is legal in your area. Emissions regulations may not allow certain changes.

To do a complete blueprint job on an engine would be cost-prohibitive and in some aspects may not provide any noticeable benefit. However, real benefits are obtainable in many areas of engine rebuilding if precise blueprinting tolerances are achieved. Many of these tolerances are easily checked with common engine rebuilding tools and equipment. Tolerances such as these are also usually published in the engine manufacturer's shop manual.

- Piston-to-cylinder wall clearance
- Piston ring-to-land clearance
- Piston ring end gap
- Piston pin fit in piston and connecting rod
- Connecting rod center-to-center length
- Piston pin-to-crown height
- Crank pin indexing
- Connecting rod side clearance
- Piston deck height

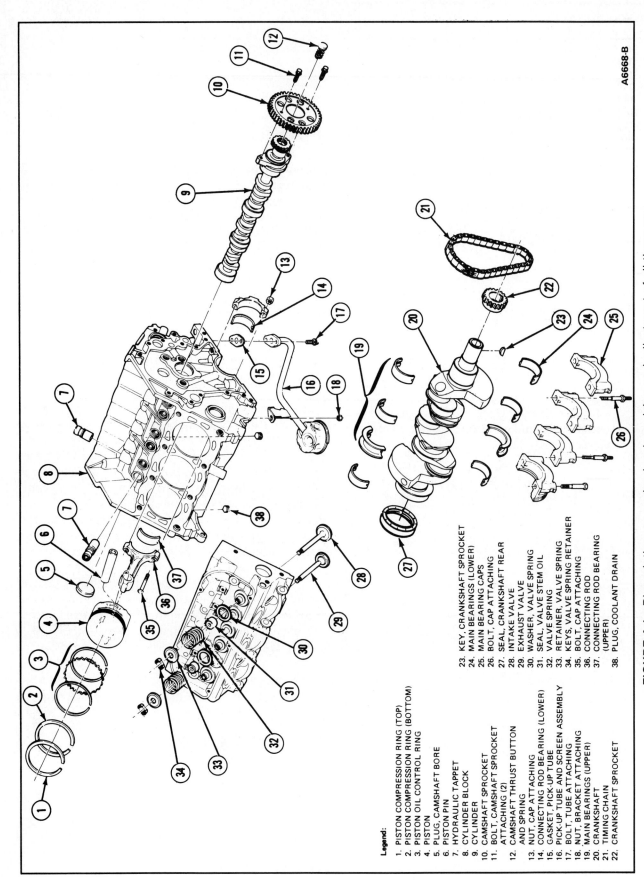

Legend:

1. PISTON COMPRESSION RING (TOP)
2. PISTON COMPRESSION RING (BOTTOM)
3. PISTON OIL CONTROL RING
4. PISTON
5. PLUG, CAMSHAFT BORE
6. PISTON PIN
7. HYDRAULIC TAPPET
8. CYLINDER BLOCK
9. CYLINDER
10. CAMSHAFT SPROCKET
11. BOLT, CAMSHAFT SPROCKET ATTACHING (2)
12. CAMSHAFT THRUST BUTTON AND SPRING
13. NUT, CAP ATTACHING
14. CONNECTING ROD BEARING (LOWER)
15. GASKET, PICK-UP TUBE
16. PICK-UP TUBE AND SCREEN ASSEMBLY
17. BOLT, TUBE ATTACHING
18. NUT, BRACKET ATTACHING
19. MAIN BEARINGS (UPPER)
20. CRANKSHAFT
21. TIMING CHAIN
22. CRANKSHAFT SPROCKET
23. KEY, CRANKSHAFT SPROCKET
24. MAIN BEARINGS (LOWER)
25. MAIN BEARING CAPS
26. BOLT, CAP ATTACHING
27. SEAL, CRANKSHAFT REAR
28. INTAKE VALVE
29. EXHAUST VALVE
30. WASHER, VALVE SPRING
31. SEAL, VALVE STEM OIL
32. VALVE SPRING
33. RETAINER, VALVE SPRING
34. KEYS, VALVE SPRING RETAINER
35. BOLT, CAP ATTACHING
36. CONNECTING ROD
37. CONNECTING ROD BEARING (UPPER)
38. PLUG, COOLANT DRAIN

FIGURE 4-1 Blueprinting an engine involves sizing and adjustment of all these engine parts and more. (Courtesy of Ford of Canada.)

A6668-B

44

3.0L ENGINE

	Standard
Type	60° V SOHC (Per Bank)
Number of Cylinders	6
Bore	91.1mm (3.59 Inch)
Stroke	76mm (2.99 Inch)
Compression Ratio	8.85:1
Displacement	3.0 Liters (181 Cu. In.)
Firing Order	1-2-3-4-5-6
Basic Ignition Timing	Refer to Emission Control Information on Label in vehicle
Valve Timing	
Intake—open	19° BTDC
—close	57° ABDC
Exhaust—open	57° BBDC
—close	19° ATDC

3.0L ENGINE SPECIFICATIONS

Description	Standard Dimension	Service Limit
Compression Pressure	178 psi @ 250 RPM	—
Maximum Variation Between Cylinders	1.0 Kg/cm^2 (14 PSI)	25%
Valve Clearance—Hot Engine	Hydraulic Lash Adjusters	—
Cylinder Head		
Flatness of Gasket Surface	0.05mm (.002 in.)	—
Grinding Limit of Gasket Surface	—	0.2mm (0.008 in)
Manifold—Flatness of Installing Surface		
Intake	0.10mm (0.004 in.)	0.2mm (0.0008 in.)
Exhaust	0.15mm (0.006 in.)	0.3mm (0.001 in.)
Valves		
Thickness of Valve Head (Margin)		
Intake	1.2mm (0.047 in.)	0.7mm (0.027 in.)
Exhaust	2.0mm (0.079 in.)	1.5mm (0.059 in.)
Valve stem to guide clearance		
Intake	0.03 to 0.06mm (0.001 to 0.002 in.)	0.10 (0.004 in.)
Exhaust	0.05 to 0.09mm (0.0019 to 0.003 in.)	0.15mm (0.006 in.)
Valve Face Angle	45° to 45°30′	—
Valve Overall Length		
Intake	103.0mm (4.055 in.)	—
Exhaust	102.7mm (4.043 in.)	—
Valve Stem Diameter		
Intake	7.960 to 7.975mm (0.313 to 0.314 in.)	—
Exhaust	7.930 to 7.950mm (0.312 to 0.3125 in.)	—
Valve Guide		
Overall Length		
Intake	44mm (1.732 in.)	—
Exhaust	48mm (1.889 in.)	—
O.D.	13.055 to 13.065mm (0.514 to 0.5143 in.)	—
I.D.	8.000 to 8.018mm (0.314 to 0.315 in.)	—
Valve Seat		
Seat Surface Angle	44° to 44°.3′	—
Contact Width	0.9 to 1.3mm (0.035 to 0.051 in.)	—
Sinkage	—	0.2mm (0.078 in.)

FIGURE 4-2 (Courtesy of Chrysler Canada Ltd.)

Description	Standard Dimension	Service Limit
Valve Spring		
Free Height	50.5mm (1.988 in.)	—
Loaded Height	40.4mm at 33 kg (1.59 in. at 73 lbs.)	—
Perpendicularty Intake and Exhaust	2° Maximum	4° Maximum
Piston		
O.D.	91.08 to 91.10mm (3.585 to 3.586 in.)	—
Piston to Cylinder Clearance	0.02 to 0.04mm (.0008 to .00015 in.)	—
Ring End Gap		
No. 1	0.30 to 0.45mm (.012 to .018 in.)	0.8mm
No. 2	0.25 to 0.40mm (.010 to .016 in.)	0.8mm
Oil	0.30 to 0.90mm (.012 to .035 in.)	1.0mm
Ring Side Clearance		
No. 1	0.05 to 0.09mm (.002 to .0035 in.)	0.1mm (.004 in.)
No. 2	0.02 to 0.06mm (.0008 to .002 in.)	0.1mm (.0039 in.)
O.S. (Service Pistons)	0.25-0.50-0.75-1.00mm	—
Connecting Rod		
Length—Center to Center	140.9 to 141.0mm (5.547 to 5.551 in.)	—
Parallelism—Twist	.05mm (.0019 in.)	—
Torsion—	0.1mm (.0039 in.)	—
Big End Thrust Clearance	0.10 to 0.25mm (.004 to .010 in.)	0.4mm (.016 in.)
Crankshaft		
End Play	0.05 to 0.25mm (.002 to .010 in.)	0.3mm (.012 in.)
Main Journal Diameter	59.980 to 60.000mm (2.361 to 2.362 in.)	—
Pin Diameter	49.980 to 50.000mm (1.968 to 1.969 in.)	—
Bearing Surface Out-of-Round	0.03mm Max. (.001 in.) Max.	—
Bearing Surface Taper	0.005mm Max. (.0002 in.) Max.	—
Bearing Oil Clearance	0.016 to 0.046mm (.0006 to .002 in.)	—
Cylinder Block		
I.D. (Bore)	91.10 to 91.13mm (3.586 to 3.587 in.)	—
Flatness of Top Surface	0.05mm (.002 in.)	0.1mm (.0039 in.)
Grinding Limit of Top Surface .... (.008 in.)*	0.2mm*	0.2mm*

*Includes/Combined with Cylinder Head Grinding

Oil Pump		
Relief Valve Opening Pressure	5.0 to 6.0 kg/cm² (71.45 to 85.75 psi)	—
Outer Rotor To Case	0.10 to 0.18mm (.004 to .007 in.)	0.3mm (.001 in.)
Rotor Side Clearance	0.04 to 0.09mm (.0015 to .0035 in.)	0.15mm (.006 in.)
Inner Rotor To Case	(.001 to .002 in.)	(.006 in.)

Figure 4-2 *(Continued)*

- Connecting rod and main bearing clearance
- Camshaft bearing clearances
- Camshaft timing
- Valve spring pressure
- Valve spring installed height
- Valve stem-to-guide clearance
- Combustion chamber volume

Engines coming off the production line are manufactured to specifications that are precise but range between minimum and maximum limits. If connecting rod length and piston crown height dimensions are stacked unfavorably, it can result in reducing the compression ratio in that cylinder by as much as half a number, from 9.0:1 down to 8.5:1, for example. This results from a connecting rod that is at the short end of the length tolerance and a piston with a crown height also at the low end of specs. Other combinations of tolerances stacked unfavorably have similar negative effects. Here are some guidelines that should help explain the blueprinting process. Measuring, machining, and reconditioning procedures are described in other chapters of this book.

PREPARING THE BLOCK

1. Be sure that the block is absolutely clean. Remove the oil gallery plugs and clean the oil passages thoroughly with a brush.

2. Check the lifter valley for the presence of casting slag. Use a small rotary grinding stone to smooth the rough edges. This prevents pieces of slag from breaking off and causing engine damage. Do the same in the entire crankcase interior.

3. Install the main bearing bore caps and torque to specifications. Check bore alignment and out-of-round. If necessary, grind the caps and align-bore to correct both out-of-round and misalignment. Out-of-round bores cause bearings to be out-of-round. Bore misalignment puts pressures on the crankshaft, reducing the oil film around the main bearing journal and causing unwanted friction. (See Chapters 3 and 7 for details).

4. Set the block up on a block resurfacing machine and check that cylinder bores are 90° to the crank centerline and that the deck surfaces are parallel to the crank centerline. Remove only enough deck material to correct any error, while maintaining equal deck-to-crank centerline height at all cylinders. (See Chapter 7 for detailed procedures).

5. If the block requires reboring, select a set of pistons sufficiently oversized to allow reboring to correct any cylinder damage. Mark each piston for a specific cylinder. Rebore and hone each cylinder to fit the designated piston with the specified piston-to-cylinder wall clearance.

6. Select the proper oversize set of piston rings to fit the rebored cylinders. Check the ring gap of each ring in its cylinder. Correct, if necessary, by using a ring gapping tool or file. Remove any resulting burrs with a hard whetstone. Burred ring ends can cause serious cylinder damage. (See Chapter 5).

7. Make sure that all connecting rods are absolutely clean. Install the caps (bearing removed), making sure that the caps are installed correctly (number on cap and yoke to same side) and torque the bolts or nuts to specifications while the rod is clamped in the appropriate holding tool. If a holding fixture is not used, torquing the nuts can easily twist the rod. Magnaflux to check for cracks. Replace cracked rods.

8. Measure the big-end bore of the rod for out-of-round or taper. Correct, if necessary, by grinding one or two thousandths off the cap mating surface. Reinstall the cap and resize the big-end bore on the appropriate machine. Perform the same procedure on all the rods. Always use new rod bolts and nuts. (See Chapters 6 and 9).

9. Measure the center-to-center length of each rod. The rod length may vary as much as 0.005 to 0.015 in. (0.127 to 0.381 mm). To ensure equal compression ratios in all cylinders, all rods should be of equal length. To correct rod length, the small end is bored and an unfinished bushing is installed. The bushing is then reamed or honed to provide the correct rod length.

10. Remove the forging lines on the connecting rods by using fine emery cloth by hand or by a belt sander for the purpose. Remove all lines, grooves, or marks that could cause cracks.

11. Shot peen the outer surfaces of the rod to further reduce the possibility of cracks and to slightly harden the surface metal. Shot peening involves bombarding the rods with tiny steel balls, called "shot," over the entire nonmachined rod surface.

12. If the rods and caps were ground, make sure the bearing locating notches are not reduced too much. If they are not deep enough, the bearings will not seat properly.

13. Make sure that all rods and caps are of exactly equal weight. If they are not, grind metal from the balance pads of the heavier rods until they are all equal in weight at both ends.

14. Using the new main bearings, check to make sure that the oil holes in the block line up with those in the bearings. To correct any misalignment, chamfer the hole in the block. Do not try to correct by changing the hole in the bearing.

15. Chase all the threaded holes in the block with the appropriate size and type of tap. Inspect all threads to make sure that they are in good condition. Chamfer threaded holes in the block deck. (see Chapter 7).

16. Check all lifter bores to make sure that they are not damaged. Light scoring can be removed with a small hone with extra-fine grit stones. Do not enlarge the lifter bores.

17. Check the crankshaft for proper indexing of crank throws. Throws on some crankshafts may be ahead or behind by as much as 0.020 in. (0.50 mm). Ignition will be late on a crank throw that is ahead and early on a crank throw that is behind its desired indexed position. If indexing is not out by more than 0.010 in. (2.5 mm), regrind all the throws to 0.010 in. (2.5 mm) undersized to correct the throw indexing. If the error is excessive, replace the crankshaft. During regrinding the journals may also be reduced in size from 0.001 to 0.003 in. (0.0254 to 0.076 mm) to provide the bearing clearance at the specifications desired. This improves lubrication. Cross-drill the crankshaft journals to provide lubrication every 180° of rotation instead of every 360°. Chamfer the holes to prevent bearings from being "shaved" by sharp edges. Have the journals "radiused" where the journal and crank cheek join. Stock crankshafts are sometimes not radiused. Providing a radius reduces the chance of stress cracks forming at this point. Have the crankshaft hard surfaced, such as by nitriding. This increases the wear resistance of the journals (See Chapter 8).

18. Paint the inside of the block crankcase area to seal any porosity and to deter the buildup of sludge and carbon.

PREPARING THE HEADS

1. Clean and inspect the cylinder heads. Check for any warpage, cracks, or damage. Resurface the head, if necessary. Deburr any sharp edges in the combustion chambers and chamfer the holes on the resurfaced side of the head. See Chapter 15.
Repair or replace the valve guides to provide the desired clearance. Make sure that valve seats have not been worn down too far. (See Chapter 15). Grind all valve seats to the angle and width specified. Use the three-angle method of seat reconditioning and face contact centering. (See Chapter 17).

2. Reface all valves to the angle specified. Check the installed valve stem height and correct as necessary to ensure equal height. Select and use valve springs with the tension designed for the type of camshaft and lifters being used and for the type of service the engine is expected to perform. Stronger springs are required for higher-speed engines to ensure that the valve train follows the cam profile. Stronger springs increase the load pressure on the valve train, camshaft, and valve seating. Special pushrods may be required to avoid bending. Test all the valve springs to ensure correct balanced spring pressures. Install the valves and springs, making sure that the spring installed height is as specified. Use positive-type valve stem seals. (See Chapter 17).

3. Check the volume of all combustion chambers after the valves are installed. Enlarge all chambers to equal the volume of the largest chamber, then reduce the volume of all chambers equally to the desired volume by carefully machining the head surface. On V engines, make sure that the intake manifold-to-head port alignment is also corrected.

4. Check to ensure that the required clearance is present between the rocker arms and valve retainers in both valve open and valve closed positions.

5. Retap and clean the spark plug holes.

Porting the Cylinder Heads

Cylinder head porting involves the removal of metal to enlarge and reshape the cross-sectional area of the ports to improve the flow of intake and exhaust gases. Porting is done by hand with an electric drill motor and suitable grinders. Porting benefits only those engines to be used for all-out performance. Porting does not benefit cars used for normal everyday driving. Many of today's engines are designed with special swirl ports as well as tuned intake and exhaust ports. Porting may well be harmful to these engines.

The degree of success achieved in porting is largely dependent on the skill and experience of the one doing the job. The procedure is time consuming and expensive. A porting service will usually carefully cut up a head first to determine where the thin spots are in the casting before working on the heads to be ported. Inexperience or lack of skill can quickly destroy a head by grinding through to the water jacket.

The porting procedure usually includes the following.

1. Enlarging the size of the ports to the size of the openings in the gasket.
2. Keeping the cross-sectional area as uniform as possible throughout the length of each port.
3. Removing any harmful bumps or restrictions that can cause the desired air flow to be disturbed. Some ports may have restrictions designed to improve the air flow and these should not be removed. An air flow meter is used to measure air flow and locate harmful restrictions.
4. Obtaining a surface finish that will provide the best possible results. This may require a slightly rough surface on the intake ports and a polished surface on the exhaust ports. Flow testing the heads with special equipment designed for the purpose determines the results achieved.
5. Matching the port openings in the heads and manifolds. Protrusions causing mismatching are carefully ground down. Specially designed templates are used to check port configuration from the guide area to the manifold gasket surface (Figure 4-3).

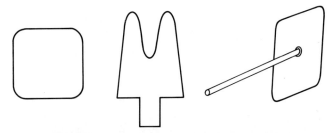

FIGURE 4-3 Typical porting templates. Handle aids in positioning template in port. Slotted template is used in valve guide area.

CALCULATING A SPECIFIC COMPRESSION RATIO

To obtain a specific compression ratio for a particular type of performance, the displacement, compression ratio, and compressed volumes must be calculated as follows.

Displacement

The displacement of an engine is the volume displaced by piston movement from BDC to TDC, multiplied by the number of engine cylinders. To calculate the displacement of an engine use either of the following formulas.

$$bore \times stroke \times 0.7854$$

or

$$bore \times 3.1416 \div 4$$

For example, a four-cylinder engine with a 3.750-in. bore and a stroke of 3.650 in. would have a displacement of 11.04 in^3, calculated as follows:

$$3.750 \times 3.750 \times 3.650 \times 0.7854 = 11.04 \text{ in}^3$$

or

$$3.750 \times 3.750 \times 3.1416 \div 4 = 11.04 \text{ in}^3$$

Compression Volume

When a specific compression ratio is desired, the required compression volume (combustion chamber volume) can be calculated from the bore and stroke dimensions as follows:

$$CV = DV \div (CR - 1)$$

For example, if the displacement volume is 45 in^3 and the desired compression ratio is 8.5:1, calculate the required compressed volume as follows:

$$45 \div (8.5 - 1) = 6 \text{ in}^3$$

Compression Ratio

The compression ratio of an engine is the comparison of the displacement volume of a cylinder to its compressed volume. The compression ratio of an engine may be calculated by using the following formula:

$$CR = (DV + CV) \div CV$$

where

$$CR = \text{compression ratio}$$

$$DV = \text{displacement volume}$$

$$CV = \text{compressed volume}$$

For example, an engine with a cylinder volume of 45 in^3 and a compressed volume of 6 in^3 has a compression ratio of 8.5:1 calculated as follows:

$$(45 + 6) \div 6 = 8.5$$

MEASURING COMBUSTION CHAMBER VOLUME

To check the volume of a combustion chamber requires a burette (mounted on a stand (graduated glass tube with a petcock as used by chemists) a rigid plexiglass plate with a small countersunk hole, and some Vaseline (Figure 4-4). With the spark plug and valves installed, the plexiglass plate is installed over the combustion chamber and sealed with the Vaseline. A clamp may be used to hold it in place. The burette is filled with test liquid (such as 10w motor oil, automatic transmission fluid, or cleaning solvent) to a precise level. The liquid is then carefully metered from the burette into the covered combustion chamber until it is full (Figure 4-5). The volume of liquid used is then easily determined from the level of fluid left in the burette. Since the test liquid creates a negative meniscus in the burette, to be accurate, the readings

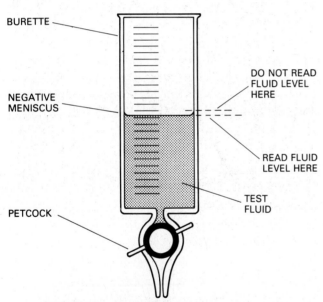

FIGURE 4-4 Burette is used to measure volume of combustion chamber.

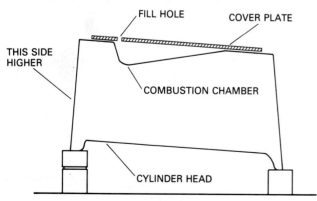

FIGURE 4-5 Setup required to measure combustion chamber volume.

should always be taken at the bottom of the meniscus. Test fluid must not be allowed to seep past the cover plate, spark plug, or valves. The volume of each chamber is carefully measured and compared to the others.

BALANCING COMBUSTION CHAMBER VOLUMES

It is important in performance engines that combustion chamber volumes be equalized. This aids in balancing output between engine cylinders and reduces vibration. Combustion chamber volumes can be balanced by removing metal from the smaller volume of the largest chamber. After balancing the volumes, the head can be milled to reduce chamber volume to the desired size and compression ratio. The increased compression ratio resulting from cylinder reboring must also be taken into consideration. The compression ratio will be raised by the same percentage as the percentage increase in displacement resulting from reboring.

Metal can be removed from the combustion chamber with a rotary file. An oval-shaped rotary file (see Figures 7-49 and 7-50 in Chapter 7) is the most useful for this. Before grinding the combustion chambers, determine where metal can safely be removed. First coat the head surface with machinist's dye and allow to dry. Use an old head gasket as a template and position it accurately on the head surface by inserting several head bolts to hold it in place. Do not use a new head gasket since it has not been compressed and would have larger cylinder bore holes. Scribe mark the head around the inside of each combustion chamber hole in the gasket. Metal should not be removed from the chamber outside this line since this would allow the head gasket to protrude into the combustion chamber. Keep a close check on the wall thickness wherever this could be a problem. Do not reduce wall thickness excessively. Only a very small amount of metal removal is normally required since modern engine production technology produces combustion chambers to very close tolerances. An alternate method may be used which involves "sinking" the valves by grinding the seats. This method, however, alters the valve stem height and the installed spring height which then must be corrected.

HIGH PERFORMANCE CAMSHAFTS

Camshafts are often described as being stock, street, or high performance. These designations indicate their intended usage. Stock camshafts are used in vehicles designed for normal everyday driving. A street cam can be used in a stock engine to increase engine performance at higher engine rpm while still maintaining a relatively smooth engine idle and reasonable low speed performance. This type of cam is considered to be "mild." A cam designed for use in competition on the track sacrifices low-end performance for improved top-end all-out performance at top engine rpm. This kind of cam is called "wild" and makes a low engine idle speed impossible.

In general high performance camshafts provide increased valve "lift" and "duration." The increased lift improves the flow of gases by increasing the size of the opening between the valve and the valve seat. This increased lift may require machining "eyebrows" in the top of the pistons to maintain the required valve to piston clearance. It may also require the use of special valve springs with wider coil spacing to prevent the spring coils from bottoming out against each other when the valve is open. Increasing the duration (valve open time in crankshaft degrees) improves engine breathing by taking advantage of the inertia of the intake gases but hampers engine breathing at low and intermediate speeds. A longer duration cam works well on the track but not so well on the street where you need some low-end power.

A longer duration cam also affects valve overlap. If valve overlap is above approximately 85° engine idle may be very rough and low speed performance very poor. An all out high performance cam could have as much as 120° of overlap which would make it impractical for use on the street.

Selecting a performance camshaft should be done by consulting the pros, the people experienced in the designing and testing of performance cams and engines. It is most important when selecting a performance cam that an entire camshaft kit be purchased. This includes the camshaft, camshaft drive, lifters, pushrods, rocker arms, valves, springs, spring retainers, valve keepers or split locks, spring dampers, and cam lubricant that are designed to work together to provide the desired performance while minimizing failure due to rapid wear or breakage. Changing only the camshaft could easily result in early failure of any of the other valve operating parts as well as the valves and seats.

DEGREEING IN A CAMSHAFT

"Degreeing in" a cam refers to the very precise timing of the camshaft-to-piston position in crankshaft degrees. This requires careful attention to detail and some special equipment. Although all engines have a TDC indicator for the number 1 cylinder, it may not be accurate. The true TDC position of the number 1 piston must therefore be checked if optimum performance is desired.

The TDC position can be checked using a degree wheel, a degree harmonic balancer, or a degree tape.

FIGURE 4-6 Typical degree wheel used to degree a cam. (Courtesy of Chrysler Canada Ltd.)

FIGURE 4-7 Close-up of degree tape. (Courtesy of Chrysler Canada Ltd.)

A degree wheel is a flat metal disc with accurate 1° divisions scribed around its outer edge. Zero, "0," on the degree wheel represents TDC (Figures 4-6 and 4-7). The degree wheel is mounted with a bolt that screws into the front of the crankshaft. It must be securely tightened to prevent even the slightest turning of the degree wheel in relation to the crankshaft. Turning the crankshaft must therefore be done from the flywheel and never from the front of the crankshaft while degreeing in a cam. On crankshafts that are not threaded at the front, the degree wheel is mounted by bolting it to the puller holes in the harmonic balancer.

Special harmonic balancers with precise degree markings are available that do not require the use of a degree wheel. Degree tape is also available and can be used to convert a harmonic balancer to a degree wheel. The balancer is first cleaned thoroughly, after which the tape is applied. The tape must be applied only after the TDC position has been established.

Locating Top Dead Center

There are two methods used to establish the top dead center (TDC) position accurately: the positive-stop method and the dial indicator method. With some variation in procedure, the positive-stop method can be used with the cylinder heads installed or with the heads removed. The dial indicator method is used with the heads removed. A positive-stop device that can be installed in the spark plug hole is used when the heads are in place (Figure 4-8). With the heads removed, a bolt-on positive-stop device is bolted in position over the number 1 piston.

FIGURE 4-8 Positive stop tool for establishing TDC position of piston through spark plug hole. (Courtesy of Chrysler Canada Ltd.)

Dial Indicator Method. The most accurate method is to use a dial indicator, as follows.

1. Mount the dial indicator on the engine block so that the stem contacts the center of the piston. This will eliminate any inaccuracy that could occur due to the piston rocking in the cylinder as it moves through the TDC position.
2. With the piston at or near TDC and the dial indicator stem in contact with the piston at all times, turn the crankshaft clockwise until the piston is .250 in. down from its uppermost position.
3. Mount the degree wheel to the crankshaft. Position the pointer (from the dial indicator set) on the block and adjust it to the zero or TDC position on the degree wheel.
4. Turn the crankshaft counterclockwise until the piston is down 0.250 in. in the cylinder.
5. Note the number of degrees the crankshaft has turned on the degree wheel. Turn the crankshaft clockwise exactly half the distance indicated on the degree wheel. That is the TDC position.
6. Repeat the procedure several times to ensure accuracy. Do not try to establish the TDC position using only the dial indicator and observing piston movement. This method is highly inaccurate since the crankshaft will turn quite a number of degrees before any piston movement is registered on the dial indicator as the crankshaft turns through the TDC area.

Positive-Stop Method. The following procedure can be used with the cylinder head removed or installed. The only difference is in the type of positive stop used. With the cylinder head in place, remove all the spark plugs and install the positive-stop device in the number 1 spark plug hole. With the head removed, the positive stop is bolted to the top of the number 1 cylinder.

1. Install the degree wheel on the crankshaft (or use a timing tape).

2. Position the number 1 piston near the bottom of the stroke.
3. Install the positive-stop device.
4. Turn the crankshaft clockwise and bring the piston carefully up against the positive stop.
5. Note the reading on the degree wheel that aligns with the pointer.
6. Turn the crankshaft counterclockwise and bring the piston carefully up to the positive stop.
7. Note the reading on the timing wheel at the pointer.
8. Divide the total number of degrees the crankshaft has turned from stop to stop and divide by 2 to establish the TDC position. This is the midpoint between the two positive-stop positions. Repeat the procedure several times to ensure accuracy.

Degreeing the Camshaft

Degreeing the camshaft is the precise and accurate timing of the opening and closing of the valves relative to the TDC and BDC positions of the piston. The following guidelines can be used for this procedure.

1. Calibrate the TDC position as outlined in the preceding section.
2. Install the lifters in the lifter bores for the number 1 cylinder. Use the lifters that are to be used with the assembled engine since using other lifters may alter the timing slightly.
3. Place a flat metal disc in each lifter to provide solid contact for the dial indicator. If these discs are not used, inaccuracy will result due to pushrod seat movement.
4. Turn the camshaft until maximum lift occurs on the intake cam. From this point turn the crankshaft 360° to position the lifter on the midpoint of the cam base circle.
5. Mount a dial indicator with a range of at least 0.500 in. plunger travel on the intake lifter. Be sure that the dial indicator plunger travel is in line with the lifter travel. An angled dial indicator mounting results in inaccuracy. Reposition the dial indicator to preload the plunger (plunger depressed at least 0.500 in.). Zero the indicator reading with the plunger in the preloaded position (Figure 4-9).
6. Turn the crankshaft clockwise through several complete revolutions of the camshaft and observe the dial indicator reading. It should return to zero each time the lifter reaches the midpoint on the base circle. If

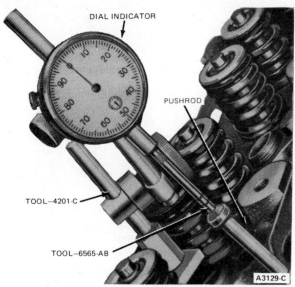

FIGURE 4-9 Dial indicator mounted in preparation for degreeing a cam. (Courtesy of Ford of Canada.)

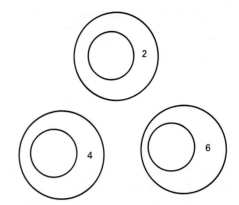

FIGURE 4-10 Offset bushings are used to advance or retard camshaft timing on camshaft drive sprockets located by a dowel pin.

FIGURE 4-11 Offset bushing in place in modified camshaft drive sprocket. This bushing is installed to retard cam timing. To advance cam timing with the same bushing install it 180° from this position.

this does not occur, readjust and/or tighten the dial indicator mounting. The lifters must move freely up and down in their bores to ensure accuracy.

7. To check the cam duration, turn the crankshaft clockwise and note the points on the degree wheel where each valve opens and closes. Never turn the crankshaft backward to recheck a reading because of clearances in engine components. If available, use the camshaft supplier's figures for checking the duration at specified checking heights since these will improve the accuracy of the checking procedure, particularly on slow initial lift cams. Repeat the checks several times to ensure accuracy while always turning the crankshaft clockwise. You should be able to obtain repeat readings within $\frac{1}{2}$°.

8. Compare the cam opening and closing points with the specifications supplied with the camshaft. If actual readings occur sooner than those called for in the specifications, the cam is advanced. If they occur later than called for, the cam is retarded.

9. Advance or retard the camshaft to achieve the cam timing specified.

CORRECTING CAMSHAFT TIMING

A number of devices are available to change camshaft timing. The method used varies with the type of camshaft drive of the engine. Among the more common devices are offset bushings, offset keys, and a crankshaft gear with multiple keyways.

On camshaft sprockets located with a dowel pin, an offset bushing can be used to change the valve timing. Offset bushings are available in 2, 4, 6, or 8° steps (Figures 4-10 and 4-11). The bushing can be installed in two ways; one way it will advance the timing, and the other way it will retard the timing. To utilize the offset bushing, the dowel hole in the cam sprocket must first be accurately drilled to the OD size of the bushing. If the sprocket mounting does not clamp the bushing in place, the sprocket should be staked around the bushing with a center punch.

On camshaft gears or sprockets located with a key, an offset key can be used to advance or retard

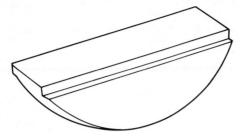

FIGURE 4-12 Offset key is used on key located cam or crank drive sprockets to advance or retard cam timing. (Courtesy of Chrysler Canada Ltd.)

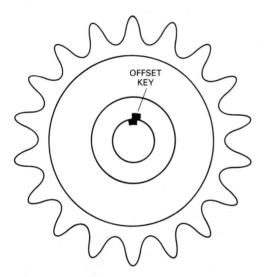

CAMSHAFT SPROCKET OR GEAR

FIGURE 4-13 Offset key in place in cam drive sprocket. This key position retards the cam on chain drives and advances the cam on gear drives.

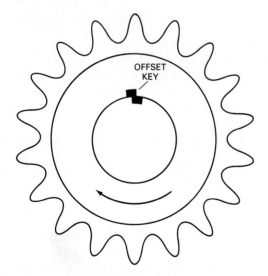

CRANKSHAFT SPROCKET OR GEAR

FIGURE 4-14 Offset key in place in crank drive sprocket. This key position retards cam timing in both gear and chain drive systems.

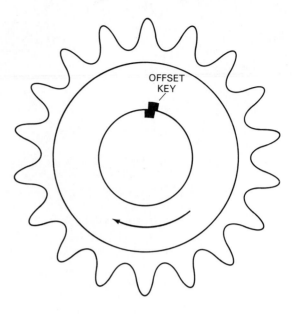

CRANKSHAFT SPROCKET OR GEAR

FIGURE 4-15 This offset key position on the crankshaft gear or sprocket advances cam timing on both gear and chain drives.

FIGURE 4-16 Special crankshaft sprocket provides for several cam timing positions. (Courtesy of Ford of Canada.)

camshaft timing (Figures 4-12 to 4-15). The alternative is to use an offset key in the crankshaft gear or sprocket. In one instance (Ford) a crankshaft sprocket is available with several keyways, which allow the sprocket to be advanced or retarded 2, 4, or 8° (Figure 4-16). More sophisticated variable and vernier cam drives are also available for some applications.

CAM CENTERLINE METHOD OF DETERMINING CORRECT CAM INSTALLATION POSITION

The cam centerline is the maximum lift position of the intake cam lobe in relation to the TDC position of the piston. The following procedure can be used to check camshaft timing by this method when camshaft specifications are stated in terms of the cam centerline.

1. Install the intake valve lifter for the number 1 cylinder. Install a degree wheel on the crankshaft. Position the piston at TDC and align the pointer to indicate TDC, as described earlier.

2. Align the camshaft timing marks with the camshaft installed.

3. Mount a dial indicator (with adequate range) and position the stem to contact the lifter top (use the flat disc on top of the hydraulic lifter). If the head is on, position the dial indicator to contact the top of the pushrod (an old pushrod ground flat will do the job). Make sure that the dial indicator is positioned to ensure that plunger travel is in line with lifter travel. An angled dial indicator position will cause inaccurate readings.

4. Turn the crankshaft clockwise until the maximum lift point has been reached. Never turn the crankshaft back to regain the position if it was turned too far. Turn the crankshaft clockwise again carefully until the maximum lift point is reached. Now zero the dial indicator.

5. Turn the crankshaft clockwise until the dial indicator reads 0.050 in. before the maximum lift point. Note the reading on the degree wheel. Turn the crankshaft carefully until the dial indicator reads 0.050 in. past the maximum lift point. Note the reading on the degree wheel. Add the two degree readings together and divide by 2 to determine the cam centerline position on the degree wheel. Repeat the procedure several times to ensure accuracy.

Example: Specifications call for a camshaft installation on a 110° centerline rotating clockwise and stopping at 0.050 in. of lift would result in a degree wheel reading of 70°. Turning the crankshaft further until 0.50 in. down from maximum lift would result in a degree wheel reading of 150°. Adding the two readings together yields 220° (70 + 150 = 220). Divide this figure by 2 and you get 110, which indicates the correct installation position. If the readings taken do not meet specifications, the camshaft timing must be advanced or retarded accordingly.

CHECKING PISTON-TO-VALVE CLEARANCE

Caution: Changing the camshaft timing alters the position of the valves in relation to the piston position during the valve overlap period. Valve and piston contact may occur as the result of valve timing changes. Advancing the camshaft timing will cause the intake valve to be open farther than normal as the piston approaches TDC on the exhaust stroke. Retarding camshaft timing results in later closing of the exhaust valve, which brings the piston closer to the valve on the exhaust stroke.

Piston-to-valve clearance should be checked during engine assembly if valve timing is changed from standard production specifications. This can be done as follows.

1. Lightly oil the top of the piston to prevent sticking of the gauging material.

2. Place a piece of strip caulking on the piston in the area of possible contact, or use a piece of modeling clay about $\frac{3}{4}$ in. thick and the size of a quarter.

3. Install the assembled cylinder head using a clean, used head gasket. Install and adjust the valve train components.

4. Carefully turn the crankshaft. If piston-to-valve contact causes turning resistance, do not force the crankshaft past this point, since this will bend the valves. If actual contact is not made, turn the crankshaft through two complete revolutions.

5. Remove the cylinder head.

6. Measure the gauging material with a pin-type depth gauge by pushing the pin into the depressed area of the material. Clearance ranges are about $\frac{3}{32}$ in. for the intake valve and $\frac{7}{64}$ in. for the exhaust valve. Follow the specifications for the specific engine being serviced. If clearance is insufficient, the pistons must be notched to provide clearance.

BALANCING VALVE SPRING PRESSURES

The objective in balancing spring pressures is to obtain equal static spring pressure for all springs, taking into consideration the difference in valve-open spring heights due to variations in cam lobe, lifter, pushrod, rocker arm, and valve stem tip dimensions. To balance valve springs, proceed as follows.

1. With the engine assembled, measure the total amount of spring deflection from the valve-closed to the valve-open position for each spring and record the results. Use a dial indicator mounted with the plunger contacting the top of the spring retainer.

2. Remove the valve springs using a spring compressor that does not require cylinder head removal. Valves must be retained to prevent

them from falling into the cylinder when springs are removed. Positively identify each spring as to its location on the head and keep them in that order.

3. Use a spring tester, test each spring at its valve-open height as recorded in step 1. Record the results for each spring.

4. Using the highest figure obtained during step 3, determine the shim thickness required to obtain the same spring tension for each of the other springs.

Example: Use the average figure of a 4-lb pressure increase per 0.015 in. of shim thickness required to achieve the pressure desired. For example, if the highest pressure obtained in step 1 was 180 lb and the spring being tested is only 172 lb, a 0.030-in. shim would be required to increase the pressure to 180 lb. The spring should be retested with the shims under the spring to ensure accuracy. Perform the same procedure on all the lower-pressure springs to achieve balanced spring pressures and assemble the springs and shims accordingly.

BALANCING THE ENGINE

An engine balancer is used to balance the rotating and reciprocating parts. Balancing should be done with all the rings, pistons, rods, and bearings and with the crankshaft, flywheel, vibration damper, and pulley. Bob weights identical in weight to the pistons, rings,

rods, and bearings are bolted to the crankpins (Figure 4-17). As the engine balancer rotates the assembly, it indicates where weight must be removed to achieve proper balance. This may require drilling the crankshaft counterweights or flywheel or grinding the balance pads on the pistons or rods.

ASSEMBLING THE ENGINE

Absolute cleanliness during assembly is always a necessity. All moving parts and parts subject to friction must be properly lubricated during assembly. Use only new connecting rod bolts and nuts. Use new cylinder head bolts. All bolts should be carefully tightened to the specified torque in steps and in proper sequence. Rotate the crankshaft at each step as the crankshaft and pistons are installed. Check all bearing clearances carefully with plastic gauge. If desired, a support "girdle" can be used to strengthen the lower end of the block, especially for the center mains. Use only good-quality gaskets and sealing materials to seal the engine and ensure against leakage.

After the heads are installed, make sure that there is no interference between the rocker arms and valve spring retainers and check to make sure that springs do not bottom out. Rocker arms may have to be relieved to provide clearance. Springs that bottom out must be replaced. After assembly and installation, be sure to prime the lubrication system of the engine before startup. This will ensure proper lubrication immediately upon starting and will prevent damage due to dry friction.

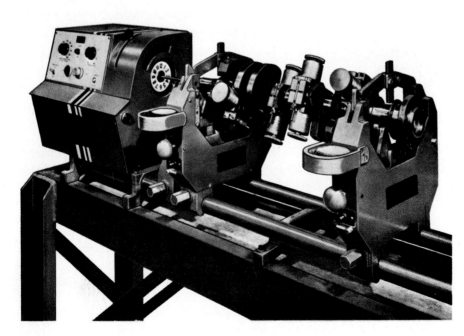

FIGURE 4-17 Engine balancer with crankshaft and bob weights in position. (Courtesy of Bear Service Equipment, Applied Power, Inc.)

REVIEW QUESTIONS

1. What is meant by "blueprinting" an engine?
2. List at least 10 engine components that are affected by blueprinting.
3. What is the purpose of removing casting slag from the inside of an engine block?
4. What negative effects result from main bearing bore misalignment and out-of-round?
5. What benefit is obtained by shot peening connecting rods?
6. Technician A says that if connecting rods are not equal in length, compression ratios between cylinders will vary. Technician B says that deck clearance will vary between cylinders if connecting rod lengths are not equal. Who is right?
   (a) technician A
   (b) technician B
   (c) both are right
   (d) both are wrong
7. The chance of cracks developing in connecting rods can be reduced by:
   (a) shot peening the rods
   (b) removing forging lines from the rods
   (c) avoiding nicks and scratches on the rod
   (d) all of the above
8. Why should all connecting rods in an engine be of equal weight?
9. How can connecting rod weights be balanced?
10. What effect does a crank throw that is indexed behind the others have on ignition timing for that cylinder?
11. What beneficial results are obtained by painting the inside of the block crankcase?
12. Engines that run at higher speeds should have valve springs that have _____ spring tension.

13. Technician A says that combustion chamber volume should be checked after the valves are installed. Technician B says that grinding the valves and seats has no effect on combustion chamber volume. Who is right?
   (a) technician A
   (b) technician B
   (c) both are right
   (d) both are wrong
14. What effect does decreasing the volume of a combustion chamber have on the compression ratio?
15. When reading the liquid level with a negative meniscus in a burette, take the reading at the (bottom/top) of the meniscus.
16. Balancing the volume of combustion chambers in an engine:
   (a) helps balance the compression ratio
   (b) helps balance the ignition timing requirements between cylinders
   (c) helps reduce vibration
   (d) all the above
17. After mounting a degree wheel on a crankshaft, the crankshaft should be turned only from the flywheel end while degreeing a cam. Why?
18. Name two methods used for accurate determination of the TDC position.
19. Name three different devices used to alter camshaft timing.
20. Define "cam centerline."
21. Advancing cam timing will cause the intake valve to be open farther than normal as the piston approaches TDC on the _____ stoke.
22. What method is used to balance valve spring pressures?

CHAPTER 5

Parts Failure Analysis

OBJECTIVES

To develop the ability to:

1. Analyze failed engine parts and determine the cause of failure.
2. Determine the correction required to avoid repeat failure.

INSPECTING PARTS FOR CAUSES OF EXCESSIVE OIL CONSUMPTION

The causes of excessive oil consumption can often be determined by careful inspection of engine parts during and after disassembly. Incomplete or inaccurate measurements may indicate that engine components are in good condition when in fact they are not. All engine components to be used over again must be measured completely and accurately to avoid the use of faulty parts. Incorrect installation of parts can also cause increased oil consumption and early engine failure. Checking parts for incorrect installation during and after engine disassembly is part of the inspection procedure. The following checklist can be used to aid in this process.

1. *Incorrect ring set*
   Check for oversized cylinders and use of standard rings. Try several rings in the cylinder; if the ring gap is excessive, the rings are the incorrect size (Figure 5-1).

FIGURE 5-1 Use of standard-size rings in oversized cylinder resulted in this wear pattern on the rings. Note lack of ring contact with cylinder near ring ends. (Courtesy of Hastings Manufacturing Company.)

2. *Rings not seated*
   Cylinders show areas where rings did not contact cylinder wall.
   a. Cylinders distorted from heat or improper torquing.
   b. Failure to deglaze cylinders properly.
3. *Rings installed incorrectly*
   a. Compression rings not installed according to instructions
   b. Rings not installed in proper groove
   c. Rings incorrect for groove width
   d. Rings installed upside down (Figure 5-2)
   e. Rings bent (Figure 5-3).

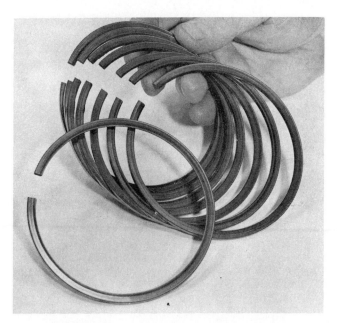

FIGURE 5-2 Compression rings installed right side up show a wear pattern on the bottom side. This results from the ring contacting the bottom of the groove on three of the four engine strokes. A ring installed up-side down shows a wear pattern on the wrong side of the ring as shown here. (Courtesy of McQuay Norris, Inc.)

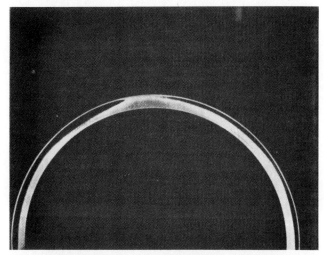

FIGURE 5-3 This wear pattern indicates that the ring was bent (twisted) during installation (Courtesy of Hastings Manufacturing Company.)

4. *Rings spinning in the groove*
   a. Sides of compression rings usually highly polished (Figure 5-4)
   b. Inside bore of scraper not showing contact points of inner spring crimps
   c. Excessive blowby
   d. Check for too much piston clearance
   e. Twisted or bent connecting rod
   f. Too much end play in the crankshaft

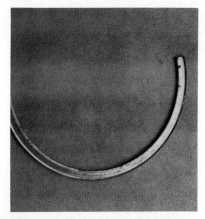

FIGURE 5-4 Highly polished sides of rings indicate ring spinning. (Courtesy of TRW, Inc., Automotive Aftermarket Division.)

FIGURE 5-5 Oil ring stuck in groove due to carbon buildup. (Courtesy of McQuay Norris, Inc.)

   g. Excessive vibration
   h. Cylinder walls highly polished, cylinders not deglazed
5. *Rings stuck in groove* (Figure 5-5)
   a. Improper side clearance
   b. Coolant seepage into cylinders
   c. Faulty cylinder head and block surface
   d. Faulty head gaskets
   e. Cracks
6. *Fractured or broken rings*
   a. Detonation, due to lugging, low-grade fuel, improper ignition timing
   b. Overheating (Figure 5-6)
   c. Careless installation when installing rings and piston in the cylinders, rings distorted
   d. Failure to remove cylinder ridge
7. *Side wear on rings—Top groove wear—Rings badly worn*
   a. Abrasive

FIGURE 5-6 Broken piston rings caused by overheating and detonation. (Courtesy of Hastings Manufacturing Company.)

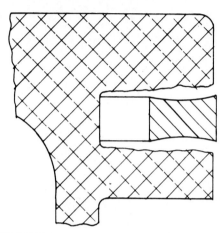

FIGURE 5-7 Worn ring groove allows ring to "hammer," resulting in excessive ring side wear and ring breakage. (Courtesy of TRW, Inc., Automotive Aftermarket Division.)

FIGURE 5-8 Scuffed piston rings caused by abrasives and overheating. (Courtesy of McQuay Norris, Inc.)

FIGURE 5-9 Broken ring lands due to pre-ignition. (Courtesy of McQuay Norris, Inc.)

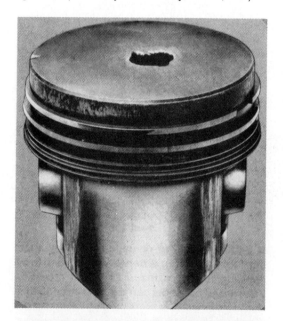

FIGURE 5-10 Piston damage due to detonation. (Courtesy of McQuay Norris, Inc.)

b. Fuel wash of cylinders—overly rich air-fuel ratio
c. Coolant seeping into cylinders
d. Detonation
e. Worn groove, allows ring to hammer (Figure 5-7)

8. *Rings scuffed* (Figure 5-8)
   a. Lack of lubrication

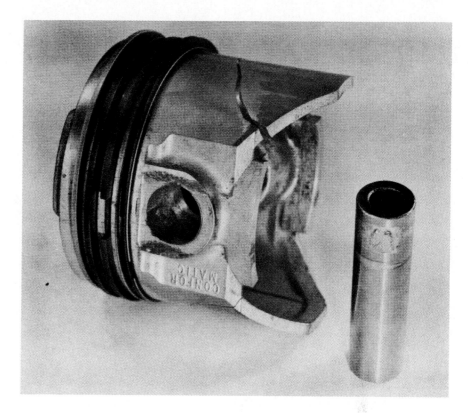

FIGURE 5-11 Piston failure caused by pin failure. (Courtesy of McQuay Norris, Inc.)

  b. Low oil pressure
  c. Idle speed too slow (especially during break-in)
  d. Engine overheated, check cooling system
  e. Check for coolant leaking into cylinders and destroying lubrication
  f. Failure to clean carbon from corners of ring grooves
  g. Distorted cylinders
  h. Improper torquing of cylinder head
 9. *Cracked or broken ring lands*
  a. Detonation
  b. Preignition (Figure 5-9)
  c. Failure to remove all the ridge before removing pistons
10. *Cracked pistons*
  a. Extreme pressure due to carbon deposit in combustion chamber
  b. Detonation (Figure 5-10)
11. *Tight piston pins*
  Affects the free action of the piston, resulting in rapid failure, and piston and cylinder damage (Figure 5-11).
12. *Cylinder taper*
  Check the cylinders at the top of ring travel Many engines will have severe taper and out-of-round in the top ½ in. (Figure 5-12).
13. *Piston clearance*
  Mike cylinders and pistons. Excessive piston clearance will shorten the life of the

FIGURE 5-12 Excessive cylinder taper causes oil consumption. (Courtesy of Sunnen Products Company.)

rings, due to the rocking action of the piston.
14. *Valves*
  Check under the head of the intake valves and the valve ports for oil or carbon deposits. Deposits indicate that oil is being drawn into the combustion chamber due to a faulty turbocharger or defective PCV system, or past the valve stems and guides.
15. *Crankshaft*
  Mike the crankshaft rod journals for size, out-of-round, and taper. When bearing clearance is increased from 0.0015 to 0.004

in., oil throw-off will increase sixfold (Figures 5-13 and 5-14).

16. *Bearing inserts*
Check the inserts for fractures, wear, and scoring.

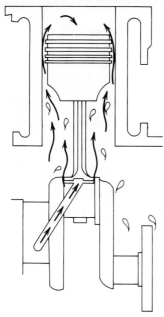

FIGURE 5-13 Excessive bearing clearance results in excessive oil throw-off and oil consumption. (Courtesy of McQuay Norris, Inc.)

FLYWHEEL PROBLEMS

The flywheel clutch friction surface must be flat and true and must be perpendicular to the crankshaft centerline to avoid face runout. The friction surface must be flat, free from heat discoloration, heat checks (small, very shallow surface cracks), cracks, or grooved wear. The ring gear must be tight on the flywheel and the teeth must be in good condition. The following flywheel problems may be encountered.

Ring Gear Damage

The ring gear teeth should be inspected for wear and chipping. Worn ring gear teeth cause drive pinion damage and poor starter engagement. The metal loss also affects flywheel balance (Figures 5-15 and 5-16).

Overheating

Overheating due to clutch slippage creates discoloration, hard spots, heat checks, and cracks on the friction surface of the flywheel. Discoloration, hard spots, and minor heat checks can be removed by resurfacing. In most cases hard spots are best removed by grinding since they are usually too hard for milling. Resurfacing

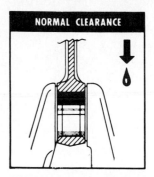

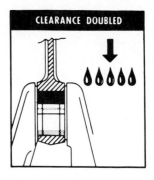

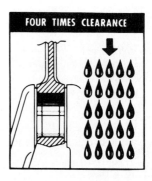

FIGURE 5-14 Oil throw-off increases drastically as clearance is doubled and quadrupled. (Courtesy of Hastings Manufacturing Company.)

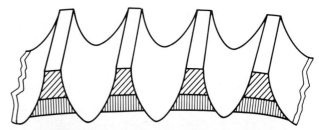

FIGURE 5-15 Damaged flywheel ring gear teeth require ring gear replacement. (Courtesy of FT Enterprises.)

FIGURE 5-16 Removing a ring gear from a flywheel. (Courtesy of Ford of Canada.)

TOOL—4201-C

FIGURE 5-17 Checking flywheel face runout. (Courtesy of Ford of Canada.)

must maintain the parallelism of the mounting flange and the friction surface. A badly cracked flywheel must be replaced.

Runout

Runout must be checked with a dial indicator while the flywheel is mounted on the crankshaft (Figure 5-17). While turning the flywheel, make sure that crankshaft end play is not a factor in the measuring procedure. Flywheel face runout should not exceed 0.0005 in. per inch of diameter. Maximum runout for a 10-in. clutch would be 0.005 in. Runout may be caused by a bent mounting flange on the flywheel or on the crankshaft, by foreign material between the mounting flanges, by raised metal on either of the mounting surfaces, or by improper resurfacing procedure. Flywheel resurfacing must be referenced from the mounting flange surface.

Dished surface

The friction surface of the flywheel may become dished, especially on lighter flywheels. Check with a straightedge and feeler gauge. A dished flywheel must be resurfaced or replaced.

Grooved Wear

Grooved wear results from hard spots on the clutch friction disc. These hard spots may develop as a result of contamination and embedded foreign material. Grooved wear can also result from a severely worn clutch disc. A grooved flywheel must be resurfaced or replaced.

VIBRATION DAMPER PROBLEMS

The vibration damper (harmonic balancer) hub must be inspected for grooved wear caused by the timing cover seal. Grooved wear can be repaired by installing a repair sleeve on the hub. This should only be done if the damper is otherwise in good condition (Figure

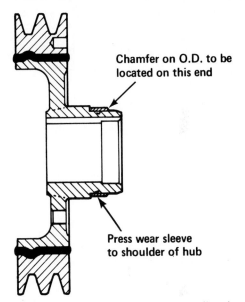

Chamfer on O.D. to be located on this end

Press wear sleeve to shoulder of hub

FIGURE 5-18 Grooved wear on vibration damper hub can be repaired by installing a repair sleeve. (Courtesy of Ford of Canada.)

FIGURE 5-19 Check vibration damper rubber condition for cracks, resilience, or missing pieces. (Courtesy of Chrysler Canada Ltd.)

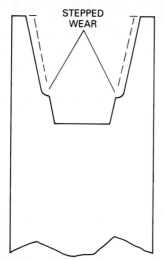

FIGURE 5-20 Pulley with stepped wear on sides of belt groove should be replaced. (Courtesy of FT Enterprises.)

FIGURE 5-21 Scored crankshaft journal due to lack of lubrication. (Courtesy of Red River Community College.)

FIGURE 5-22 Severely worn crankshaft journal caused by running engine for a long period after initial bearing failure. (Courtesy of Red River Community College.)

5-18). Check the resilience of the rubber ring by pressing a screwdriver tip against it. It should feel resilient. If it is badly cracked, is loose, or there are pieces missing, the damper must be replaced (Figure 5-19). If the pulley is part of the balancer, inspect the pulley groove for stepped wear on the sides. A badly worn pulley should be replaced (Figure 5-20).

CRANKSHAFT FAILURE

Crankshaft failure can result from wear, torsional stresses, and bending stresses.

Crankshaft Journal Wear

Main and connecting rod journal wear can be the result of normal loads and friction, inadequate lubrication, using the incorrect type or grade of engine oil, excessive oil temperatures, abrasives, detonation and preignition, or insufficient or excessive bearing clearances. Any of these factors will cause wear and increase bearing clearances, oil throw-off, and oil consumption (Figures 5-21 to 5-24).

Torsional Stress Failure

Torsional stresses on the crankshaft are a result of normal engine operation. Each crankpin is speeded up on the power stroke and slowed down on the compression stroke. This creates torque reversal stress on the crankshaft. The crankshaft must twist and untwist at a rate of 12,000 times per minute in a six-cylinder engine at 4000 rpm. In spite of this, crankshafts generally are able to withstand severe punishment over a long period of time before torsional fatigue failure occurs. Torsional stresses are reduced by the use of a vibration

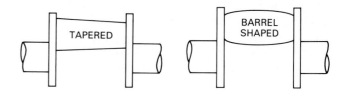

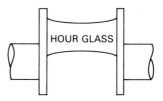

FIGURE 5-23 Crankshaft bearing journal wear can result in tapered barrel-shaped or hourglass-shaped journals. Bent connecting rods and tapered connecting rod big-end bore are causes. (Courtesy of FT Enterprises.)

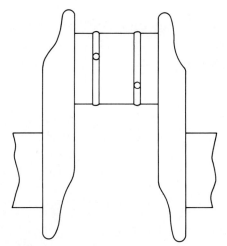

FIGURE 5-24 Journal ridging results from grooved bearing leaving part of the journal unworn. (Courtesy of FT Enterprises.)

FIGURE 5-25 Torsional fatigue failure of crankshaft. (Courtesy of Red River Community College.)

damper (harmonic balancer). Vibration dampers are designed to match the particular torsional characteristics of different engines.

Figure 5-25 shows a torsional fatigue failure between two crankpin journals across the drilled oil passage. The possibility of torsional fatigue failure is increased by the following conditions:

- Faulty vibration damper
- Using the wrong type of vibration damper
- Loose flywheel
- Using the wrong torque converter
- Unbalanced torque converter
- Unbalanced flywheel
- Unbalanced clutch assembly
- Unbalanced drive pulleys or accessories
- Drive belt whip
- Excessive slack in the camshaft drive

Bending Stress Failure

The crankshaft is subjected to bending stresses imposed on it by the power impulses of alternately firing cylinders under normal engine operation. The main bearing journals supported by the main bearings and caps support the crankshaft, while every cylinder that fires tries to force the crankshaft down. The alternately firing cylinders produce a cyclic bending stress pattern, which can result in eventual crankshaft bending stress fatigue failure (Figure 5-26). Conditions that increase the potential for bending stress failure include the following.

- Excessive bearing clearances
- Loose bearing caps
- Overheating of the crank journals
- Main bearing bore misalignment
- Bent crankshaft
- Incorrect fillet radius
- Bent connecting rod
- Improper bearing journal finish

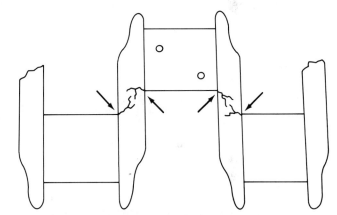

FIGURE 5-26 Cracks resulting from bending stress. (Courtesy of FT Enterprises.)

BEARING FAILURE

Engine bearings are designed to absorb tremendous loads over long periods of time in most engine operating conditions. When bearings fail, the cause can usually be attributed to any of the following, according to Federal-Mogul Corporation, manufacturers of engine bearings.

Dirt	43.4%
Insufficient lubrication	16.6%
Misassembly	12.2%
Misalignment	11.7%
Overloading	6.7%

| Corrosion | 4.0% |
| Other causes | 5.4% |

Dirt

"Dirt" found in engine bearings is defined as any foreign material circulating internally through the engine and the lubrication system that is not produced as a result of normal engine wear. Normal engine wear produces very fine particles which are easily suspended in the engine oil. Consequently, most of this very fine dirt is removed by the engine oil filter. However, if this dirt is allowed to accumulate in the engine, it will act as an abrasive, increasing wear of engine parts such as piston rings, cylinder walls, oil pumps, rocker arms, pushrods, lifters, and crankshafts. If not cleaned properly, material from cleaning and machining engine parts can add to the dirt problem. This includes grit from grinding and honing, sand and pellets from shot blasting, and iron and steel particles from machining. Figure 5-27 shows a bearing that has entrapped iron and steel particles. Entrapped dirt in bearings is shown in Figures 5-28 and 5-29. Dirt particles between the bearing back and the bearing bore also cause bearing damage, as shown in Figure 5-30.

FIGURE 5-28 Dirt embedded in bearing material. (Courtesy of Federal-Mogul Corporation.)

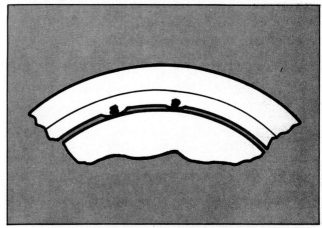

FIGURE 5-29 How dirt embeds in bearing lining material and raises metal surrounding dirt particle. (Courtesy of Federal-Mogul Corporation.)

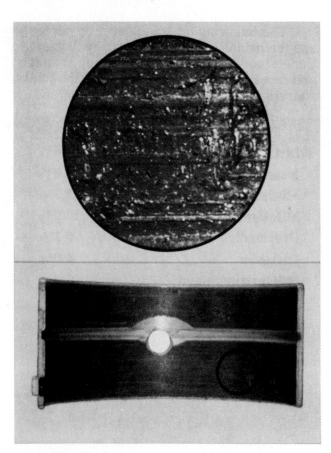

FIGURE 5-27 Iron and steel particles entrapped in bearing liner. (Courtesy of Federal-Mogul Corporation.)

FIGURE 5-30 Dirt trapped between bearing back and bearing bore caused this damage. (Courtesy of Federal-Mogul Corporation.)

Insufficient Lubrication

Insufficient lubrication may affect only one engine bearing or it may affect several or even all the engine bearings. If engine oil pressure is good and the right type of engine oil is being used, a bearing can still fail if the oil passage to the bearing is restricted (Figure 5-31). Poor oil pressure, a low oil level, or the use of inferior or incorrect engine oil will affect all engine bearings as well as other engine parts that rely on lubrication to reduce friction. Figure 5-32 shows a main bearing with the lining material wiped away due to insufficient lubrication. Other causes of insufficient lubrication include the following.

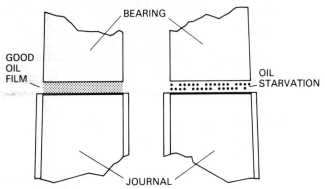

FIGURE 5-31 Difference between a good oil film and oil starvation. (Courtesy of FT Enterprises.)

FIGURE 5-32 Oil starvation bearing damage. (Courtesy of Federal-Mogul Corporation.)

Dry Start. A dry start results from starting an engine after an engine overhaul without first priming the engine's lubrication system. The light film of oil put on the bearings during assembly does not provide adequate lubrication during the time it takes the engine's oil pump to fill all the lubrication passages and the oil filter. It is during this time that severe damage to the bearings can occur (wiping or seizure) if the engine lubrication system is not primed prior to startup. Figure 5-33 shows bearing failure caused by dry start.

Insufficient Clearance. Too little clearance between the bearing and bearing journal can result in wiping or seizure of the bearing. Too little clearance does not allow a good oil film to be established and maintained between the bearing and bearing journal. Figure 5-34 shows bearing failure caused by insufficient clearance.

Oil Dilution. The lubricating quality of engine oil is reduced when it is diluted with gasoline. Excessive choking during engine starting and an overly rich air/fuel ratio will cause oil dilution. Figure 5-35 shows bearing damage caused by oil dilution.

FIGURE 5-33 Bearing damage caused by dry start. (Courtesy of Federal-Mogul Corporation.)

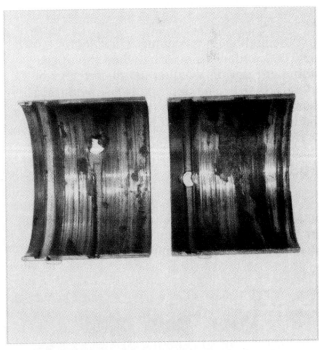

FIGURE 5-34 Insufficient oil clearance caused this bearing damage. (Courtesy of Federal-Mogul Corporation.)

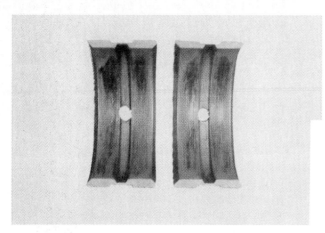

FIGURE 5-35 Bearing damage caused by oil dilution. (Courtesy of Federal-Mogul Corporation.)

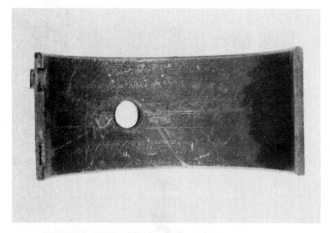

FIGURE 5-36 When a bearing cap is reversed the two halves are offset, resulting in bearing wear shown here. (Courtesy of Federal-Mogul Corporation.)

Misassembly and Misalignment

Bearings Reversed. When bearing halves are identical, either insert can be installed in the upper position. Where only one main bearing insert has an oil hole, it must be installed in the upper position. Installing the lower insert in the upper position blocks off the oil hole in the saddle bore, thereby preventing oil from reaching the bearing. Obviously, this will cause almost immediate bearing failure.

Bearing Caps Reversed or Mixed. When bearing caps are installed in the reverse position, the saddle bore or connecting rod bore halves are not in alignment. The upper half and the lower half of the bore will be slightly offset, causing extreme pressure in those areas. Figure 5-36 shows bearing damage caused by caps assembled in reverse position.

Bearing Insert Reversed. The bearing insert must be installed with the locating lug positioned in

the notch. If the insert is reversed, the locating lug and the immediate area next to it will be distorted as the cap is tightened. This will cause high unit loading in the area and complete bearing failure.

Overloading

Overloading results from lugging the engine, overspeeding preignition, detonation, and excessive bear-

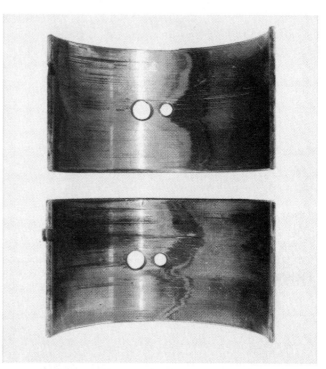

FIGURE 5-37 Pressure pattern caused by lugging. (Courtesy of Federal-Mogul Corporation.)

FIGURE 5-38 Bearing failure due to corrosion. (Courtesy of McQuay Norris, Inc.)

ing clearances. "Lugging" the engine is defined as the condition when increasing the throttle opening does not result in increasing the engine speed. Lugging causes fatigue failure in the upper half of the connecting rod bearings and in the lower half of the main bearings. Figure 5-37 shows bearing failure caused by lugging. Overspeeding, preignition, detonation, and excessive bearing clearance also result in bearing overload since the bearings are literally being hammered under those conditions. Bearing failure will result.

Corrosion

Following proper oil and filter change recommendations generally eliminates any possibility of bearing corrosion. Corrosion damage can result from infrequent oil changes, excessively hot temperatures, excessive blowby, and the use of low-quality engine oils (Figure 5-38).

Other Causes

A variety of other engine conditions, some of which are shown in Figures 5-39 to 5-41, can also cause bearing damage.

FIGURE 5-40 Incorrect crankshaft end play caused this thrust bearing damage. (Courtesy of Federal-Mogul Corporation.)

FIGURE 5-41 A bent connecting rod caused this wear pattern. (Courtesy of Federal-Mogul Corporation.)

PISTON FAILURE

Pistons are designed to withstand the force and stress of engine operation. They can withstand considerably greater-than-normal engine operating conditions without failure. However, there are limits that the piston cannot exceed without failure. Piston failure is usually the result of severe abnormal operation, excessive heat or load, inadequate lubrication, or mishandling. The following are examples of this type of failure.

Preignition Damage

Sharp edges, carbon deposits, or spark plugs with too high a heat range can cause the air–fuel mixture in the combustion chamber to ignite before the spark at the spark plug occurs. The resulting collision of two flame

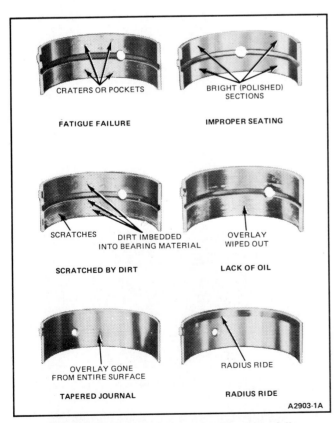

FIGURE 5-39 Other causes of bearing failure (Courtesy of Ford of Canada.)

FIGURE 5-42 Preignition damage above top ring. (Courtesy of Federal-Mogul Corporation.)

FIGURE 5-43 The hole in this piston was caused by detonation. (Courtesy of McQuay Norris, Inc.)

fronts creates tremendous heat and pressure increases. Continued engine operation under these conditions can cause damage as shown in Figures 5-42 and 5-43.

Detonation Damage

Detonation occurs after the spark plug has ignited the air–fuel mixture in the combustion chamber. Excessive engine temperatures and the expanding flame front of combustion can result in enough heat to cause the remaining air–fuel mixture in the cylinder to detonate or explode. The colliding flame fronts result in reverberations that can damage pistons as shown in Figure 5-43.

Lack of Lubrication

Improper lubrication of the cylinder walls and pistons results in piston scuffing as shown in Figure 5-44. Operating with insufficient engine oil, improper oil viscosity for prevailing ambient temperatures, plugged oil squirt hole in connecting rod, dirty oil, an excessively rich air–fuel mixture, or insufficient piston-to-cylinder wall clearance.

Pin Fit Too Tight

If the piston pin fits too tightly in the piston, scoring can result (Figure 5-45). The piston can seize on the pin and normal piston cam action is lost. The piston cannot expand or contract along the pin axis, causing expansion and contraction to occur across the thrust faces. This results in either too much expansion and excessive skirt-to-cylinder pressure or too much contraction and excessive piston-to-skirt clearance, depending on whether the piston seizes on the pin when

FIGURE 5-44 Piston scuffing and scoring caused by too little clearance or lack of lubrication. (Courtesy of TRW, Inc., Automotive Aftermarket Division.)

FIGURE 5-45 This cutaway shows the damage caused by a pin fit that was too tight. (Courtesy of Sunnen Products Company.)

hot or when suddenly cooled. The result can be scoring of the piston, piston slap, or both.

Misaligned Connecting Rod

A diagonal wear pattern will develop on the piston skirt by a bent connecting rod. Extreme misalignment can also cause the skirts to collapse (Figure 5-46) and even crack (Figure 5-47).

FIGURE 5-46 Wear pattern caused by bent connecting rod. (Courtesy of McQuay Norris, Inc.)

FIGURE 5-47 This piston cracked because of a bent connecting rod. (Courtesy of Red River Community College.)

Galled Piston Pin

Piston pin galling occurs most commonly while assembling the piston and connecting rod. If the piston pin bosses are not properly supported and aligned, the piston will be distorted, which results in the misalignment of the two pin holes. This results in excessive pin pres-

sure against one side of the pin bosses and consequent galling. Piston scuffing and cracking can occur. Only proper pin fit and assembly procedures can avoid pin galling.

Deck Height Too Low

When too much material is removed from the cylinder block deck or cylinder head and this has not been compensated for by a thicker head gasket, the piston can strike an overhanging part of the cylinder head. The piston becomes overstressed and cracks (Figure 5-48).

Broken Valve

When a valve head breaks off or the stem breaks, allowing the valve to slide into the combustion chamber, extensive piston damage occurs (Figures 5-49 and 5-50).

FIGURE 5-48 Too low a deck height caused this piston to crack from striking the head. (Courtesy of Federal-Mogul Corporation.)

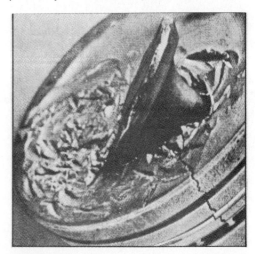

FIGURE 5-49 Broken valve embedded in piston. (Courtesy of Federal-Mogul Corporation.)

FIGURE 5-50 A broken valve caused this piston destruction. (Courtesy of Red River Community College.)

FIGURE 5-51 Piston pin lock ring failure. (Courtesy of Sunnen Products Company.)

FIGURE 5-52 Cylinder damage caused by piston pin lock ring failure. (Courtesy of Ford of Canada.)

Coolant in Cylinder

If engine coolant leaks into a cylinder in sufficient quantity, the piston head can crack and force an opening through the piston for coolant to escape into the crankcase. Antifreeze in the engine oil can cause serious bearing damage and engine seizure.

Loss of Piston Pin Lock Ring

A faulty or improperly installed piston pin retaining ring can come loose and become lodged between the piston and the cylinder wall. Extensive piston and cylinder wall damage can result (Figures 5-51 and 5-52). Only new retaining rings should be used in pistons that have ring grooves in good condition. Rings must not be distorted during installation.

PISTON RING FAILURE

Piston ring failures may be caused by (1) abrasives in the engine, (2) excessive engine temperatures, (3) detonation, (4) improper ring installation, (5) excessive ring rotation, and (6) blasting bead damage.

Scored Piston Rings

Scored piston rings result in the loss of compression and power, increased blowby and oil consumption, increased engine temperature, and hard starting when hot. Dust and dirt entering the engine cylinders and lubrication system increase friction and cause abrasive action to occur between the piston rings and cylinder walls. Rapid ring and cylinder wear are the result. Piston rings show rough vertical scratches on the ring face similar to those shown in Figure 5-53. Cylinder wall surfaces will show similar vertical wear patterns.

FIGURE 5-53 Scored piston rings. (Courtesy of TRW, Inc., Automotive Aftermarket Division.)

Scuffed Piston Rings

Piston rings become scuffed (light vertical scratches on the face of the ring) due to the breakdown of the oil film between the rings and cylinder wall. The increased friction generates excessive heat, momentarily causing the ring to weld itself to the cylinder wall at the TDC position. As the piston moves down the weld is broken, causing scuffing and scoring the face of the ring (see Figure 5-8).

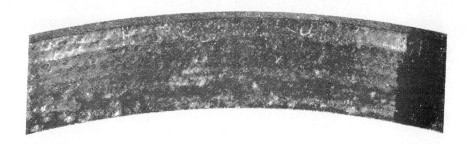

FIGURE 5-54 Residue from bead blasting left in engine caused these "peen" marks on side of ring. (Courtesy of Hastings Manufacturing Company.)

Detonation Damage

The high-pressure shock waves and reverberations of detonation can cause serious ring damage, break ring lands, and piston rings (see Figure 5-10).

Improper Installation

Improper ring installation includes installing compression rings upside down, spiraling a compression ring onto the piston, and using a ring size smaller in diameter than the cylinder bore size. A torsional twist or taper-faced compression ring installed upside down will scrape the cylinder wall going up instead of down. This pumps oil into the combustion chamber. Spiraling a compression ring into place bends the ring, resulting in poor sealing on the sides of the ring. Figure 5-3 shows the wear pattern resulting from a bent ring. Using a ring size smaller than the cylinder bore size results in poor ring-to-cylinder contact and causes a wear pattern as shown in Figure 5-1. Note the dark unworn part of the rings, which have not sealed against the cylinder wall.

Excessive Ring Rotation

Piston rings that rotate excessively on the piston cause ring and groove wear as shown in Figure 5-4. Excessive rotation is caused by too rough a cylinder bore finish, a unidirectional honing pattern, or a bent connecting rod.

Blasting Bead Damage

Bead blasting material left in the piston ring grooves after cleaning will result in peen marks on the sides of the piston rings (Figure 5-54). Many ring and piston manufacturers do not recommend cleaning pistons with bead blasting equipment for this reason.

CONNECTING ROD FAILURE

Connecting rods are well designed and built to withstand heavy loads over long periods. Despite this, failure can occur. There are several reasons for connecting rod failure.

Improper Installation Procedure

This includes failure to tighten rod bolts or nuts to specific torque (overtightening or undertightening), using faulty bolts or nuts and reversing the rod offset. Overtightening can cause threads to shear partially or cause a bolt to overstretch and break at the thread base. Undertightening can cause rod bolts to stretch and break or allow the nuts to loosen, resulting in severe impact loading, hinge stressing of the cap, and breakage. The threads in connecting rod bolts are heavily stressed the first time they are tightened. This deforms the threads and if used again will result in increased friction during tightening. This affects the torque on the nut and the tension on the bolt. Tightening a used nut to specified torque results in less-than-required bolt tension and eventual loosening and breakage. Always use new rod bolt nuts of good quality to avoid this problem. Reversing an offset rod causes rod-to-crank cheek contact as well as rod-to-pin boss contact with resulting rapid wear (Figure 5-55).

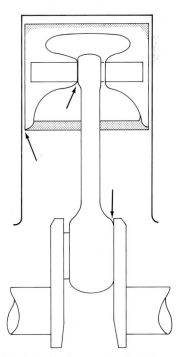

FIGURE 5-55 Offset rod installed in reverse causes contact and rapid wear at points indicated. (Courtesy of FT Enterprises.)

Big-End Bore Out-of-Round

If big-end bore out-of-round is not corrected during the rebuilding process, the new bearings will assume the shape of the bore. High engine speeds and heavy loading create increased stress and high-impact loading, which can result in rod breakage.

Pin Fit Too Tight in Piston

A pin fit that is too tight in the piston can result in pin seizure and rod breakage on rods where the pin is a press-fit in the rod.

Excessive Bearing Clearance

This results in reduced oil pressure at the bearing and eventual bearing knock. Continued knocking stretches the bearing cap and breakage at the cap results (Figure 5-56).

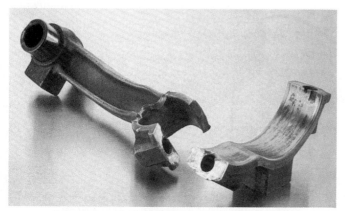

FIGURE 5-56 Rod breakage caused by excessive bearing clearance. (Courtesy of TRW, Inc., Automotive Aftermarket Division.)

Mishandling

Connecting rods that are nicked, scratched, or clamped in a vise (without using soft jaws) are weakened at that point. Under high speeds or heavy loads, breakage can result. Always handle rods carefully to avoid possible damage. When clamping rods in a vise, protect the rod by using soft jaws, or use the special clamp designed for holding connecting rods. When loosening or tightening connecting rod bolts, clamp the rod across the yoke. Clamping the shank of the rod while loosening or tightening bolts can cause the rod to twist.

Bent Rod

Bent connecting rods result in bearing wear (Figure 5-41), piston wear (Figure 5-46), and barrel-shaped journal wear.

PISTON PIN FAILURE

The two most common causes of pin failure and the subsequent damage are pin seizure in the piston and piston pin lock ring failure.

Piston Pin Seizure

After a sustained period of heavy engine loading with very light loading immediately thereafter, a condition may develop that causes the pin to seize in the piston. When this occurs on pins that are a press-fit in the connecting rod, oscillation is no longer possible and the piston breaks at the pin bosses. This happens because engine parts are heated to a very high temperature during heavy loading. When under a period of light load immediately thereafter, the piston pin bosses cool more rapidly than the piston pin. The cooling shrinks the pin holes in the piston, which results in seizure. Full-floating piston pins can seize in the rod bushing or piston if fitted too tightly or if the bushing was not properly expanded after replacement (Figure 5-57).

FIGURE 5-57 Full-floating piston pin seized in piston caused this damage. (Courtesy of Sunnen Products Company.)

Piston Pin Lock Ring Failure

When a piston pin lock ring comes out of its groove, the ring and piston pin can cause severe cylinder wall damage (Figure 5-52). Figure 5-58 shows piston lock ring groove damage. Improper lock ring installation, the use of used lock rings, or a damaged lock ring groove can cause lock ring failure. Lock rings must not be distorted by twisting or overcompressing during installation since tension will be lost and the ring may come out of the groove. Use only new lock rings and install them with the gap down. The heaviest loading during engine operation occurs when the piston passes

FIGURE 5-58 Cutaway of piston, showing damaged lock ring groove at left side. (Courtesy of Sunnen Products Company.)

TDC and when the cylinder fires. Tanged lock rings must be installed with the gap down and the tang pointing away from the pin.

VALVE FAILURE

Exhaust Valve Guttering

Figure 5-59 shows severe valve damage caused by pre-ignition. Extremely high temperatures and pressures resulting from preignition literally blew away the valve

material with a cutting-torch effect. The metal erosion starts with a slight leakage past the valve. As the hot gases blow past the valve during combustion, the exposed edges become molten and are blown away. Causes of preignition include:

- Lean fuel mixture
- Excessive engine operating temperatures
- Insufficient valve margin
- Spark plug heat range too high
- Spark timing too far advanced
- Fuel octane too low

The only correction for this type of failure is valve replacement and a complete valve guide and valve seat reconditioning job.

Exhaust Valve Leakage Failure

This type of valve failure is similar to that caused by preignition. The valve in Figure 5-60 shows failure resulting from exhaust valve leakage. Valve leakage can be caused by a small deposit on the valve face or seat, insufficient valve lash, excessive stem-to-guide clearance, a warped valve head, or a distorted valve spring. Hot combustion gases under extreme pressure are forced out past the leakage area. This overheats the area, causing guttering and cracking. This type of failure requires replacing the valve, reconditioning the guides and seats, and possibly replacing the springs.

FIGURE 5-59 Severe valve damage caused by high temperatures resulting from preignition. (Courtesy of TRW, Inc., Automotive Aftermarket Division.)

FIGURE 5-60 This valve failure started with minor exhaust leakage past the seated valve. (Courtesy of McQuay Norris, Inc.)

Intake Valve Tulipped (Cupped)

A tulipped intake valve is shown in Figure 5-61. The head of the valve (Figure 5-62) is extensively distorted and has become cupped. Valve cupping causes reduced valve lash, reduced valve spring closing pressures, and a possible high engine rpm miss. The excessively high temperatures of preignition overheat the intake valve, causing stretching in the valve head.

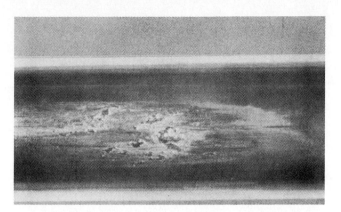

FIGURE 5-63 Scuffed valve stem caused by lack of lubrication or insufficient clearance. (Courtesy of McQuay Norris, Inc.)

- Insufficient lubrication
- Misalignment between valve stem and rocker arm
- Too little clearance between valve stem and guide
- Too much clearance between valve stem and guide
- Lack of prelubrication of valve stem during assembly

Make sure that sufficient lubrication reaches the valve stems by ensuring that hollow oil feed pushrods or rocker arm feed holes are not clogged. When doing a valve job, make sure that the valve stem tip and the rocker arm stem contact area are flat and parallel. Stem-to-guide clearance must not be excessive, to ensure valve alignment with the guide and seat. To prevent sticking from thermal expansion, make sure that clearance is not less than the minimum specified. Always prelube valve stems during assembly. If assembled dry, the stems will scuff before any oil gets there for lubrication.

FIGURE 5-61 Tulipped intake valve. (Courtesy of McQuay Norris, Inc.)

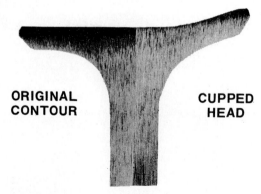

ORIGINAL CONTOUR   CUPPED HEAD

FIGURE 5-62 New valve contour (left) and cupped valve (right.) (Courtesy of TRW, Inc., Automotive Aftermarket Division.)

Scuffed Valve Stems

When valve stem scuffing has progressed far enough, the valves tend to stick in the guides, resulting in a missing condition (Figure 5-63), and it may cause guide wear or damage. Valve stems can become scuffed due to:

Valve Necking

Valve necking (erosion of metal below the valve head) is caused by poor valve seating. This may result from a distorted valve, distorted seat, misaligned valve guide, or excessive stem-to-guide clearance. The valve shown in Figure 5-64 is severely necked and would soon break, allowing the head to be pounded by the piston, with resulting piston and combustion chamber damage. Hot combustion gases blowing past the poorly seated valve have eroded the metal from the valve.

Valve Face Deposits

Figure 5-65 shows a valve burned, pitted, and eroded due to face deposits being repeatedly built up on the valve. The valve does not seat properly because of the

FIGURE 5-64 Severely necked valve. (Courtesy of TRW, Inc., Automotive Aftermarket Division.)

FIGURE 5-65 Burned and eroded valve caused by face deposits. (Courtesy of TRW, Inc., Automotive Aftermarket Division.)

FIGURE 5-66 Cracked valve caused by heat stress. (Courtesy of McQuay Norris, Inc.)

face deposits when it is closed. Hot combustion gases literally burn away the metal.

Heat-Stressed Valves

Heat stress results in thermal fatigue. Excessive heat may cause a valve to crack before it is eroded. Heating of the valve beyond its design rating is the cause. As the valve cools more rapidly around its edges than it does in the center, metal contraction causes the valve to crack (Figure 5-66).

Worn Valve Guides

The valve in Figure 5-67 shows the results of a worn valve guide. Carbon buildup projecting into the guide causes abrasive wear on the valve stem. Cocking of the valve in the guide prevents proper seating of the valve, with valve face burning being the result.

FIGURE 5-67 Effects of worn valve guide on valve stem. (Courtesy of McQuay Norris, Inc.)

Excessive Dynamic Forces

A mechanical problem in the valve train allows the valve to be pounded open and closed. Usually beginning with valve train noise, complete valve failure and breakage can result, as shown in Figure 5-68. Excessive valve lash (clearance between the rocker arm and valve stem tip) reduces the effective range of the opening and closing ramps on the camshaft lobes, causing the valves to be slammed open and closed. Weak valve springs do not keep the valve train in contact with the cam profile at higher engine speeds. Worn cam lobes can destroy the opening and closing ramp action from its original design. Excessive engine speeds may result in the inability of the valve train to follow the cam

FIGURE 5-68 This valve failure began at the keeper groove. (Courtesy of McQuay Norris, Inc.)

FIGURE 5-69 Scuffed keeper groove can lead to complete valve failure. (Courtesy of McQuay Norris, Inc.)

FIGURE 5-70 Fatigue failure of valve stem with magnified view of break. (Courtesy of TRW, Inc., Automotive Aftermarket Division.)

FIGURE 5-71 Impact failure of valve stem with magnified view of break. (Courtesy of TRW, Inc., Automotive Aftermarket Division.)

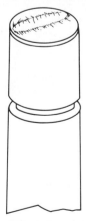

FIGURE 5-72 Off-center wear pattern on valve stem tip indicates rocker arm misalignment. (Courtesy of FT Enterprises.)

profile. The opening and closing ramps of the cam are designed to lift and close the valves as gently as possible.

The valve in Figure 5-68 first broke at the keeper grooves, allowing the valve to drop into the combustion chamber, where it broke further when pounded by the piston. Piston and combustion chamber damage may also occur. Figure 5-69 shows a scuffed keeper groove, which could lead to eventual valve breakage, caused by excessive dynamic forces. Figures 5-70 and 5-71 show fatigue failure and impact failure.

Rocker Arm Misalignment

A misaligned rocker arm causes an off-center wear pattern on the valve stem tip (Figure 5-72). Rocker arms may be misaligned due to stud misalignment, fulcrum wear, or improper assembly. Rocker arm misalignment can also cause rocker-to-stud contact and cause rocker arm stud wear.

HYDRAULIC LIFTER FAILURE

Lifter Noise

Lifter noise is not necessarily an indication of lifter failure. There are several reasons why lifters can become noisy. Some of the more common causes are:

- Engine oil level too low
- Engine oil pressure too low
- Use of incorrect viscosity engine oil
- Excessive lifter leakdown
- Seized lifter plunger

Operating the engine for an extended period with any of these conditions present will cause lifter wear and scoring as well as other engine problems. Other valve train noises which are very similar to lifter noises include the following:

- Broken valve spring
- Badly worn cam lobes
- Incorrect valve adjustment
- Worn valve stem tips and rocker arm pads
- Excessive clearance between lifter and lifter bore
- Excessive valve stem to guide clearance
- Sticking valves
- Loose valve seat insert
- Loose rocker arm shaft mounting
- Insufficient lubrication between rocker arm and pushrod
- Insufficient lubrication between rocker arm pad and valve stem tip
- Badly worn lifter base

Isolating the Noise

This can be done with a stethoscope. Place the sensing tip alternately on each end of the rocker arm with the engine running at idle. This will indicate whether the noise is at the valve end or at the pushrod end. This can also be done by using a piece of heater hose (or garden hose) about 3 ft long. Place one end of the hose to your ear and the other end alternately to each end of each rocker arm.

A feeler gauge may be used to check whether excessive valve lash is causing the noise. Place a 0.010- to 0.015-in. feeler gauge between the rocker arm and valve stem tip. If this eliminates the noise, it indicates excessive lash or excessive lifter leakage, or a seized lifter plunger. A preliminary check of excessive lifter leakdown can be made as follows. With the engine stopped, push down firmly on the pushrod end of each rocker arm and compare the time it takes to bleed down each lifter. Any lifter that "bottoms" easily has excessive leakdown.

Seized Lifter Plunger

A lifter plunger can seize due to excessive varnish buildup between the plunger and the lifter body. Any foreign matter such as carbon particles or metal chips lodging between the plunger and lifter body can also cause the plunger to seize.

Excessive Lifter Leakdown

Excessive lifter leakdown will cause valve train noise when the engine is warm. Excessive leakdown is caused by excessive clearance between the lifter plunger and the lifter body or by a leaking check valve. Excessive plunger-to-body clearance is the result of wear. Foreign matter such as carbon or varnish can cause both the disc-type and the ball-type check valve to fail to seal properly. The check valve must seal in order to trap the oil below the plunger.

Lifter and Cam Lobe Wear

Wear on the lifter base and the cam lobe is normal. Tremendous pressure is brought to bear between the lifter base and the cam lobe. This pressure can reach 100,000 psi. Initially, the interface between the lifter base and cam lobe wear together until the mating surfaces are matched, and then the wear virtually stops as long as good lubrication and normal operating conditions exist. Wear is accelerated by poor lubrication, abrasives, and extreme loads resulting from abnormal engine operation. The cam lobe becomes rounded and the lifter base becomes dished (concave; Figures 5-73 and 5-74). Originally, the lifter base is slightly convex

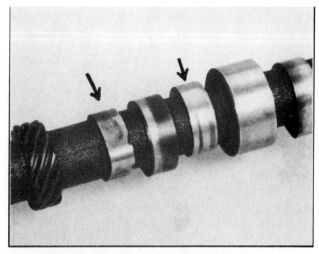

FIGURE 5-73 Worn cam lobes are rounded across the nose. (Courtesy of McQuay Norris, Inc.)

FIGURE 5-74 Badly worn cam lobes and lifters. Lifter base is worn right through on lifter second from left. (Courtesy of TRW, Inc., Automotive Aftermarket Division.)

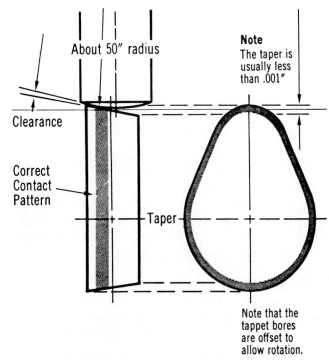

FIGURE 5-75 Convex lifter base and tapered cam nose on new lifter and cam (exaggerated). (Courtesy of Sealed Power Corporation.)

FIGURE 5-76 Channeled wear pattern on lifter base is caused by lack of lifter rotation. (Courtesy of TRW, Inc., Automotive Aftermarket Division.)

FIGURE 5-77 Wear on OD of lifters caused by using poor-quality engine oil. (Courtesy of McQuay Norris, Inc.)

The only way to achieve proper lifter and cam action is to replace the camshaft and the lifters at the same time. A good high-pressure prelubricant must also be used on the cam lobes during assembly to prevent scoring of cam lobes and lifters after initial engine startup. Figure 5-77 shows wear on the lifter OD caused by use of poor-quality engine oil.

CAMSHAFT DRIVE MECHANISM FAILURE

Accurate valve timing is critical to efficient engine operation. The valves must open and close at the appropriate time in relation to piston position. Excessive wear of camshaft drive mechanisms results in the disruption of this valve action/piston position relationship. Incorrect assembly of camshaft drive gears, chains and sprockets, or timing belts and sprockets also disrupts this relationship, with a resultant lack of performance and fuel efficiency.

Timing Gear Failure

Timing gear teeth wear excessively due to lack of lubrication, abrasives, and excessive loads (Figure 5-78). Incorrect valve train clearance adjustment can result in severely shock-loading the gear teeth. The constant torque load reversal on the camshaft caused by valve

and the cam nose is slightly tapered (Figure 5-75). This ensures even wear distribution on the lifter base, cam lobe, and between the lifter and lifter bore. A channeled wear pattern develops on the lifter base if the lifter does not turn in its bore (Figure 5-76).

Using new lifters on a worn camshaft results in a mismatch of mating surfaces between the lifter base and cam lobe. The result is poor or no lifter rotation and rapid wear in the center of the lifter base from the rounded nose of the worn cam. Using old lifters with a new cam causes similar problems. Poor lifter rotation, edge loading of the lifters and cam lobe, cam lobe chipping, and rapid wear are the result.

opening and closing action also contributes to gear tooth wear. As the gear teeth wear, backlash increases further, increasing the shock loads and causing eventual gear tooth breakage (Figure 5-79). Insufficient backlash between gear teeth results in increased friction, reduced lubrication, and rapid wear. Insufficient backlash results from mismatched timing gears. Excessive gear runout can cause rapid wear due to increased friction and insufficient backlash.

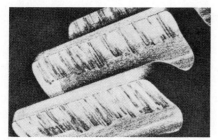

FIGURE 5-78 Worn timing gear teeth. (Courtesy of McQuay Norris, Inc.)

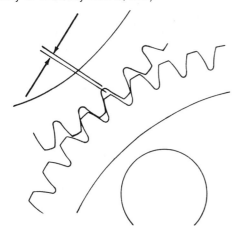

FIGURE 5-79 Timing gear backlash. (Courtesy of Ford of Canada.)

Timing Chain and Sprocket Failure

The reasons for timing chain and sprocket wear are the same as for timing gear wear. The timing chain has a great many timing pins, each of which has its own bearing surfaces. All of these pins and bearings are subject to camshaft drive loads, which eventually result in increased slack in the chain. Timing chains can wear to the point where they can jump a cog on the sprocket, severely affecting valve timing. Timing chain slack can also cause the chain to contact and wear through the timing chain cover (Figure 5-80). As timing chains wear, so do the sprocket teeth (Figure 5-81).

To check for excessive timing chain wear, turn the cam sprocket to get all the slack on one side of the chain. If the slack side of the chain on an engine with an in-block camshaft can be moved in and out more

FIGURE 5-80 Slack in worn timing chain caused chain to rub against timing cover. (Courtesy of Red River Community College.)

FIGURE 5-81 Badly worn and damaged camshaft sprocket. (Courtesy of McQuay Norris, Inc.)

than $\frac{1}{2}$ in., the chain should be replaced. (Figure 5-82). To check for sprocket wear, run your fingernail across both sides of the sprocket teeth. If there is any evidence of grooving or stepped wear on the teeth the sprockets should be replaced. Normal practice is to replace both sprockets and the chain at the same time. Excessive chain wear results from using a new chain on worn sprockets.

If the camshaft and crankshaft sprockets are misaligned, the chain must operate with excessive side loading. This causes rapid wear on the sides of sprocket teeth as well as on the chain itself. Excessive sprocket runout due to improper installation methods

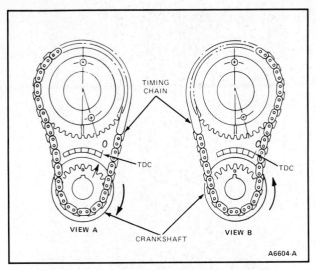

FIGURE 5-82 Checking slack (wear) in timing chain. (Courtesy of Ford of Canada.)

FIGURE 5-83 Broken timing chain. (Courtesy of TRW, Inc., Automotive Aftermarket Division.)

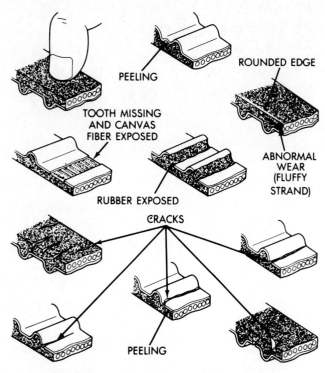

FIGURE 5-84 Timing belt inspection. (Courtesy of Chrysler Canada Ltd.)

can also cause rapid chain and sprocket wear. High stress loads imposed on the chain due to misalignment and excessive sprocket runout can cause the chain to break (Figure 5-83).

Improper installation methods include hammering directly on the sprocket, which can cause distortion, fractures, and cracking. Foreign matter between a sprocket flange and the camshaft flange causes misalignment. Mating surfaces must be clean and smooth for proper alignment.

Timing Belt Failure

Cogged or toothed camshaft timing belt failures include fabric cracking or peeling, missing or worn teeth, and edge wear (Figure 5-84). Misalignment of shields can cause rubbing of the belt against the shield. Prying against the belt with a screwdriver or pry bar will invariably damage the fabric or belt teeth. Flying gravel or other foreign matter that may get between the belt and sprockets will also damage the belt.

VALVE SPRING FAILURE

Valve springs are designed to keep the valve lifter in contact with the cam lobe at all times during opening and closing of the valves. Insufficient valve spring pressure will fail to perform this vital task. Insufficient valve spring pressure may be the result of weak springs or excessive installed spring height. Springs that are too weak must be replaced. Incorrect installed spring height can be corrected by the use of appropriate shims under the valve spring. See Chapter 18 for this procedure.

Valve springs should be inspected for rust, corrosion, pitting, acid etching, collapse, squareness, breakage, and wear at the ends. Damaged springs should be replaced. The free length of valve springs should be within $\frac{1}{16}$ in. of each other and within length specifications (Figure 5-85). Springs between 2 and $2\frac{1}{2}$ in. in height should be no more than $\frac{1}{16}$ in. out of square

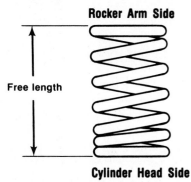

FIGURE 5-85 Valve spring free length. (Courtesy of Chrysler Canada Ltd.)

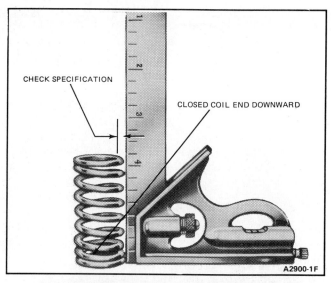

FIGURE 5-86 Checking valve spring squareness. (Courtesy of Ford of Canada.)

FIGURE 5-87 Shiny base of valve spring indicates valve spring flutter. (Courtesy of TRW, Inc., Automotive Aftermarket Division.)

FIGURE 5-88 Collapsed spring on left does not meet spring free length dimensions. (Courtesy of TRW, Inc., Automotive Aftermarket Division.)

(Figure 5-86). Valve springs can break due to acid etching, pitting, corrosion, rust, or excessive surge (Figures 5-87 to 5-89). Valve spring pressure must be checked on a spring tester. See Chapter 16 for details.

FIGURE 5-89 Broken valve spring caused by excessive flutter or surge. (Courtesy of TRW, Inc., Automotive Aftermarket Division.)

CHECKING PUSHRODS

All pushrods should be checked for wear or ridging on the ends. Worn pushrods must be replaced. Check whether the pushrod is bent by rolling it on a flat machined surface and checking for clearance anywhere between the pushrod and the flat surface. Bent pushrods should be replaced (Figure 5-90).

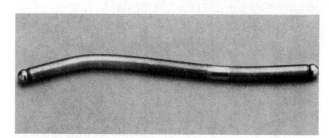

FIGURE 5-90 Bent pushrod. (Courtesy of FT Enterprises.)

CHECKING ROCKER ARMS

Inspect the rocker arms for wear at the pushrod end and at the valve stem end. If any obvious wear is present, the rocker arm must be replaced. In some cases the valve stem end of the cast-type rocker arm can be reground to restore the wear surface. Check the service manual to determine whether this can be done. Stamped steel rocker arms that are worn must be replaced.

Inspect the pivot area of the rocker arm for wear and scoring. Check for galling of the metal on the bottom of the rocker shaft hole on cast rocker arms. Check the ball seat area of stamped steel rocker arms for wear as well as the pivot ball itself. Replace both the rocker arm and ball if either show noticeable wear. New rocker arms should not be used with worn ball pivots, nor should worn rocker arms be used with new ball pivots. Rocker arms and ball pivots are wear mated and mismatching can result in excessive wear and high unit loading (Figures 5-91 to 5-94).

FIGURE 5-91 Rocker arm worn at valve stem end. (Courtesy of TRW, Inc., Automotive Aftermarket Division.)

FIGURE 5-92 Rocker arm worn at pushrod end. Note worn pushrod ball. (Courtesy of TRW, Inc., automotive Aftermarket Division.)

FIGURE 5-93 Broken rocker arms. (Courtesy of TRW, Inc., Automotive Aftermarket Division.)

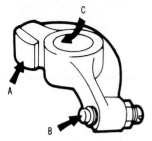

FIGURE 5-94 Wear points on cast rocker arm, shaft pivoted. (Courtesy of Chrysler Canada Ltd.)

CHECKING ROCKER ARM STUDS

Check each rocker arm stud for possible rocker arm-to-stud contact. The stud may show wear on its sides if the rocker arm was misaligned or on the side next to the valve if excessive valve float or valve lift have occurred. Continued rocker-to-stud contact eventually results in severe stud wear. A worn stud is prone to breakage and should be replaced. The reason for rocker-to-stud contact should also be corrected. This might be a bent pushrod, incorrect rocker arm, or bent rocker stud (Figure 5-95).

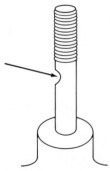

FIGURE 5-95 Rocker arm misalignment can wear into side of rocker arm stud. (Courtesy of FT Enterprises.)

CHECKING ROCKER ARM SHAFTS

Carefully inspect the bottom of the rocker arm shaft for wear at each of the rocker arm contact areas. Upward pressure of the rocker arm on the shaft can cause wear. Look for scoring, galling, or ridging. If wear is smooth and shiny, measure the worn area with a micrometer to determine the wear. If stepped wear, scoring, or galling is present, replace the rocker shaft (Figure 5-96).

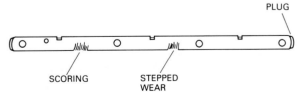

FIGURE 5-96 Rocker arm shaft wear. (Courtesy of FT Enterprises.)

1. State one reason for each of the following piston ring problems:
   (a) rings not seated properly;
   (b) rings spinning in groove;
   (c) rings stuck in groove;
   (d) rings broken;
   (e) side wear on rings;
   (f) scuffed rings.

2. List four causes of excessive oil consumption other than those listed in Question 1.

3. Worn ring gear teeth cause:
   (a) poor starter engagement
   (b) flywheel imbalance
   (c) drive pinion damage
   (d) all of the above

4. Flywheel runout is best checked with:
   (a) a straightedge and feeler gauge
   (b) a dial indicator
   (c) either (a) or (b)
   (d) neither (a) nor (b)

5. List five causes of crankshaft journal wear.

6. In a six-cylinder engine running at 4000 rpm there are how many torsional stresses per minute?
   (a) 120
   (b) 1200
   (c) 12,000
   (d) 120,000

7. Torsional fatigue failure of a crankshaft may be caused by:
   (a) a faulty vibration damper
   (b) a loose flywheel
   (c) either (a) or (b)
   (d) neither (a) nor (b)

8. List four causes of crankshaft bending stress failure.

9. Insufficient bearing lubrication may be the result of:
   (a) dry startup
   (b) insufficient bearing clearance
   (c) oil dilution
   (d) all of the above

10. List three types of main bearing misassembly.

11. Technician A says that detonation is ignition occurring in the cylinder before the spark plug fires. Technician B says that detonation can create a hole in a piston. Who is right?
    (a) technician A
    (b) technician B
    (c) both are right
    (d) both are wrong

12. List four causes of piston damage.

13. Exhaust valve guttering may be caused by preignition. Preignition may be caused by a fuel mixture that is too lean or excessive engine operating temperatures. Which of these statements is correct?
    (a) the first
    (b) the second
    (c) both are right
    (d) both are wrong

14. List three reasons why valve stem scuffing occurs.

15. What is valve "necking"?

16. List four reasons for excessive valve dynamic forces.

17. Hydraulic lifters may become noisy due to:
    (a) engine oil level being too low
    (b) engine oil pressure being too low
    (c) an excessive lifter leakdown rate
    (d) any of the above

18. List six reasons for a noisy valve train.

19. Why should a worn camshaft not be used with new valve lifters?

20. Why is the base of new lifters convex?

21. Severe shock loading of camshaft drive gears can result from excessive valve train ————.

22. Camshaft and crankshaft sprocket ———— can cause severe side loading of the chain.

23. Insufficient valve spring pressure may be the result of weak ———— or excessive spring installed ————.

24. How can a pushrod be checked for straightness?

25. Rocker arm misalignment can cause side wear on the rocker arm ————.

CHAPTER 6

Crack and Thread Repair

OBJECTIVES

To develop the ability to:

1. Detect cracks by:
   - Magnetic particle inspection
   - Wet magnetic flux inspection
   - Dye penetrant inspection
2. Repair cracks in cast iron blocks and heads by:
   - Using epoxy
   - Using threaded taper pins
   - Welding
3. Use ceramic sealing equipment.
4. Repair damaged threads with the use of thread inserts.
5. Remove broken bolts.

INTRODUCTION

Many cracked cylinder blocks and heads can be repaired successfully by relatively low cost methods. However, before any kind of rebuilding work is done on the engine, the block and head must be inspected for cracks to determine the location, extent, and repairability of the crack. There is no point in wasting time and expense on a head or block that is cracked beyond repair.

CAUSES OF CRACKS

Most cracks are very small at the beginning but develop into larger cracks and eventual part failure over a period of time. Cracks may start as the result of minor casting process imperfections, such as porosity or incomplete alloy dispersion. Over time the stresses and strain of engine operation cause the metal to break down at the weak point, resulting at the start in a small crack. Continued operation causes the crack to extend, resulting eventually in complete failure. Other causes of cracks include the following:

1. *Fatigue*. Fatigue is the result of continued repeating stress cycles of loading and unloading of a part. Fatigue failures begin as a tiny crack that continues to extend as the stress cycles are repeated, resulting eventually in breakage.
2. *Thermal shock*. Sudden chilling of hot metal by a rush of cold air or cold water. The forced sudden contraction of metal can result in cracks.
3. *Overheating*. Cracks resulting from engine overheating generally show up in the exhaust valve seat area of the cylinder heads.
4. *Incorrect tightening of bolts*. Excessive tightening, tightening in the wrong sequence, and misalignment causes metal to be put under excessive stresses, which can result in cracks developing.
5. *Freeze expansion*. In rare cases where water is used in a cooling system and the engine is left exposed to freezing temperatures, the head or block may crack due to expansion of the water as it turns to ice.

CRACK DETECTION

There are several commonly used methods of crack detection in the engine rebuilding industry. These include magnetic particle inspection, wet magnetic flux inspection, dye penetrant inspection, and pressure testing. Before using any of these methods, the area to be inspected must be thoroughly cleaned. Ultraviolet absorbing glasses must be used for eye protection.

Magnetic Particle Inspection

A powerful electromagnet is used for magnetic inspection methods. Place the magnet over the suspected area before applying the magnetic powder or liquid

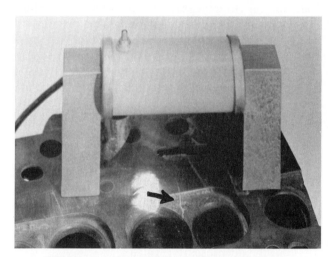

FIGURE 6-1 Magnetic particle crack detection of cylinder head. Electromagnet causes magnetic particles sprayed on suspected area to align in the cracked area exposing the crack. (Courtesy of Irontite Products Co., Inc.)

FIGURE 6-2 Magnetic particle inspection of cylinder block shows cracks. (Courtesy of Jasper Engine and Transmission Exchange.)

(Figures 6-1 to 6-4). Spray the powder over the suspected area from about 6 in. away. This allows the powder to align with the magnetic poles. Reposition the magnet by raising it and turning it in several dif-

FIGURE 6-3 Magnetic particle inspection of crankshaft. A swivel coil electromagnet and black light are used. (Courtesy of Kwik-Way Manufacturing Company.)

FIGURE 6-4 Magnetic inspection station with curtained hood and black light. (Courtesy of Kwik-Way Manufacturing Company.)

ferent positions from the first position. This ensures that the two poles will straddle any crack in at least one position. If the magnet does not straddle the crack, the opposing magnetic fields may not develop sufficiently to cause the particles to align and expose the crack.

Magnetic particle crack detection is used on cast-iron cylinder heads and blocks. It does not work on metals that cannot be magnetized, such as aluminum or bronze. The procedure is also restricted to exposed surfaces such as combustion chambers, machined surfaces, block decks, and main bearing areas. Internal cracks cannot be detected by this method. Magnaflux is a typical example of this procedure.

Wet Magnetic Flux Inspection

The wet magnetic flux inspection method is used on connecting rods and crankshafts. It uses fluorescent magnetic particles suspended in oil or water. The solution is sprayed on the suspected area and then inspected with black light. Any cracks will show up as white lines. Magnaglo is a typical example of this method.

Dye Penetrant Inspection

The dye penetrant inspection method can be used on all magnetic or nonmagnetic materials. For this reason it can be used to check aluminum cylinder heads or blocks as well. The area to be inspected must be cleaned, degreased, and decarbonized. Depending on the product being used, spray or paint the suspected area with the penetrant. Allow the penetrant to dry for a few seconds. Wash off any excess penetrant with a dye remover. Rinse the remover off with water and wipe the area dry. Spray the developer over the surface and allow a few minutes for it to dry. As the developer dries it absorbs the penetrant from the crack and the crack shows up as a line through the developer. Figure 6-5 shows a dye penetrant kit.

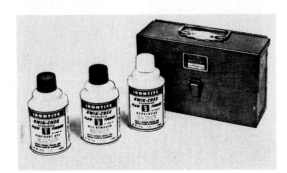

FIGURE 6-5 Dye penetrant kit includes dye penetrant, dye remover, and dye developer. (Courtesy of Irontite Products Co., Inc.)

Fluorescent Dye Penetrant

In another method, a fluorescent penetrant is sprayed over the suspected area, then dried by blotting with paper towel. Next a thin coat of developer is sprayed on and allowed to dry to a white color. The area is then inspected with black light, under which any crack will be visible. Zyglo is an example of this inspection method.

Pressure Testing

The pressure testing method of crack detection is commonly used by many engine rebuilders. This method detects both internal and external cracks in areas inaccessible by other methods. The procedure requires some special equipment, including a variety of plugs and clamps used to close coolant passage openings in heads and blocks. Also required is a method of pressurizing the coolant passages in the head or block. Figure 6-6 is one example of this equipment. The procedures for pressure testing a cylinder head or block are generally quite similar.

1. Place the head or block on the pressure testing stand and mount it securely.
2. Close all the water ports with the proper size closure pads and clamp them securely in place.
3. Connect the air pressure line to the cylinder head and pressurize to about 50 to 60 psi with the air pressure regulator.
4. Apply a bubble solution (a soap or shampoo solution can be used) to the head or block surfaces. If any leaks are present, bubbles will appear.
5. Clearly mark the entire length of any cracks that are discovered with chalk.
6. Repair the crack and again pressure test the head or block to make sure that the repair does not leak.

CRACK REPAIR

Crack Repair with Epoxy

Some minor cracks and imperfections such as porosity can be repaired with an epoxy sealer or a metallic plastic material. This type of repair is restricted to certain areas of the cylinder block which are not subject to high pressures or stresses (Figures 6-7 and 6-8). The following procedure is typical.

1. Clean the repair area thoroughly by grinding or rotary filing and follow by cleaning with lacquer thinner.

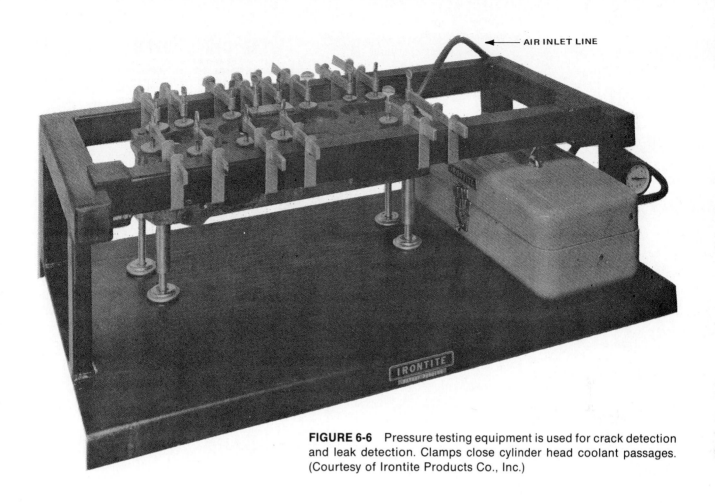

FIGURE 6-6 Pressure testing equipment is used for crack detection and leak detection. Clamps close cylinder head coolant passages. (Courtesy of Irontite Products Co., Inc.)

AIR INLET LINE

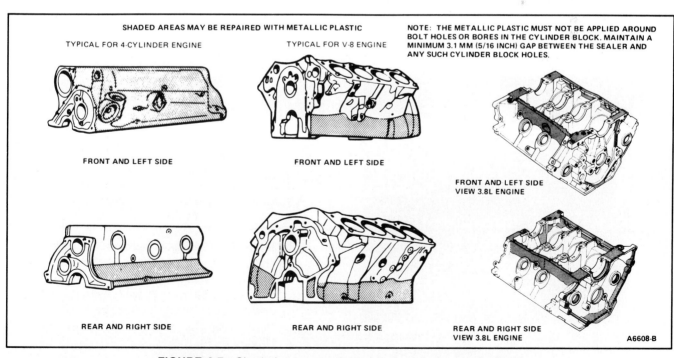

SHADED AREAS MAY BE REPAIRED WITH METALLIC PLASTIC

NOTE: THE METALLIC PLASTIC MUST NOT BE APPLIED AROUND BOLT HOLES OR BORES IN THE CYLINDER BLOCK. MAINTAIN A MINIMUM 3.1 MM (5/16 INCH) GAP BETWEEN THE SEALER AND ANY SUCH CYLINDER BLOCK HOLES.

TYPICAL FOR 4-CYLINDER ENGINE

TYPICAL FOR V-8 ENGINE

FRONT AND LEFT SIDE

FRONT AND LEFT SIDE

FRONT AND LEFT SIDE VIEW 3.8L ENGINE

REAR AND RIGHT SIDE

REAR AND RIGHT SIDE

REAR AND RIGHT SIDE VIEW 3.8L ENGINE

A6608-B

FIGURE 6-7 Shaded areas on these blocks may be repaired with epoxy or metallic plastic. (Courtesy of Ford of Canada.)

FIGURE 6-8 High-temperature crack repair material for use on cast iron or aluminum. (Courtesy of Silver Seal Products Company, Inc.)

2. Prepare the epoxy or metallic plastic as outlined on the container.

3. With a putty knife, apply the sealer to the area to be repaired. Press the material firmly into place to get maximum penetration and adhesion.

4. Allow the repair to dry thoroughly as directed on the container, usually 12 hours or more.

Crack Repair by Pinning

Threaded tapered plugs can be used to repair cracks in cast-iron cylinder heads and blocks as long as the crack is accessible for drilling reaming and tapping. The plugs are available in a range of sizes to suit the particular repair. A good sense of drilling, reaming, and taper tapping is needed for a successful repair. The following procedure is typical.

1. Locate the crack with one of the foregoing methods of crack detection. Be sure to mark its entire length.

FIGURE 6-9 Drilling both sides of a crack in a block in preparation for pinning to squeeze crack shut. (Courtesy of Irontite Products Co., Inc.)

TAP DRILL SIZES
Recommended for
AMERICAN NATIONAL SCREW THREAD PITCHES

COARSE STANDARD THREAD (N. C.) Formerly U. S. Standard Thread					SPECIAL THREAD (N. S.)				
Sizes	Threads Per Inch	Outside Diameter at Screw	Tap Drill Sizes	Decimal Equivalent of Drill	Sizes	Threads Per Inch	Outside Diameter at Screw	Tap Drill Sizes	Decimal Equivalent of Drill
1	64	.073	53	0.0595	1	56	.0730	54	0.0550
2	56	.086	50	0.0700	4	32	.1120	45	0.0820
3	48	.099	47	0.0785	4	36	.1120	44	0.0860
4	40	.112	43	0.0890	6	36	.1380	34	0.1110
5	40	.125	38	0.1015	8	40	.1640	28	0.1405
6	32	.138	36	0.1065	10	30	.1900	22	0.1570
8	32	.164	29	0.1360	12	32	.2160	13	0.1850
10	24	.190	25	0.1495	14	20	.2420	10	0.1935
12	24	.216	16	0.1770	14	24	.2420	7	0.2010
1/4	20	.250	7	0.2010	1/16	64	.0625	3/64	0.0469
5/16	18	.3125	F	0.2570	3/32	48	.0938	49	0.0730
3/8	16	.375	5/16	0.3125	1/8	40	.1250	38	0.1015
7/16	14	.4375	U	0.3680	5/32	32	.1563	1/8	0.1250
1/2	13	.500	27/64	0.4219	5/32	36	.1563	30	0.1285
9/16	12	.5625	31/64	0.4843	3/16	24	.1875	26	0.1470
5/8	11	.625	17/32	0.5312	3/16	32	.1875	22	0.1570
3/4	10	.750	21/32	0.6562	7/32	24	.2188	16	0.1770
7/8	9	.875	49/64	0.7656	7/32	32	.2188	12	0.1890
1	8	1.000	7/8	0.875	1/4	24	.250	4	0.2090
1 1/8	7	1.125	63/64	0.9843	1/4	27	.250	3	0.2130
1 1/4	7	1.250	1 7/64	1.1093	1/4	32	.250	7/32	0.2187
					5/16	20	.3125	17/64	0.2656
FINE STANDARD THREAD (N. F.) Formerly S.A.E. Thread					5/16	27	.3125	J	0.2770
Sizes	Threads Per Inch	Outside Diameter at Screw	Tap Drill Sizes	Decimal Equivalent of Drill	5/16	32	.3125	9/32	0.2812
0	80	.060	3/64	0.0469	3/8	20	.375	21/64	0.3281
1	72	.073	53	0.0595	3/8	27	.375	R	0.3390
2	64	.086	50	0.0700	7/16	24	.4375	X	0.3970
3	56	.099	45	0.0820	7/16	27	.4375	Y	0.4040
4	48	.112	42	0.0935	1/2	12	.500	27/64	0.4219
5	44	.125	37	0.1040	1/2	24	.500	29/64	0.4531
6	40	.138	33	0.1130	1/2	27	.500	15/32	0.4687
8	36	.164	29	0.1360	9/16	27	.5625	17/32	0.5312
10	32	.190	21	0.1590	5/8	12	.625	35/64	0.5469
12	28	.216	14	0.1820	5/8	27	.625	19/32	0.5937
1/4	28	.250	3	0.2130	11/16	11	.6875	19/32	0.5937
5/16	24	.3125	I	0.2720	11/16	16	.6875	5/8	0.6250
3/8	24	.375	Q	0.3320	3/4	12	.750	43/64	0.6719
7/16	20	.4375	25/64	0.3906	3/4	27	.750	23/32	0.7187
1/2	20	.500	29/64	0.4531	7/8	12	.875	51/64	0.7969
9/16	18	.5625	0.5062	0.5062	7/8	18	.875	53/64	0.8281
5/8	18	.625	0.5687	0.5687	7/8	27	.875	27/32	0.8437
3/4	16	.750	11/16	0.6875	1	12	1.000	59/64	0.9219
7/8	14	.875	0.8020	0.8020	1	27	1.000	31/32	0.9687
1	14	1.000	0.9274	0.9274					
1 1/8	12	1.125	1 3/64	1.0468					
1 1/4	12	1.250	1 11/64	1.1718					

TAP DRILL SIZES (METRIC)

Bolt Diameter (in mm)	Distance Between Threads (in mm)	Diameter of Drill (in mm)	Bolt Diameter (in mm)	Distance Between Threads (in mm)	Diameter of Drill (in mm)
M 2	0.4	1.6	M 22	2.5	19.5
M 2.2	0.45	1.75	M 24	3	21
M 2.5	0.45	2.05	M 27	3	24
			M 30	3.5	26.5
M 3	0.5	2.5			
M 3.5	0.6	2.9	M 33	3.5	29.5
M 4	0.7	3.3	M 36	4	32
M 4.5	0.75	3.7	M 39	4	35
			M 42	4.5	37.5
M 5	0.8	4.2			
M 6	1	5	M 45	4.5	40.5
M 8	1.25	6.8	M 48	5	43
M 10	1.5	8.5	M 52	5	47
M 12	1.75	10.2	M 56	5.5	50.5
M 14	2	12	M 60	5.5	55
M 16	2	14	M 64	6	58
M 18	2.5	15.5	M 68	6	62
M 20	2.5	17.5			

FIGURE 6-10 Tap and drill size chart. (Courtesy of Mac Tools, Inc.)

2. Drill a hole 1/8 in. past each end of the crack (Figure 6-9). This prevents any furthering of the crack during the repair procedure. Use the same drill size as will be used for the crack repair. The drill size is determined by the size of plug best suited for the repair (Figure 6-10).

3. Starting at one end of the crack, ream the hole with the proper size tapered reamer

FIGURE 6-11 Tapered reamers are used to ream drilled holes prior to tapping. (Courtesy of Irontite Products Co., Inc.)

FIGURE 6-12 Taper taps are used for tapping crack pinning holes. (Courtesy of Irontite Products Co., Inc.)

(Figure 6-11). Do not ream excessively. (Reaming makes it easier to tap the hole.)

4. Tap the hole with the proper size taper tap for the plug to be used. Use the appropriate cutting lubricant (Figures 6-12 and 6-13).

5. Coat the taper plug threads with a good sealer and torque the plug into place (Figure 6-14).

6. Remove the protruding end of the plug. If accessible, it can be cut off with a saw (Figure 6-15). Another method is to cross-drill

FIGURE 6-14 Typical taper plugs used for crack repair. (Courtesy of Irontite Products Co., Inc.)

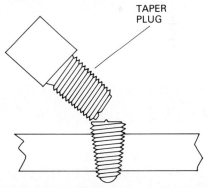

FIGURE 6-15 Taper plug installed. Plug is cut with hacksaw and broken off. Enough plug extends above surface for peening. (Courtesy of FT Enterprises.)

FIGURE 6-13 Lubricants for cutting tools. (Courtesy of The L. S. Starrett Company.)

Lubricants for Cutting Tools

Material	Turning	Chucking	Drilling Milling	Reaming	Tapping
Tool Steel	Dry or Oil	Oil or Soda Water	Oil	Lard Oil	Oil
Soft Steel	Dry or Soda Water	Soda Water	Oil or Soda Water	Lard Oil	Oil
Wrought Iron	Dry or Soda Water	Soda Water	Oil or Soda Water	Lard Oil	Oil
Cast Iron	Dry	Dry	Dry	Dry	Oil
Brass	Dry	Dry	Dry	Dry	Oil
Copper	Dry	Dry	Oil	Mixture	Oil
Babbitt	Dry	Dry	Dry	Dry	Oil
Glass				Turpentine or Kerosene	

Mixture is ⅓ Crude Petroleum, ⅔ Lard Oil. When two lubricants are mentioned the first is preferable.

the plug just above the casting surface and break the plug off after drilling. Leave just enough plug material above the casting surface to allow peening of the pin.

7. After installing the plug, drill the next hole so that the edge of the hole just meets the edge of the previous pin. This results in an overlap of plugs, which locks them in place. Repeat the drilling until the repair is completed. Figures 6-16 and 6-17.

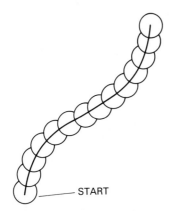

FIGURE 6-16 Overlapping method of crack pinning. (Courtesy of FT Enterprises.)

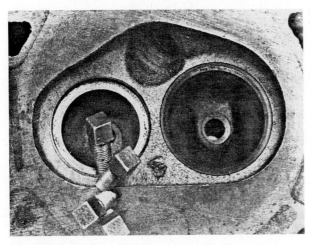

FIGURE 6-17 Cracked cylinder head properly pinned. (Courtesy of Irontite Products Co., Inc.)

8. Peen the exposed ends of the plugs to expand them (Figure 6-18).
9. After peening grind the repair down to make it blend with the surrounding surface. Use a rotary file or grinder (Figure 6-19).
10. Inspect the work with a magnetic crack detector to ensure that the entire crack has been repaired.
11. To ensure that any imperfections in the repair are sealed, use the ceramic sealer circulator described later to seal the repair.

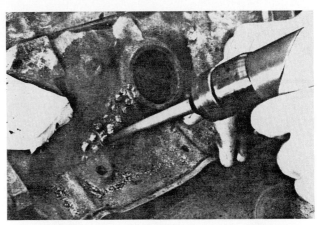

FIGURE 6-18 Peening with an air hammer and peening tool. (Courtesy of Irontite Products Co., Inc.)

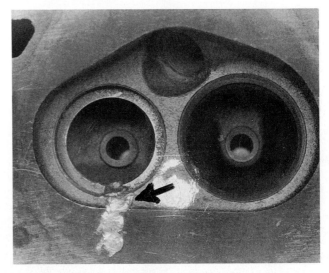

FIGURE 6-19 Plugs cut off and peened. (Courtesy of Irontite Products Co., Inc.)

Crack Repair by Welding

Crack repair by welding requires carefully controlled preheating of the casting to be repaired prior to welding. This prevents stress cracks from forming due to extreme differences in temperature of the casting during the welding process and during the cooling-off period. The cooling of the casting should also be controlled for the same reason. If allowed to cool too rapidly, stress cracks may occur. The rate of cooling should be about 100°F per hour. Typically, the procedure is as follows.

1. Clean the area to be repaired by grinding or rotary filing.
2. Place the casting in a furnace for the purpose and heat to about 1400°F or until a dull cherry red.
3. Keep the casting near that temperature during the welding process.

4. Weld the crack with a neutral acetylene flame using a cast-iron welding rod and a good-quality glass flux.

5. After the repair, heat the casting again to 1400°F, then allow to cool at about 100°F per hour to minimize any stresses.

CERAMIC SEALING

Ceramic sealing can be done after crack repairs have been made to seal any minute cracks or porosity that may remain. The same equipment is used as in pressure testing, but a circulator is used to circulate the ceramic solution through the water jackets. Figures 6-20 and 6-21). The circulator has a thermostatically controlled heating element which heats the ceramic solution since the solution is more effective when hot.

The circulator hoses are equipped with pressure caps at the outer ends which are connected to opposite ends of the cylinder head. Valves at the tank end of the hoses can be opened or closed to control the flow of the solution. The hot solution is circulated through the head for about 15 minutes. After the circulation procedure the head can be pressurized with air to help force the solution into any imperfections in the casting or the repair. The circulation valves must be closed before pressurizing the head with air.

After pressurizing the head for about half a minute, shut off the air pressure, open the circulator valves for a few minutes, then close the valve at the circulation outlet hose and force the solution out of the head. Remove the closure fittings from the head. Remove the head from the test stand and set it aside to allow the ceramic sealer to dry and set. Wash off any ceramic solution spills immediately. If left to dry on other surfaces, it sets and can only be removed by abrasion. Avoid spilling the solution on unwanted surfaces.

REPAIRING THREADS WITH A THREAD INSERT

A thread insert is a coil of steel threads used to replace damaged threads. Thread inserts are available in various sizes and thread types to suit the needed repair. Thread inserts are usually $1\frac{1}{2}$ times as long as the thread

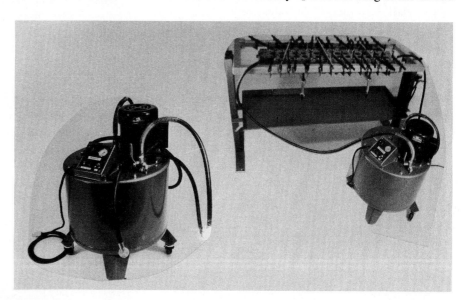

FIGURE 6-20 Circulator (left) is used with pressure testing equipment for ceramic sealing of cylinder head. (Courtesy of Irontite Products Co., Inc.)

FIGURE 6-21 Cooling system flushing and sealing materials. (Courtesy of Irontite Products Co., Inc.)

diameter to ensure proper thread strength. Helicoil thread inserts are made of stainless steel wire. A special tool is required to install the thread insert (Figures 6-22 and 6-23). The general procedure is as follows.

1. Select the drill size appropriate for the thread insert to be used as specified on the insert package.

2. Carefully drill out the damaged threads.

3. Using the tap specified on the insert package, tap the hole. Use cutting oil, especially in

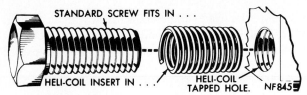

FIGURE 6-22 Typical thread repair insert. (Courtesy of Chrysler Canada Ltd.)

  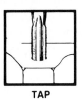

STRIPPED   DRILL   TAP

INSTALL   FIXED!

FIGURE 6-23 Steps in repairing damaged threads. (Courtesy of Chrysler Canada Ltd.)

aluminum castings. Remove all metal chips from the hole.

4. With the insert tool engaging the tang at the bottom of the thread insert, screw the insert into the threaded hole. Be careful not to destroy the insert by pushing down on the tool. If pressure is needed to start the insert, push down on the insert, not on the tool.

5. After the insert is in place, carefully break off and remove the tang by bending it back and forth with a pair of needle-nose pliers.

6. Lightly stake the exposed thread above the thread insert to prevent the insert from coming out when the cap screw is removed on the next disassembly.

REPAIRING A DAMAGED SPARK PLUG HOLE THREAD

Damaged spark plug threads in an aluminum cylinder head can be repaired using special tools and thread inserts called Taperserts, similar to the procedure described above. The repair must be performed with the cylinder head removed to prevent metal chips from entering the combustion chamber. The repair is permanent and has no negative effects. The procedure is done by hand as follows.

1. Thoroughly clean the spark plug counter bore, seat, and threads of all dirt or other foreign material.

2. Start the tap into the spark plug hole, being careful to keep it properly aligned. As the tap begins cutting new threads, apply aluminum cutting oil to the tap (Figure 6-24).

3. Continue cutting the threads and applying the oil until the stop ring bottoms against the stop collar. The stop collar should be against the spark plug seat and the stop ring. If the collar is loose, the top has not been threaded far enough.

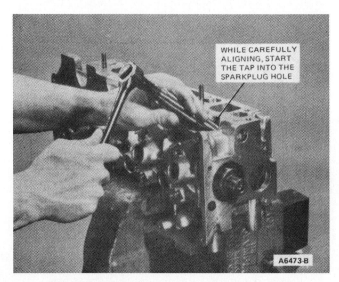

FIGURE 6-24 Tapping a damaged spark plug hole prior to installing a thread repair insert. (Courtesy of Ford of Canada.)

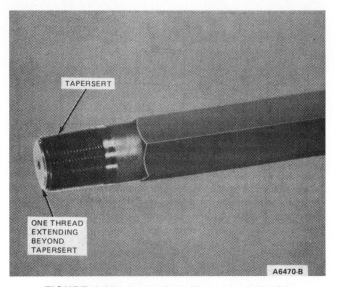

FIGURE 6-25 Insert installing mandrel with thread repair insert in place. (Courtesy of Ford of Canada.)

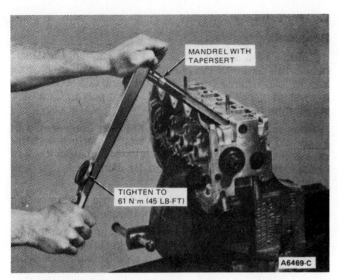

FIGURE 6-26 Installing the thread repair insert with a torque wrench. (Courtesy of Ford of Canada.)

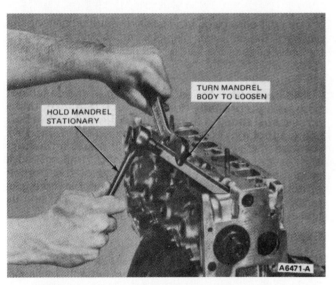

FIGURE 6-27 Removing the mandrel after insert installation. (Courtesy of Ford of Canada.)

4. Remove the tap. Remove all metal chips using compressed air.

5. Coat the threads of the mandrel with cutting oil. Thread the Tapersert onto the mandrel until one thread of the mandrel extends beyond the Tapersert (Figure 6-25).

6. Thread the Tapersert into the tapped spark plug hole using a torque wrench (Figure 6-26). Continue tightening the mandrel until the torque wrench indicates 45 ft-lb (62 N·m).

7. To loosen the mandrel for removal, hold the mandrel stationary and turn the mandrel body approximately one-half turn (Figure 6-27). Remove the mandrel.

Note: A properly installed Tapersert will be flush to 1 mm below the spark plug gasket seat.

CUTTING TOOL LUBRICANTS

Whether machining, drilling, milling, reaming, or tapping metals, using the proper lubricant will help to produce good work. Figure 6-13 shows common applications of cutting tool lubricants.

REMOVING BROKEN BOLTS

Repairing engines often includes the need to remove broken bolts or studs. Bolts and studs can break, leaving the threaded part of the bolt in the engine component. The bolt may break flush with the surface or it may break above or below the surface.

The following procedures are commonly used to remove broken bolts and studs. First saturate the threads with penetrating oil. If the threads are not seized and the bolt is flush with or below the surface, use a center punch and hammer to remove the piece. Place the point of the center punch near the edge of the bolt at an angle that will unscrew the bolt as you tap the center punch with a hammer.

If the bolt extends far enough above the surface, grip the bolt with locking pliers, stud remover (Figure 6-28), or a pipe wrench and unscrew the bolt. Alternative methods include sawing a slot in the bolt top and using a screwdriver, or welding a nut to the bolt top and using a wrench on the nut. The heat from welding will often loosen seized threads in the process.

Often a bolt that is broken well below the surface and is seized in the threads must be removed by drilling and the use of a screw extractor. First carefully center punch the bolt in the exact center. Select the largest appropriate extractor and drill size for the bolt, then

FIGURE 6-28 Stud removing tool wedges serrated cam against stud for removal. (Courtesy of Easco/KD Tools, Inc.)

Removing Broken Bolts **95**

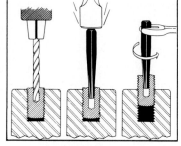

FIGURE 6-29 Tapered square screw extractors and how to use them. (Courtesy of Easco/KD Tools, Inc.)

drill the broken bolt in the center. Install the extractor securely in the hole and turn the extractor with a wrench (Figures 6-29 and 6-30). If all else fails, the broken bolt can be drilled out completely, including the threads, and a thread repair insert installed as described earlier.

FIGURE 6-30 Fluted screw extractor and drill set. (Courtesy of Easco/KD Tools, Inc.)

REVIEW QUESTIONS

1. List five causes of engine block or cylinder head cracks.

2. Magnetic particle inspection involves:
   (a) the use of an electromagnet
   (b) spraying magnetic powder over the suspected area
   (c) both (a) and (b)
   (d) neither (a) nor (b)

3. What method of crack detection is used on aluminum castings?

4. During pressure testing of a cylinder head:
   (a) a thin coat of developer is sprayed over the suspected area
   (b) dye penetrant is used
   (c) both (a) and (b)
   (d) neither (a) nor (b)

5. A soap or shampoo solution is used during:
   (a) magnetic particle crack detection
   (b) fluorescent dye penetrant crack detection
   (c) pressure testing for cracks
   (d) none of the above

6. What is the purpose of ceramic sealing of a cylinder head or block?

7. If a ceramic solution spill is not washed off immediately:
   (a) it will penetrate the metal
   (b) it will set and can only be removed by abrasion
   (c) it will evaporate as a harmful vapor
   (d) none of the above

8. The epoxy or metallic plastic crack repair must be allowed to dry for at least:
   (a) 4 hours
   (b) 6 hours
   (c) 10 hours
   (d) 12 hours

9. Epoxy and metallic plastic crack repairs can be made:
   (a) anywhere on the cylinder block or head
   (b) only in low-pressure areas of the cylinder block or head
   (c) only after welding to seal any leaks
   (d) none of the above

10. The procedure for crack pinning requires the following steps in order:
    (a) cleaning, drilling, reaming, threading, pinning, cutting, and peening

(b) cleaning, drilling, threading, reaming, pinning, peening, and cutting

(c) cleaning, drilling, reaming, threading, pinning, peening, and cutting

(d) cleaning, drilling, threading, reaming, pinning, cutting, and peening

11. What are Helicoil inserts made from?

12. True or false: Thread repair inserts can only be made on cast-iron parts.

13. Repairing damaged spark plug hole threads requires:
(a) cylinder head removal
(b) drilling and tapping the hole
(c) installing the thread insert with a special tool
(d) all of the above

14. List four methods that can be used to remove a broken stud or cap screw.

CHAPTER 7

Cylinder Block Reconditioning

OBJECTIVES

To develop the ability to perform the following reconditioning procedures to production engine rebuilders' standards:

- Measure cylinder wall thickness.
- Clean up all threaded holes by "chasing" the threads with a tap.
- Chamfer all threaded holes where threads extend to the surface.
- Remove all casting slag and burrs from the block interior by grinding.
- Replace all core hole plugs and oil gallery plugs, both soft and threaded types.
- Align bore or align hone the main bearing saddle bores.
- Hone the lifter bores.
- Resurface the block deck by grinding or milling.
- Recondition the cylinders by boring, honing, or sleeving, and fitting the pistons.
- Clean the block by scrubbing with soap and water.
- Install core plugs.
- Paint the block interior and exterior.

INTRODUCTION

After the block has been degreased and checked for cracks as described in Chapter 6, it is ready for reconditioning. Reconditioning the block includes thread cleaning and chamfering, deburring, align boring or align honing, cylinder resizing, resurfacing the deck, honing the lifter bores, and installing new core plugs and oil gallery plugs. This chapter deals with these procedures.

CYLINDER WALL THICKNESS

The thickness of the cylinder walls must be established prior to cylinder resizing. Cylinder walls must maintain minimum thickness limits after reboring. A cylinder wall that is too thin after reboring is subject to distortion and failure. Previously rebored cylinders may already be at the maximum oversize possible. In some instances the cylinder wall may be thinner on one side than it is on the other, due to core shift during production boring of the block.

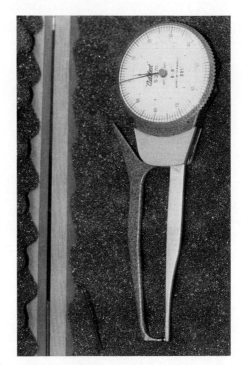

FIGURE 7-1 Cylinder wall thickness gauge. (Courtesy of Western Engine Ltd.)

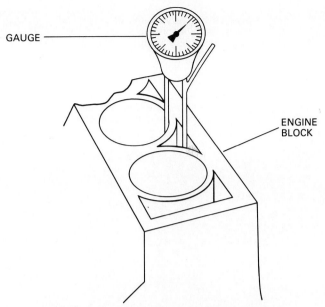

GAUGE

ENGINE BLOCK

FIGURE 7-2 Measuring cylinder wall thickness with dial gauge. (Courtesy of FT Enterprises.)

Several methods are used to check cylinder wall thickness: a dial-type thickness gauge and an ultrasonic tester. The dial gauge is operated like a spring-loaded clamp. It has one movable jaw which registers its position on the dial gauge (Figures 7-1 and 7-2). The ultrasonic tester is a meter equipped with a probing device. The probe is placed on the cylinder wall and held in place with heavy grease. Readings taken are accurate within 0.010 in.

DECKING THE BLOCK

When surface irregularities, corrosion, or erosion of the block deck are present to the extent that gasket sealing is affected, the block deck must be resurfaced. After cleaning the block, careful visual inspection of the deck surface is required to check for damage by erosion or corrosion. The flatness of the deck surface is checked with a straightedge and feeler gauge as shown in Figure 7-3.

To prepare the cylinder block for resurfacing, the dowel pins must first be removed from the deck. Keep the dowel pins for installation after resurfacing the deck. The main bearing caps must also be removed. The block is then placed on the deck resurfacing machine, leveled, and locked firmly in place.

Figures 7-4 and 7-5 illustrate a wet surface grinder used to resurface cylinder blocks, heads, and manifolds. Gauging is done by dial indicator. The block position is referenced from the main bearing saddle bores to maintain parallelism. A rollover fixture permits grinding both decks of V engines with one setup. The grinding head traverses across the work along hardened, ground ways.

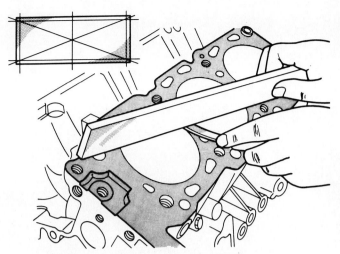

FIGURE 7-3 Checking the block deck for surface distortion. (Courtesy of Chrysler Canada Ltd.)

Figures 7-6 and 7-7 illustrate a milling machine with a cylinder block in position. The rotating milling head has a series of cutters arranged near the outer edge. The block is leveled and firmly locked in place on the machine base. The milling head traverses across the deck surface. Rotating and traverse speeds and depth of cut determine the microinch surface finish obtained. A surface finish that is too smooth or too rough can result in early head gasket failure. Follow engine and equipment manufacturers' specifications for the proper rotating and traverse speeds to use for a particular surface finish.

BLOCK DECK HEIGHT AND CLEARANCE

The deck height of a cylinder block is the distance from the main bearing saddle bore centerline to the deck or top of the block (Figure 7-8). This height is an important factor in that it affects the engine's compression ratio. The deck surface must also be parallel to the main bearing centerline to ensure that the compression ratio will be equal in all cylinders. On V engines the deck height dimension must be equal for both cylinder banks. If they are unequal, cylinder balance will be affected.

Deck clearance is the difference in height between the deck and the top of the piston when it is at the TDC position (Figure 7-9). Deck clearance can be positive—the pistons at TDC being higher than the deck, or negative—the pistons at TDC being lower than the deck (Figure 7-10). Factors that affect deck clearance are:

- Deck height
- Connecting rod length
- Piston height: pin centerline to top of piston

Block Deck Height and Clearance **99**

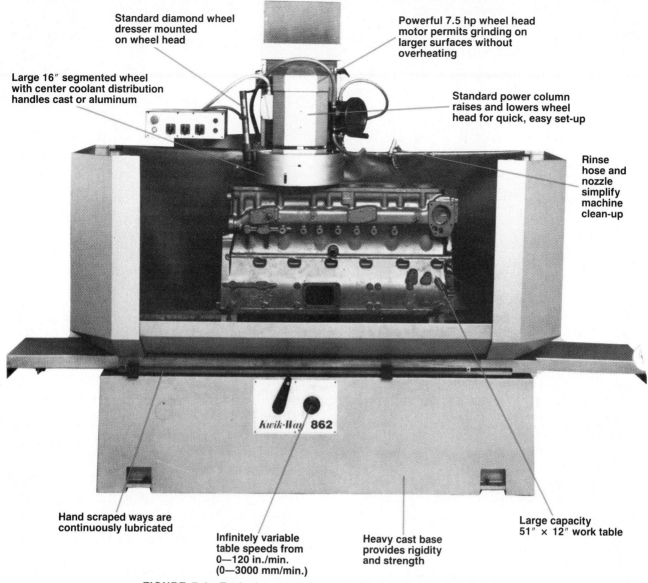

Standard diamond wheel dresser mounted on wheel head

Powerful 7.5 hp wheel head motor permits grinding on larger surfaces without overheating

Large 16″ segmented wheel with center coolant distribution handles cast or aluminum

Standard power column raises and lowers wheel head for quick, easy set-up

Rinse hose and nozzle simplify machine clean-up

Kwik-Way 862

Hand scraped ways are continuously lubricated

Infinitely variable table speeds from 0—120 in./min. (0—3000 mm/min.)

Heavy cast base provides rigidity and strength

Large capacity 51″ × 12″ work table

FIGURE 7-4 Typical wet surface grinder for resurfacing cylinder blocks, heads, and manifolds. (Courtesy of Kwik-Way Manufacturing Company.)

FIGURE 7-5 Cylinder block positioned on rollover fixture of surface grinder with dial indicator in position for leveling. (Courtesy of Kwik-Way Manufacturing Company.)

- Crankpin stroke
- Vertical positioning of crankshaft in block

Decking the block positions the cylinder heads closer to the crankshaft. This can have the following effects, depending on how much metal is removed.

- Misalignment of intake ports with manifold (V engines)
- Misalignment of bolt holes in intake manifold and cylinder head (V engines)
- Pistons striking valves
- Effective length of valve train too long
- Combustion chamber volume and compression ratio altered

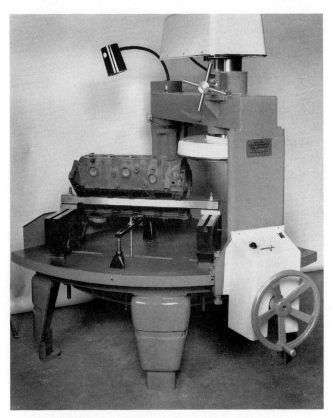

FIGURE 7-6 Block deck milling machine. (Courtesy of Storm Vulcan Co.)

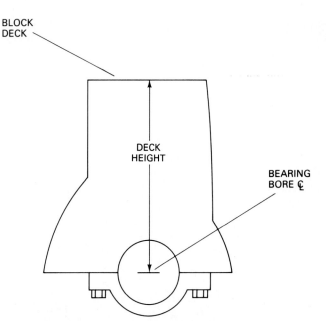

FIGURE 7-8 Block deck height dimension. (Courtesy of FT Enterprises.)

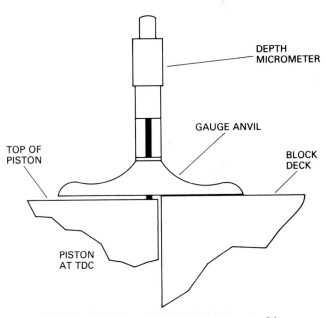

FIGURE 7-9 Measuring deck clearance with a depth micrometer. (Courtesy of FT Enterprises.)

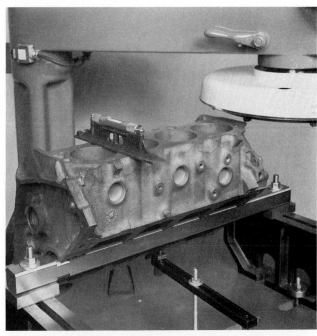

FIGURE 7-7 Leveling the block on a deck milling machine. (Courtesy of Storm Vulcan Co.)

On V engines, if more than 0.010 in. of stock is removed from either the block or the head (or both combined), stock must also be removed from the intake side of the head. (See Chapter 15 for this procedure.) On engines where the intake manifold serves as the lifter chamber cover, the bottom of the manifold gasket surface may also have to be machined. If there is any danger of the pistons striking the valves, the

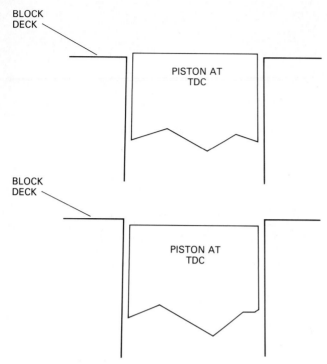

FIGURE 7-10 Positive (top) and negative (bottom) deck clearance. (Courtesy of FT Enterprises.)

MAIN BEARING BORE ALIGNMENT

The main bearing saddle bores must be in vertical and horizontal alignment to prevent the new crankshaft and bearings from binding. This binding results in rapid bearing wear and overheating of the bearings (Figure 7-11). The stresses of heating and cooling of the engine block over thousands of miles eventually cause block warpage. The process is slow, allowing the bearings to compensate for the misalignment through gradual wear. However, when a new or reground crankshaft and bearings are installed, binding and rapid wear will occur.

MAIN BEARING BORE STRETCH

The continuous pounding absorbed by the main bearing caps as a result of engine power impulses eventually results in a vertical stretch of the caps. At the same time, they tend to "pinch in" at the cap parting line, resulting in a slightly oval bore opening. Heavy loading of the engine increases this tendency. Since this process happens slowly over a period of time, the bearings compensate for this change by wear. Installing new bearing inserts in stretched, out-of-round bearing bores results in crankshaft bind, rapid bearing wear, or complete failure of both (Figure 7-12).

BEARING SPIN

If the engine is overloaded and the crankshaft subjected to extreme loads and overheating, the bearing inserts may seize to the crankshaft and be forced to

pistons may have to be notched to provide clearance. Selective length pushrods are available for some engines with nonadjustable valve trains to ensure centering of the hydraulic lifter plunger. On shaft-mounted rocker arms the pedestals can be shimmed. To correct the combustion chamber volume and compression ratio, see Chapter 4.

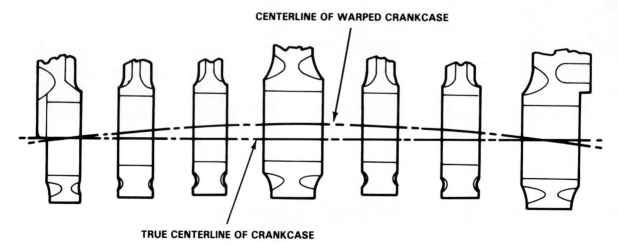

FIGURE 7-11 Main bearing saddle bore misalignment. (Courtesy of Federal-Mogul Corporation.)

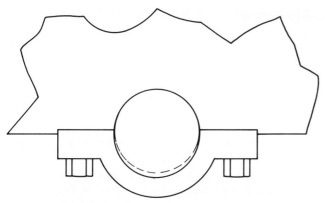

FIGURE 7-12 Main bearing cap stretch results in out-of-round bearing bore. (Courtesy of FT Enterprises.)

FIGURE 7-13 Measuring main bearing saddle bore out-of-round with a dial bore gauge. (Courtesy of Sunnen Products Company.)

turn with the journal. This results in a scored main bearing bore, a ruined bearing, and a damaged crankshaft.

MEASURING MAIN BEARING BORES (SADDLE BORES)

Main bearing bores (saddle bores) must be measured to determine their roundness and alignment. To do this, the main bearing caps are torqued in place in their proper position with the bearing inserts removed.

Saddle bore out-of-round can be checked with s dial bore gauge, inside micrometer or telescoping gauge, and outside micrometer. Measurements are taken vertically and horizontally. The difference between these two measurements is the amount of out-of-round. Each saddle bore must be measured separately (Figure 7-13). A maximum of 0.001 in. of out-of-round is normally allowed. Record the measurements for reference when grinding main bearing caps prior to align boring or honing.

To measure saddle bore alignment, either of two methods may be used. One is to use a round precision arbor 0.001 in. less in diameter than the standard saddle bore diameter and long enough to extend through all the saddle bores. With the main bearing caps re-

FIGURE 7-14 Measuring saddle bore alignment with a precision arbor. (Courtesy of Federal-Mogul Corporation.)

moved, place the arbor in the saddle bores and torque the bearing caps into place. If the arbor can be turned with a 12-in.-long handle, the saddle bores are considered to be within alignment limits. If the arbor will not turn, the bores are out of alignment (Figure 7-14).

Another method is to use a precision straightedge and a 0.0015-in. feeler gauge. While holding the straightedge firmly against the sides of the bores, try to insert the feeler gauge at each bore. This checks the lateral alignment. Do the same while holding the straightedge against the top of the bores (Figure 7-15). This checks the vertical alignment of the saddle bores. If the feeler gauge cannot be inserted at any saddle bore in both straightedge positions, the saddle bores are in acceptable alignment limits. If they are not the caps must be shaved and the bores align bored or honed.

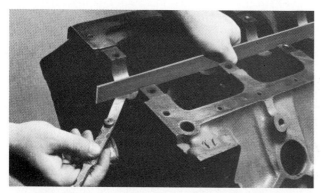

FIGURE 7-15 Measuring saddle bore alignment with a straightedge and feeler gauge. (Courtesy of TRW, Inc., Automotive Aftermarket Division.)

CORRECTING BORE ALIGNMENT

There are three ways to correct main bearing bore misalignment and cap stretch: (1) align boring the saddle bores, (2) align honing the saddle bores, and (3) align boring semifinished main bearings. The first two methods are preferred.

Align Boring

The general procedure for align boring is as follows (Figures 7-16 to 7-18).

1. Remove stock from the parting face of the main bearing caps to reduce the bore diameter. Use only a precision cap and rod grinder to ensure accuracy. Remove 0.001 in. more than the maximum out-of-round measurement recorded earlier. Debur the ground edges if necessary.

2. Make certain that the parting surfaces on the caps and block are clean. Position the caps front side forward on the block. Make sure that they are in order front to rear.

3. Install the cleaned cap bolts and torque them alternately in steps to specified torque.

4. Position the boring equipment on the block as shown in Figure 7-17. Set the cutting bits to the desired cutting diameter Figure 7-18 (original bore diameter) and lock them into place. Select the desired cutting speed and engage the drive to complete the boring. The cutting depth, cutter rotating speed, and rate of feed must be set to produce a 60- to 90-microinch (min.) surface finish to ensure good heat transfer from the bearings.

Caution: The maximum cut into the block should not exceed 0.005 in. The camshaft-to-crankshaft centers must not be decreased by more than 0.005 in.

Align Honing

Align honing is a popular method of correcting main bearing bore alignment. With this method a minimum amount of stock is removed from the cap parting surface, after which the caps are torqued into place as for

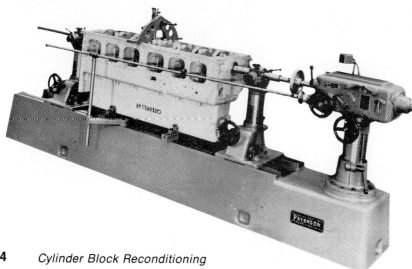

FIGURE 7-16 Align boring a large engine block. (Courtesy of Peterson Machine Tool, Inc.)

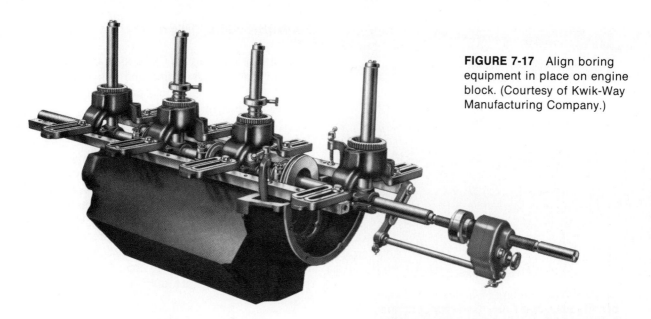

FIGURE 7-17 Align boring equipment in place on engine block. (Courtesy of Kwik-Way Manufacturing Company.)

FIGURE 7-18 A special micrometer is used to set the cutting tool shown in Figure 7-17. (Courtesy of Kwik-Way Manufacturing Company.)

align boring. The block is mounted on the cradle (Figures 7-19 and 7-20) and adjusted to the proper height. The honing unit is inserted and adjusted. Next the drive arm is swung into position and coupled to the hone. The honing stops are adjusted to limit stroke length. Actual honing time is only approximately 1 minute. An oil control bar directs honing oil to the main bearing cap area. Various size honing units are available for different engines.

Semifinished Bearings

Another method of correcting bearing bore misalignment is to use semifinished bearings. These bearings have sufficient extra wall thickness to allow align boring after installation. Semifinished cam bearings are

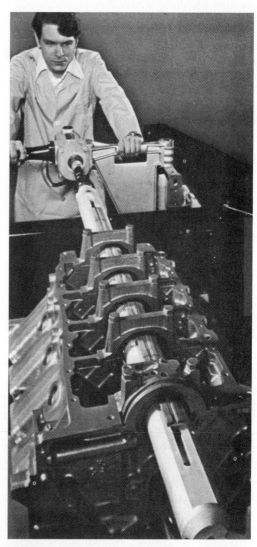

FIGURE 7-19 Align honing main bearing bores. (Courtesy of Sunnen Products Company.)

FIGURE 7-20 Honing oil is directed at each bearing bore by a T bar. (Courtesy of Sunnen Products Company.)

also available for installation and align boring to correct cam bearing bore misalignment. Finishing semi-finished bearings to the correct ID and surface finish requires particularly careful attention when using align boring equipment.

CYLINDER RECONDITIONING

Several different methods of cylinder reconditioning are employed by engine rebuilders, as follows:

1. *Cylinder boring.* Cylinders are bored to within about 0.003 in., then finish honed to fit the piston.
2. *Cylinder honing.* Cylinders are precision honed to exact preset size on an automatic honing machine.
3. *Cylinder sleeving.* Cylinder is resized to accommodate a repair sleeve, which is finished to the desired bore size after installation.
4. *Replaceable sleeves.* Some engines are designed with removable cylinder sleeves which are replaced when worn.
5. *Cylinder deglazing.* Engines with relatively little cylinder wear are often repaired by a minor overhaul. This includes replacing the piston rings. To ensure good ring seating against the cylinder wall, the cylinders are deglazed with a glaze breaker type of hone.

Before discussing these, a few facts about cylinder wall surface finish must be considered.

CYLINDER WALL FINISH

The cylinder wall surface finish must meet certain criteria for proper ring seating and cylinder wall lubrication. The following requirements should be met when reconditioning engine cylinders.

Crosshatch Pattern

A uniform crosshatch honing pattern is required. The included angle of this pattern should be approximately 45°. This means that the angle of the honing pattern is to be about 20 to 23° from horizontal (Figure 7-21). The crosshatch angle is a product of the rotating speed of the hone in revolutions per minute and the stroking rate in cycles per minute. A hone speed of about 400 rpm and a stroking rate of 30 cycles per minute are typical.

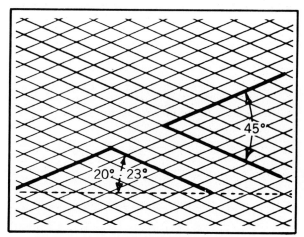

FIGURE 7-21 Ideal crosshatch pattern produced by honing. (Courtesy of Ford of Canada.)

Surface Finish

The cylinder walls should be free of glazed or burnished sections, embedded particles, tooling marks left by boring, tears, folds, and voids. There should be a plateaued area of from 60 to 80%. The plateaued area provides the ring seating surface and the valleys provide oil retention (Figure 7-22). The smoothness of the surface finish is measured in microinches [1 microinch is one-millionth of an inch (0.000001 in.)] using a highly sophisticated measuring instrument called a profilometer. A hand-held stylus is passed back and forth over the area being checked. The stylus is connected to the profilometer by an electrical cord. As the stylus passes over the surface, the profilometer automatically computes the average groove depth. The result of this is displayed on the profilometer scale as an rms (root-mean-square) figure. Master blocks of various surface

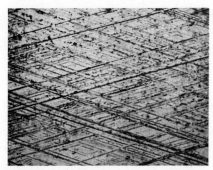

FIGURE 7-22 Faxfilm of recommended cylinder bore finish magnified 50 times. (Courtesy of Muskegon Piston Ring Company.)

smoothness are used for comparison purposes. Cylinder smoothness ranges from approximately 5 to 30 rms. Typical deviations from a good bore finish are shown in Figure 7-23.

Piston Ring Design and Cylinder Finish

The type of piston ring to be used in an engine determines the surface finish required on the cylinder walls. The almost mirror-smooth finish of a glazed cylinder wall is about 5 rms. This is too smooth for ring seating to occur. A 30-μ in. finish is close to the honed finish of a production engine. The following examples of the required surface finish for different ring types can be

COMMON DEVIATIONS FROM A GOOD BORE FINISH

ITEM	PHOTOGRAPH	EFFECT ON ENGINE PERFORMANCE	COMMON MAJOR CAUSES
1. Cross hatch grooves irregularly spaced	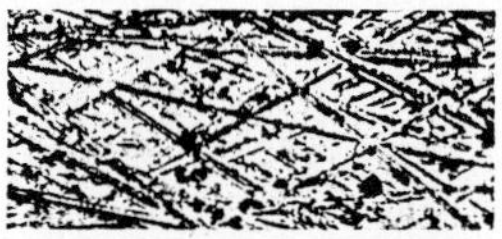	Poor oil distribution, erratic break-in oil economy.	Varying reciprocation rate, poor stone breakdown with alternate loading and breakdown, stone grade too hard and/or grit too coarse.
2. Wide deep cross hatch grooves		Causes abnormal wear, excessive oil consumption, poor and variable break-in period.	Stone grit too coarse, excessive stone pressure, poor stone breakdown, insufficient lubricant.
3. One directional cut		Causes rapid wear, poor seating, ring rotation, lowers oil control.	Excessive play in hone components.
4. Low cross hatch angle		High impact forces causing excessive wear, poor oil distribution, slower break-in of rings.	Low reciprocation rate compared to RPM used.
5. Cross hatch grooves folded and fragmented		Causes scratching, high wear, raises ring temperatures, high and erratic oil consumption.	Stone grit too coarse and/or grade too soft, insufficient lubricant.
6. Too little plateau		Excessive ring wear, high temperatures.	Stone grit too coarse, insufficient stock removal per cut.
7. Glazed plateau		Slows down seating-in and lowers economy.	Excessive stone pressures, loaded stones (too hard a grade) insufficient lubricant.
8. Excessive voids in cylinder wall		Lowers economy, rapid wear, reduces uniformity from engine to engine.	Insufficient stock removal, excessive stone pressure, lack of lubrication.

FIGURE 7-23 Common deviations from a good bore finish. (Courtesy of Muskegon Piston Ring Company.)

used as a guide, but the best procedure is to follow the piston ring manufacturer's recommendations for required surface finish.

Plain cast-iron rings	25–30 rms
Chrome-plated rings	20–25 rms
Stainless steel rings	15–20 rms
Moly rings	10–15 rms

Honing Stone Selection

A honing stone of a certain grit size will produce a specified microinch finish if the proper pressure and rotating speed are used. A 220-grit stone, for example, will produce a surface finish of 20 to 35 rms. Typical microinch surface finishes obtainable with different-grit-size honing stones are listed in Figure 7-24.

Grit Size	Microinch Finish	Automatic Hone CK-10 Stone Set	Hand-Operated Stone Set
70	85–105	EHU-135	
70	135–170		AN-101
150	25–40		AN-201
220	25–35	EHU-525	
220	20–25		AN-301
280	14–23	JHU-625	
280	15–20		AN-501
400	8–13	JHU-820	
400	5–10		N37-J85
600	4–8	C-30-CO3-81	
600	3–5		NN40-CO5

FIGURE 7-24 Microinch surface finish obtainable with stones of different grit sizes for automatic and hand-operated hones. (Courtesy of Sunnen Products Company.)

PISTON SIZES

Pistons are available in standard, + 0.030, + 0.040, and +0.060-in. sizes. Standard pistons can vary in size due to production tolerances by as much as 0.0006 in. Some manufacturers have as many as four or five different-size standard pistons to accommodate cylinder-size variations during production. For example, the Ford 2.3-liter four-cylinder engine has standard piston sizes available as follows:

Coded red	3.7764–3.7770 in.
Coded blue	3.7776–3.7782 in.
0.003 in. oversize.	3.7788–3.7794 in.

The Ford 351 V8 has the following sizes available:

Coded red	3.9978–3.9984 in.
Coded blue	3.9990–3.9996 in.
Coded yellow	4.0014–4.0020 in.
0.003 in. oversize.	4.0002–4.0008 in.

These figures indicate a production tolerance range of 0.0006 in. With a piston-to-bore clearance specification of from 0.0018 to 0.0026 in. it is easy to see that piston selection and finished bore size after honing are critical to maintaining proper piston-to-bore clearance. Most piston manufacturers build the operating clearance into the piston. A piston marked 0.030 in. oversize is in fact slightly smaller for this reason. This type of piston sizing allows cylinders to be rebored and finish honed to the size stated on the piston. Some piston manufacturers finish pistons to the exact size stated on the piston. In this case cylinder boring and honing must take into consideration the required operating clearance. Since piston sizes vary to such an extent, it is important that cylinders be bored and honed to fit the pistons to be used.

CYLINDER BORING

Cylinder boring machines with one or two boring bars are used in production engine remanufacturing. Portable boring bars are used in smaller shops. The boring bar utilizes a cutting tool mounted on a rotating cutter head. Several rotating speeds and feed travel rates are possible to accommodate differences in cutting depth and metal hardness. A typical speed–feed chart is shown in Figure 7-25.

Boring head rpm	Feed travel (inches per minute)			
	1 min.	2 min.	3 min.	4 min.
83 rpm	.127″	.249″	.393″	.564″
142 rpm	.217″	.426″	.670″	.966″
297 rpm	.455″	.891″	1.400″	2.000″
505 rpm	.772″	1.515″	2.380″	3.380″

FIGURE 7-25 Cylinder boring speed and feed chart. (Courtesy of Kwik-Way Manufacturing Company.)

Preparing the Block

Many remanufacturing shops use a deck plate during all boring and honing procedures (Figure 7-26). The deck plate is bolted to the top of the block and torqued to the cylinder head bolt torque specifications. This prestresses the block in the same manner as it will be after assembly. This is particularly critical on modern thin-walled block castings. If bored and honed without the deck plate, the cylinder may distort when the head

FIGURE 7-26 Typical deck plate used to prestress cylinder block during boring and honing. (Courtesy of Chrysler Canada Ltd.)

bolts are tightened. An angled deck plate must be used with blocks having an angled deck surface.

The main bearing caps must be installed and the bolts tightened to specifications prior to boring or honing the cylinders. If this is not done, the cylinders will become distorted when the caps are installed after boring and honing. Prior to boring, the deck surface must be parallel to the main bearing bore centerline. This ensures that cylinders will be perpendicular to the crankshaft since most boring machines are referenced from the block deck. The cylinders should be clean and dry.

Boring Procedure

The boring procedure includes the following basic steps.

1. Prepare the block for boring (as above).
2. Position and secure the block in the boring machine (Figure 7-27).
3. Center the boring bar (Figure 7-28).
4. Anchor the boring bar to the block (Figure 7-29).
5. Set the cutting tool to the desired depth of cut (Figure 7-30).

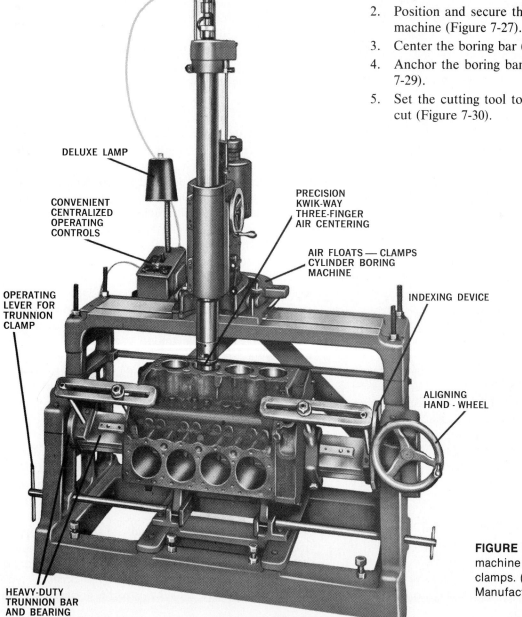

DELUXE LAMP

CONVENIENT CENTRALIZED OPERATING CONTROLS

PRECISION KWIK-WAY THREE-FINGER AIR CENTERING

AIR FLOATS — CLAMPS CYLINDER BORING MACHINE

OPERATING LEVER FOR TRUNNION CLAMP

INDEXING DEVICE

ALIGNING HAND - WHEEL

HEAVY-DUTY TRUNNION BAR AND BEARING

FIGURE 7-27 Cylinder boring machine with air float and clamps. (Courtesy of Kwik-Way Manufacturing Company.)

Cylinder Boring **109**

FIGURE 7-28 Three-finger centering is operated by a cone-shaped plunger. (Courtesy of Kwik-Way Manufacturing Company.)

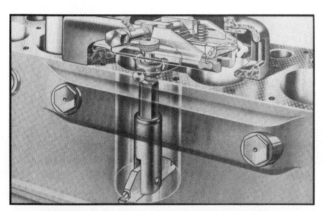

FIGURE 7-29 Clamping device anchors boring bar to cylinder block. (Courtesy of Kwik-Way Manufacturing Company.)

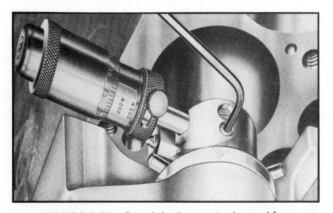

FIGURE 7-30 Special micrometer is used for setting depth of cutting tool. (Courtesy of Kwik-Way Manufacturing Company.)

6. Set the boring bar stop to the required stroke length for the cylinders being bored.
7. Select the appropriate boring speed and the rate of feed.
8. Bore each cylinder to the desired oversize (0.003 in. less than finished size).

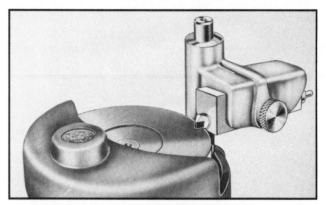

FIGURE 7-31 Dressing the cutting tool of the boring bar. (Courtesy of Kwik-Way Manufacturing Company.)

Make sure that the cutting tool is kept sharp (Figure 7-31). To ease and speed up the setting-up operation, modern boring machines are equipped with air centering, air float, and air clamping devices. An air-operated retracting loading table is also available on some machines. After boring to within 0.003 in. of the size desired, the cylinder is finish honed to fit the piston with a rigid hone.

Boring Bar Centering

Several approaches may be used to center the boring bar. The bar may be centered in the most worn part of the cylinder, in the unworn part of the cylinder, or in between. The method used is determined by individual experience and preference. If the bar is centered in the unworn part of the cylinder, all cylinders may not clean up at the preferred oversize setting. A larger bore diameter is often required when using this method since cylinders often wear more on one side than on the other. If the bar is centered in the most worn part of the cylinder, the smallest possible oversize can be used. This results in the least amount of metal removal but may cause the cylinder bore to be shifted over with respect to the original cylinder spacing. The boring bar may also be centered halfway up the bore. This results in a compromise between the other two methods.

SLEEVING A CYLINDER

If normal reboring does not clean up a damaged cylinder, it may be salvaged by installing a sleeve. The procedure consists of boring the cylinder oversize and installing a sleeve. A typical sleeve is shown in Figure 7-32.

Cast-iron sleeves are usually available in $\frac{1}{8}$- or $\frac{3}{32}$-in. wall thickness. The sleeve should be purchased prior to boring the affected cylinder. If the cylinder is bored first and the desired sleeve wall thickness is not available, the block may have to be discarded. The

FIGURE 7-32 Cutaway of cylinder sleeve. (Courtesy of Chrysler Canada Ltd.)

forming on the sleeve, place it in a pail of solvent and pack the dry ice around the outside and on the inside of the sleeve.

Place the chamfered end of the sleeve in the rebored cylinder. Position the sleeve squarely and keep it square at all times. Press or drive the sleeve in place using a driver that will not damage the sleeve. Using the boring bar, cut the top of the sleeve flush with the block deck. Bore the cylinder to the desired size and chamfer the top. Carefully check and correct the adjacent cylinders for distortion that may have occurred during sleeve installation. A 440-CID aluminum block with sleeves is shown in Figure 7-33. If there is any question as to success of the sleeving operation, the block should be pressure tested before proceeding further with rebuilding.

FIGURE 7-33 A sleeved 440-CID aluminum block. (Courtesy of Chrysler Canada Ltd.)

outside diameter of a sleeve of $\frac{3}{32}$ in. wall thickness is 0.253 in. larger than the standard bore size of the engine. The $\frac{1}{8}$-in.-wall-thickness sleeve is 0.1905 in. larger than the standard bore size. The extra 0.003 in. of stock provides for a tight press-fit in the rebored cylinder.

The cylinder should be bored 0.003 in. smaller than the OD of the sleeve to be installed. Reboring should extend to about $\frac{1}{8}$ in. from the bottom of the cylinder. This leaves a shoulder at the bottom to prevent the sleeve from moving downward in operation. On thin-walled G.M. 350-CID blocks, the ledge should be only $\frac{1}{16}$ in. high to ensure that the bottom of the sleeve has solid casting support below the bottom of the water jacket. The cylinder should be bored as smooth as possible to ensure good heat transfer.

Cut the sleeve $\frac{1}{8}$ in. longer than the rebored section of the cylinder. Make sure that you cut the sleeve at the end that is not chamfered and that you cut it square. The best method is to use a lathe. Remove any burrs. Coat the entire outer surface with a good non-insulating antiseize compound to prevent galling during installation. Use RTV sealer at the lower edge of the bore. To facilitate installation the cylinder can be carefully heated uniformly with a torch or the sleeve can be shrunk by cooling with dry ice. To prevent ice

RIGID HONING OF CYLINDERS

Finish Honing

Finish honing cylinders after reboring requires a rigid hone. The process can be done by hand or by use of an automatic honing machine with an expandable cylinder hone. Since boring cylinders leaves a relatively rough surface with torn metal, the last 0.003 in. of material must be removed with a rigid hone (Figure 7-34). Cylinders must be clean and dry prior to honing. The following honing procedure can be used (Figures 7-35 to 7-39).

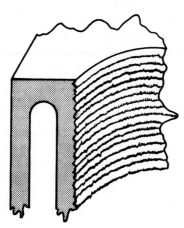

FIGURE 7-34 Cylinder boring leaves a relatively rough surface (exaggerated here) which must be honed to produce the correct surface finish. (Courtesy of TRW, Inc., Automotive Aftermarket Division.)

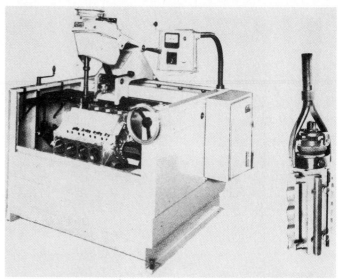

FIGURE 7-35 This honing machine controls stock removal and size automatically. More than 0.008 in. of stock can be removed in 1 minute. The average V8 block can be resized without reboring in 40 minutes. (Courtesy of Sunnen Products Company.)

FIGURE 7-36 This cylinder honing machine features a dwell mode which allows positioning the bone head at any height in the cylinder bore. Also featured are a variable stroking rate and auto retract. (Courtesy of Kwik-Way Manufacturing Company.)

FIGURE 7-37 Honing plate bolted in place on block. (Courtesy of TRW, Inc., Automotive Aftermarket Division.)

FIGURE 7-39 Cylinder hone in operation. (Courtesy of Western Engine Ltd.)

FIGURE 7-38 Hone in position in cylinder and coolant flow directed at cylinder. (Courtesy of Jasper Engine and Transmission Exchange.)

Honing Guidelines

1. Select honing stones of 180 grit to rough hone the cylinder to within 0.0005 in. Use a 280- to 320-grit stone to finish hone the cylinder.

2. Use a honing speed of about 200 to 400 rpm and a stroking speed of about 30 cycles per minute. Adjust the speed or stroking cycle as required to achieve the desired crosshatch pattern. Make sure that the hone travels the full length of the cylinder without coming in contact with the main bearing saddle bores.

3. True the stones in the cylinder with the hones expanded to provide light pressure. Stroke the hone up and down in the cylinder until the full length of the stones make contact with the cylinder wall.

4. Direct a continuous supply of honing oil to the stones. Honing oil flushes away the metal removed from the cylinder walls and the grit worn from the stones. This keeps the hones cutting and prevents the stones from loading up with foreign matter and overheating.

5. Make sure that hone guides are not too tight. With the hone expanded you should be able to wiggle the guides by hand. If they are too tight, remove the hone and file the guides until about 0.030 in. has been removed.

6. Make sure that the hone is stroking the full length of the cylinder and that adequate pressure is applied to reduce the possibility of stones becoming tapered.

7. The preferred method of honing is to use honing oil; however, honing can also be done dry. Stones that have been used with honing oil must not be used for dry honing since they will clog quickly and will not cut.

Resizing Cylinders with a Rigid Hone

Many shops use an automatic honing machine for resizing cylinders. This eliminates the need for a boring machine. The machine provides automatic stone feed, automatic stroking, and automatic shutoff. Honing is begun in the unworn part of the cylinder with short strokes to ensure original alignment throughout the honing operation. Cylinders are rough honed to within 0.0005 to 0.003 in. of finished size and then finish honed. A chamfering cone is used to chamfer the cylinders after honing.

CHAMFERING CYLINDER BORES

Cylinder bores should be chamfered after boring and honing. Piston assemblies are much easier to install in chamfered cylinders, and there is less danger of ring damage caused during installation. The chamfer should be cut at a 45° angle and should be $\frac{1}{16}$ in. wide. Insufficient chamfer can cause ring damage during piston installation. Too much chamfer will allow the oil ring rails to move outward too much during installation and allow them to be bent as they leave the ring compressor. Chamfering can be done by using a chamfering bit in the boring bar after cylinder boring is completed. Another method is to use a chamfering cone and an electric drill motor to do it by hand (Figures 7-40 and 7-41).

FIGURE 7-40 Chamfering cone has abrasive surface for chamfering cylinders after honing. (Courtesy of Sunnen Products Company.)

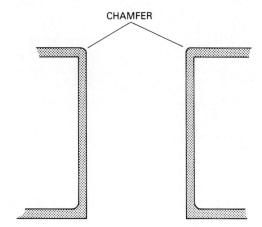

FIGURE 7-41 Chamfering cylinders aids in piston and ring installation. (Courtesy of FT Enterprises.)

DEGLAZING CYLINDERS

When piston rings are replaced in an engine without reboring or resizing the cylinders, the glaze must be removed from the cylinder walls. Cylinder walls be-

come glazed due to the polishing action of the piston rings and lubricating oils. If the glaze is not removed, the new rings will not seat properly and a good seal will not be formed between the rings and the cylinder wall.

The glaze should be removed with a flexible cylinder hone or beaded brush type of hone (Figures 7-

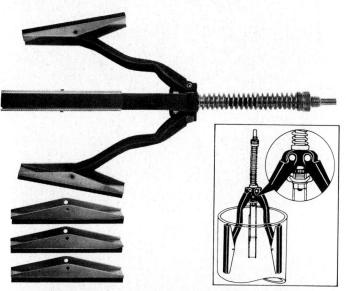

FIGURE 7-42 Flexible glaze breaker hone. (Courtesy of Easco/KD Tools, Inc.)

FIGURE 7-43 Beaded brush glaze breaker. (Courtesy of Easco/KD Tools, Inc.)

FIGURE 7-44 Deglazing with flexible cylinder hone. (Courtesy of Ford of Canada.)

FIGURE 7-45 Deglazing with bead glaze breaker. (Courtesy of Ford of Canada.)

42 to 7-45. The hone is driven by an electric drill motor at between 200 and 400 rpm. Light hone pressure against the cylinder wall is provided by spring action in the hone. Stroking speed should be adjusted to achieve the proper crosshatch pattern (Figure 7-46). A light oil should be used to prevent overheating and provide flushing action. An advantage of the beaded brush hone is that it also hones the tapered area just below the top of ring travel.

HONING THE LIFTER BORES

The lifter bores can be cleaned with an ordinary brake cylinder hone. The process is effective for removing varnish, rust, minor burrs, or scuffing. In no case should any significant amount of metal be removed. The bore diameter should not be increased by more than 0.0005 in. since excessive clearance results in reduced oil pressure.

THREADED HOLES

The threaded holes in the block should be cleaned up with the proper-size bottoming tap to remove any burrs from the threads. The holes should be checked at the surface for any sign of thread pull (Figure 7-47). Thread pull results in a slight raising of the metal surrounding the top of the hole. A fine mill file can be used to detect and remove any thread pull. The holes should be chamfered to avoid this condition when the engine is assembled (Figure 7-48).

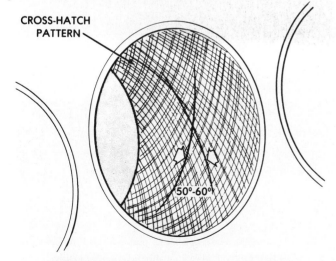

FIGURE 7-46 Crosshatch pattern should be produced by a cylinder deglazing. (Courtesy of Chrysler Canada Ltd.)

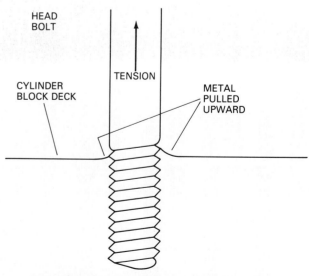

FIGURE 7-47 Thread pull around bolt holes in cylinder block. (Courtesy of FT Enterprises.)

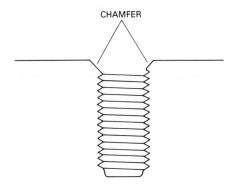

CHAMFER

FIGURE 7-48 Properly chamfered bolt hole. (Courtesy of FT Enterprises.)

DEBURRING THE BLOCK

Inspect the interior of the block for any casting slag or burrs. Remove burrs and slag with a high-speed grinder or rotary file as shown in Figures 7-49 and 7-50. Casting slag or burrs can hide dirt and sand, which can loosen and cause engine damage. A block that is properly deburred will feel smooth all over when you run your hand over the interior surface.

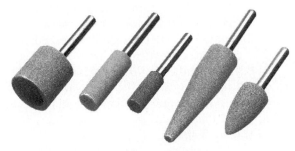

FIGURE 7-49 Rotary aluminum oxide deburring stones. (Courtesy of Mac Tools, Inc.)

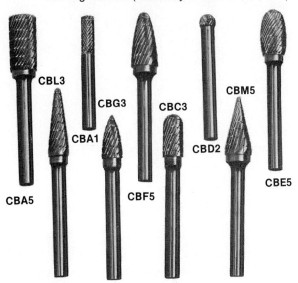

CBL3
CBG3
CBA1
CBC3
CBM5
CBD2
CBA5
CBF5
CBE5

FIGURE 7-50 Carbide rotary deburring tools. (Courtesy of Mac Tools, Inc.)

CLEANING THE BLOCK

The entire cylinder block must be cleaned after all the reconditioning and machining operations have been completed. Particular attention must be given to the engine cylinders, oil galleries, lifter bores, and to all holes and cavities. Abrasives and metal chips produced by machining must not be left in the engine, since they are highly destructive. Dirt, sludge, and carbon residues left in bolt holes results in false torque readings and can cause block damage. Foreign material left in oil galleries can restrict oil flow and cause scoring of moving parts (Figure 7-51).

The oil galleries, holes, and cavities should be cleaned with hot water and a good laundry detergent using a spray jet under pressure. Oil gallery and core plugs, and the cam bearing bore plug must be removed. This is normally done right after disassembly and before degreasing. Special attention must be given to cleaning the cylinders. Absolutely all abrasives must be removed. Use a solution of hot water and detergent and thoroughly wash each cylinder from top to bottom with a good soft-bristle brush. Keep washing until the solution remains clean. Dry the block with compressed

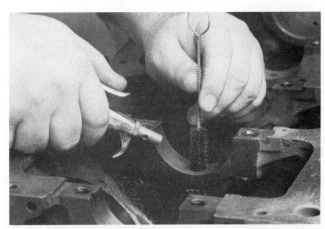

FIGURE 7-51 Cleaning the oil passages in the cylinder block. (Courtesy of TRW, Inc., Automotive Aftermarket Division.)

FIGURE 7-52 Cylinders should be absolutely clean. A white cloth should not smudge when wiping the cylinders. (Courtesy of TRW, Inc., Automotive Aftermarket Division.)

air and blow out all oil passages and bolt holes. Use a face mask for protection. The block should be clean enough not to smudge a white cloth when wiping the cylinders (Figure 7-52).

CORE PLUGS

Core Plug Dimensions

Core plugs are available in sizes with as little as 0.010 in. difference in diameter. Accurate sizing of core plug holes and core plug selection is therefore critical to leak-free operation. Cup-type plugs are also selected on the basis of cup depth and taper. Several cup depths are available in each size, as well as reverse taper plugs. Use only the type, diameter, and cup depth specified by the original-equipment manufacturer when installing core plugs (Figure 7-53).

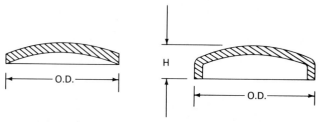

FIGURE 7-53 Dished (left) and cup (right) core plugs. (Courtesy of Ford of Canada.)

Threaded Plugs

Make sure that all threads are clean and in good condition. Use a good thread sealer on the threads of the plug and install to the specified torque.

Installing the Core Plugs

Core plugs must be installed to ensure leak-free operation. An incorrectly installed core plug may leak and cause the engine to overheat (Figures 7-54 to 7-56). The following guidelines will help eliminate core plug problems.

1. Make sure that the core holes in the cylinder block are absolutely clean. Remove all old sealer, burrs, rust, and dirt.
2. Use a good hardening type of sealer on the core plug. Coat the sealing surface with sealer just before installing the plug.
3. Make sure to use the correct type and size of core plug. There are three basic types.
   a. *Disc-type (dished) plugs*. These are used in core holes that are counterbored. The crowned side of the plug faces out. The

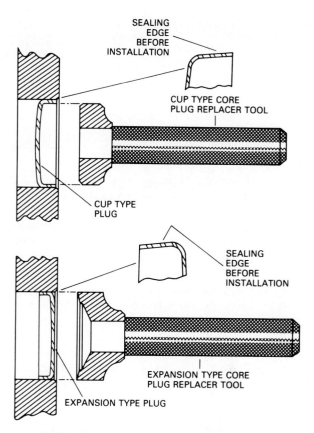

FIGURE 7-54 Cup plug with sealing edge at rim (top) is installed open side out. Plug with sealing edge near the closed side of the plug is installed with the closed side out (bottom.) (Courtesy of Ford of Canada.)

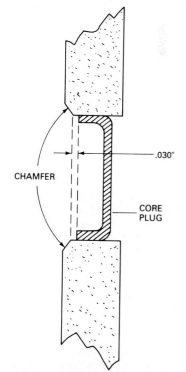

FIGURE 7-55 Typical cup plug installation. (Courtesy of FT Enterprises.)

FIGURE 7-56 Core plug installing tools. (Courtesy of Silver Seal Products Company.)

plug is installed with a driver that contacts the crowned center area of the plug. With the plug seated against the counterbore, flatten the plug enough to expand the plug tightly against the hole. Too little flattening can leave the plug loose. Flattening the plug too much or dishing it inward will loosen the plugs as well.

b. *Cup-type plug with open side out.* This plug has a tapered sealing edge. The plug is larger in diameter across the open edges than it is across the face. Use the proper-size driving tool to prevent cup distortion during installation. Keep the cup square during installation and install it to the specified depth, usually $\frac{1}{32}$ in. past the inner edge of the hole chamfer.

c. *Cup-type plug with open side in.* This plug has a reverse taper sealing edge. The open side of the cup is smaller in diameter than it is across the face. Use only the correct size of driver and install to the specified depth.

HEAT TABS

Heat-sensitive tabs are used by some engine rebuilders to ensure against false warranty claims due to overheating. The heat tab is installed on the cylinder head or behind the timing cover on the block. When the engine temperature reaches 250°F, the center melts, providing visual evidence of overheating. The heat tabs are installed with heat-resistant bonding materials, one type for cast iron and one for aluminum.

PAINTING THE BLOCK

The interior of the engine block is not normally painted by production engine rebuilders. For high performance, however, painting the interior of the block has some advantages. It enhances oil drain back, protects the surfaces from acid etching, and may help contain some casting grit that might otherwise loosen and cause problems. The areas painted include the front of the block, which is covered by the timing cover, the crankcase between and below the cylinders, and the lifter chamber. A good paint to use is GE Glyptal, available from electric motor rebuilding shops.

The exterior of the block should be painted with a good grade of heat-resistant engine enamel. Engine rebuilders usually paint the exterior of the short block or complete engine as the final operation before preparation for shipment. Areas of the engine where paint is not desired are masked before the engine is spray painted. Spray paint equipment similar to that used by body shops is used. Small shops may use spray paint in aerosol cans (Figure 7-57).

FIGURE 7-57 Spray can of exterior engine enamel. (Courtesy of Chrysler Canada Ltd.)

REVIEW QUESTIONS

1. What two methods can be used to determine the thickness of cylinder walls?

2. To prepare the cylinder block for decking:
   (a) remove the main bearing caps
   (b) remove the dowel pins
   (c) both (a) and (b)
   (d) neither (a) nor (b)

3. Name two types of equipment used to deck the cylinder block.

4. What determines whether a block needs decking?

5. Define "block deck height."

6. Define "positive and negative deck clearance."

7. Decking the block:
   (a) positions the head closer to the crankshaft
   (b) reduces block deck height
   (c) affects deck clearance
   (d) all of the above

8. Factors that affect deck clearance are:
   (a) connecting rod length
   (b) piston height
   (c) deck height
   (d) all of the above

9. Define "main bearing saddle bore alignment."

10. What effect does main bearing bore misalignment have on the main bearings?

11. What effect does main bearing cap stretch have on the main bearing?

12. To measure main bearing bore alignment:
    (a) the main bearing caps must be in place
    (b) use a dial bore gauge
    (c) both (a) and (b)
    (d) neither (a) nor (b)

13. When measuring main bearing bore alignment with a precision arbor, the arbor should be:
    (a) the same diameter as the saddle bores
    (b) 0.0001 in. less than the bore diameter
    (c) 0.001 in. less than the bore diameter
    (d) 0.01 in. less than the bore diameter

14. List the two most common methods of correcting main bearing bore misalignment.

15. The surface finish on a cylinder wall should:
    (a) have a 45° crosshatch pattern
    (b) be free of tooling marks left by boring
    (c) have a plateaued area of 60 to 80%
    (d) all of the above

16. True or false: Standard piston sizes may vary as much as 0.0006 in.

17. Define "microinch."

18. What piston oversizes are normally available?

19. What is a deck plate, and why is it used?

20. Engine cylinders are bored to within:
    (a) 0.03 in. of finished size
    (b) 0.003 in. of finished size
    (c) 0.0003 in. of finished size
    (d) 0.3 in. of finished size

21. The preferred method of centering the boring bar is in the part of the cylinder that:
    (a) is least worn
    (b) most worn
    (c) is not worn
    (d) is in between the least-worn and most-worn area

22. When sleeving a badly damaged cylinder:
    (a) purchase the sleeve before boring the cylinder
    (b) the cylinder should be bored 0.003 in. smaller than the sleeve OD
    (c) the cylinder should be bored as smooth as possible
    (d) all of the above

23. The crosshatch pattern in a cylinder is the product of the rotating speed of the hone and the ——————— speed.

24. What is the purpose of honing oil?

25. What can be done about honing guides that are too tight?

26. Why should cylinders be chamfered after honing?

27. What is meant by "cylinder deglazing"?

28. What is thread pull?

29. What is the advantage of deburring the block?

30. What are heat tabs used for?

CHAPTER 8

Crankshafts, Flywheels, and Balancers

OBJECTIVES

1. To develop an understanding of the purpose, design, construction, and operation of the crankshaft, flywheel, and balancer.
2. To develop the ability to inspect and rebuild engine crankshafts, including straightening, journal rebuilding, shot peening, machining and polishing of bearing journals, and radius grinding.
3. To develop the ability to resurface the flywheel, replace the starter ring gear and replace the harmonic balancer.

CRANKSHAFT MATERIALS

Automotive crankshafts are made of iron or steel alloys. They are cast, billeted, or forged (Figures 8-1 to 8-3). Cast crankshafts are made by heating metal to a liquid state and pouring it into a crankshaft mold. After the metal has solidified, it is removed and machined. Cast cranks have little or no grain. A billet crankshaft is produced from a solid chunk of metal by machining its shape as well as its finish. The metal billet has a grain to it, which provides greater strength. A forged crankshaft begins with a slug of metal that is formed in a heavy die under extremely high pressures. This results in a grain being created along the full length of the crankshaft.

Cast crankshafts satisfy the requirements of most standard production engines. High output and performance engines often use forged or billet cranks. One advantage of the billet crank is that it can be machined to very exact tolerances with regard to indexing of crank pins.

You can distinguish between forged and cast crankshafts by lightly striking a counterweight with a hammer. The forged crankshaft will have a ringing sound, while the cast crankshaft will have a much more dull sound to it, although some cast-steel shafts may be hard enough to have a similar-sounding ring. However, a cast crankshaft has straight, narrow casting mold parting lines visible on the crank arms, whereas forged crankshafts have wide twist lines on the crank throws, resulting from twisting the crank to index the throws.

CRANKSHAFT DESIGN

Main bearing journals are machined to a highly polished finish and are in perfect alignment with each other. The main bearing journals are mounted in the cylinder block with split-type precision bearing inserts, which are held in place by the main bearing caps. The crankshaft rotates freely in these bearings.

The connecting rod journals are offset from the crankshaft centerline, which causes the crankpin journals to orbit the crankshaft centerline as the shaft rotates. The distance from the main bearing journal center to the connecting rod bearing journal center is exactly one-half of the engine's stroke. This is sometimes called the crank throw (Figure 8-4).

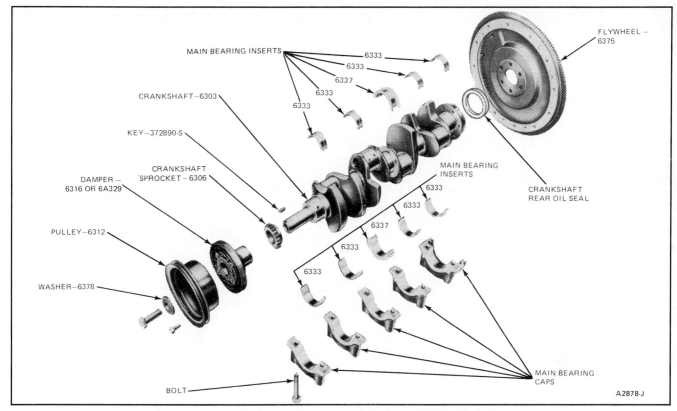

FIGURE 8-1 Typical crankshaft assembly including flywheel, damper, and pulley. (Courtesy of Ford of Canada.)

FIGURE 8-2 Cast-steel shot-peened crankshaft (top) and forged crankshaft (bottom.) Note sharp edges of cast shaft compared to forged shaft. (Courtesy of Chrysler Canada Ltd.)

FIGURE 8-3 Crankshafts awaiting reconditioning. Note mold parting lines on cast crankshafts. (Courtesy of Western Engine Ltd.)

A flange at the rear of the crankshaft provides the means for mounting the flywheel or converter drive plate. A seal journal is machined just ahead of the flange or on the OD of the flange to allow the rear main oil seal to effectively prevent oil leakage past the seal (Figures 8-5 and 8-6). An integral oil slinger is usually provided just ahead of the seal journal to deflect oil away from the seal. This increases the effectiveness of the seal.

The front of the crankshaft is machined with a keyway for the camshaft drive gear or sprocket and the vibration damper. A threaded hole is provided for secure bolting of the vibration damper to the crank (Figure 8-7). The front oil seal, located in the timing cover, seals between the cover and the vibration damper hub. An oil slinger deflects oil away from the seal.

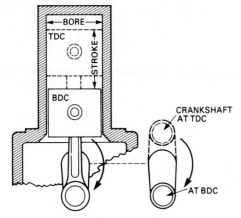

FIGURE 8-4 Crank throw dimension is exactly one-half the stroke dimension. (Courtesy of Ford of Canada.)

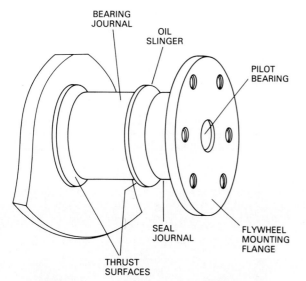

FIGURE 8-5 Flywheel end of crankshaft with oil slinger and flywheel mounting flange. (Courtesy of Ford of Canada.)

FIGURE 8-6 This crankshaft is designed with a spirally grooved seal journal on the flange OD and an oil slinger that is part of the flange. (Courtesy of Ford of Canada.)

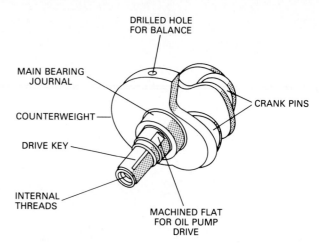

FIGURE 8-7 Typical crankshaft components at front end.

Thrust surfaces are machined on the side of one of the main bearing journals. These thrust surfaces and the flanged main bearing or thrust washers control crankshaft end play. Oil holes are drilled from the main bearing journals to the connecting rod journals to lubricate the connecting rod bearings. The main bearings receive their lubrication from the oil galleries in the engine block (Figures 8-8 ad 8-9).

Crankshafts have heavy sections of metal, or counterweights, opposite the crank throws. Counterweights offset the weight of the crank throws and connecting rods to provide crankshaft balance. Finish balancing is achieved by drilling the counterweights or sometimes also the connecting rod journals to remove some metal. This provides for a smoother-running en-

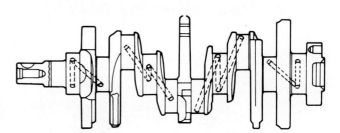

FIGURE 8-8 Dotted lines indicate drilled oil passages in crankshaft. (Courtesy of Chrysler Canada Ltd.)

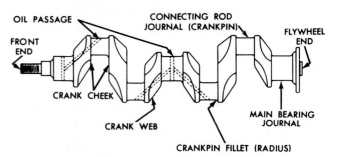

FIGURE 8-9 Crankshaft nomenclature. (Courtesy of Ford of Canada.)

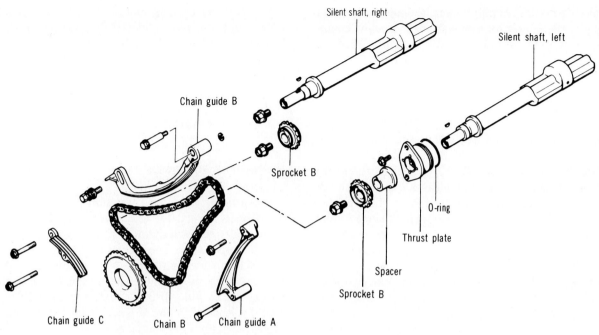

FIGURE 8-10 Balance shafts are used in many four-cylinder engines to reduce vibration. (Courtesy of Chrysler Canada Ltd.)

gine and longer crankshaft and bearing life. Some engines use balance shafts to improve engine balance (Figure 8-10).

Crankshaft configuration varies from engine to engine. Four-cylinder crankshafts have four connecting rod journals and usually three or five main bearing journals. Connecting rod journals are indexed (positioned) 180° apart to provide evenly spaced power impulses. Six-cylinder in-line crankshafts have six connecting rod journals and either four or seven main bearing journals. The four-main-bearing crankshafts must be of heavier construction than the seven-main-bearing crankshafts. Connecting rod journals are spaced 120° apart for even cylinder firing. The V6 crankshaft has four main journals and six crank throws spaced for even firing (Figure 8-11). The V8 crankshaft has five main journals and four connecting rod journals.

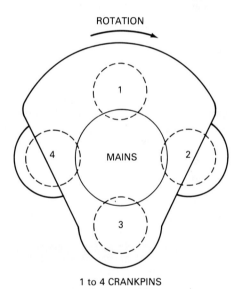

FIGURE 8-12 Four double-width crankpin journals are indexed 90° apart for even firing of a 90° V-8 engine. (Courtesy of FT Enterprises.)

Each connecting rod journal has two connecting rods attached to it. Crank throws are spaced 90° apart for even cylinder firing (Figure 8-12).

Diesel Engine Crankshafts

Diesel engine crankshafts are forged rather than cast. The forging process, although a more expensive manufacturing process, is needed to provide greater strength. Main and rod bearing journals on diesel

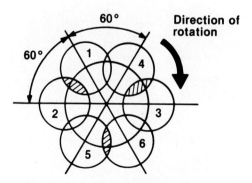

FIGURE 8-11 Six crankpin journals are indexed 60° apart for even firing of a 60° V-6 engine. (Courtesy of Chrysler Canada, Ltd.)

crankshafts are generally larger in diameter and may be wider than those for a similar-sized gasoline engine. Bearing journals are also usually induction hardened to increase wear resistance and service life.

FLYWHEEL

A flywheel is required to stabilize the speed fluctuations of the crankshaft resulting from power impulses of the engine cylinders. The flywheel stores energy during the power stroke and releases it during the non-power-producing strokes of the engine. The flywheel has an important speed-governing effect since it limits the speed increase or decrease during sudden changes in engine load. The flywheel provides a convenient means for mounting a large ring gear used for starting purposes. It also provides the frictional drive surface for a clutch-type mechanical drive. Engine timing marks are sometimes located on the flywheel. These degree markings indicate the TDC piston position usually for cylinder number 1 and are used to determine ignition or injection timing.

Flywheels are made from cast-iron alloys or cast or rolled steel. Provision for mounting is provided by the flywheel web in the center area. The precisely machined opening in the center ensures centering of the flywheel with the rotating axis of the crankshaft. The radial positioning of the flywheel is maintained by unevenly spaced mounting bolt holes or by a dowel pin. This relationship must be maintained for timing marks to be valid and to maintain balance.

Flywheel designs include the flat and recessed types and the torque converter flex plate (Figure 8-13). The recessed type reduces the overall size of the clutch and flywheel assembly. On engines equipped with automatic transmissions the torque converter serves the purpose of the flywheel and clutch assembly. The flex plate provides the connection between the crankshaft and torque converter.

HARMONIC BALANCER

The harmonic balancer (vibration damper) is needed to dampen normal torsional vibrations of the engine crankshaft. As each cylinder fires, it causes the crank throw to speed up. The inertia of the rest of the shaft causes it to stay slightly behind, resulting in a twisting action on the crankshaft. The torsional pulsations of successive cylinder firings create vibrational frequencies that vary with engine speed and with the number of engine cylinders. The vibration damper reduces the effects of these vibrations.

The vibration damper consists primarily of a hub and an inertia ring (Figure 8-14). The inertia ring is bonded to the hub through a flexible elastomer (rubber compound) insert (Figure 8-15). The inertia ring moves slightly in relation to crankshaft rotation as each cylinder fires, thereby dampening the torsional vibrations of the crankshaft over a wide range of engine speeds. Some dampers are designed with two inertia rings of different sizes for more effective control over a wide range of vibrational frequencies.

Over an extended period, the elastomer can deteriorate or the bonding can let go, rendering the damper ineffective or causing vibrations itself as a result. A damaged damper must be replaced. On damper designs where the hub is also a seal journal, the seal can wear a groove in the hub, resulting in oil leakage. A sleeve-type repair can restore the damper if it is other-

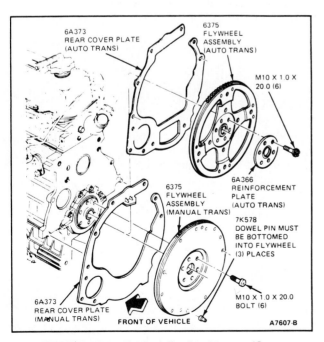

FIGURE 8-13 Typical flywheel types. (Courtesy of Ford of Canada.)

FIGURE 8-14 Typical vibration damper. (Courtesy of Red River Community College.)

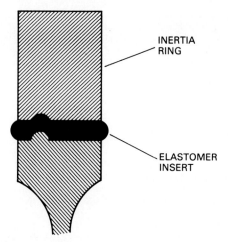

FIGURE 8-15 Cross section of vibration damper ring. (Courtesy of FT Enterprises.)

wise in good condition. The hub may require machining in some cases to accommodate the repair sleeve. See Figure 5-18 in Chapter 5.

BALANCE SHAFTS OR GEARS

In six- and eight-cylinder engines, the inertia forces of the reciprocating piston and rod assemblies can be more easily offset by the use of crankshaft counterweights. This can be done due to the closer crank indexing. On four-cylinder engines, this is not possible. To offset these forces, some four-cylinder engines use two counterweighted balance shafts or gears turning in opposite directions and turning at twice the crankshaft speed (Figure 8-16). These counterweights provide equal but opposing forces to engine vertical forces. The shafts or gears must be timed to the crank-

shaft position. Even though individual engine components such as crankshafts and flywheels are balanced, the entire rotating assembly must also be balanced. There are two forces at work that tend to create an unbalanced condition:

1. *Centrifugal force.* This results from rotary motion of the crankshaft, creating a force outward from the axis of rotation. The magnitude of this force changes with engine speed.
2. *Inertia forces.* These are caused by the reciprocating movement of pistons and connecting rods. The magnitude of these forces also changes with engine speed.

The centrifugal forces of the rotating crankshaft and the inertia forces of the reciprocating pistons can be offset by the use of crankshaft counterweights in four-cycle engines with more than four cylinders. This is possible since crank throws can be positioned to achieve the desired result. On engines with four cylinders or less, this is not possible. Even though the crankshaft as a whole can be balanced, individual crank throws remain unbalanced.

To reduce the stress caused by these forces on higher-speed engines, gear-driven balance weights or balance shafts are used. Balance weights are positioned to offset the inertia forces of the pistons. Two gears with balance weights are used in one balancer design. The gears are timed to the crankshaft gear in a manner that results in the upward force of the piston being offset by the downward force of the counterweight on the balancer. Gear-driven balance shafts with counterweights operate in a similar manner (Figures 8-17 to 8-20).

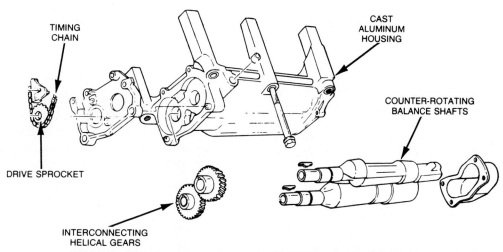

FIGURE 8-16 Counter-rotating balance shafts, mounting and drive components for Chrysler 2.5-liter, four-cylinder engine. (Courtesy of Chrysler Canada Ltd.)

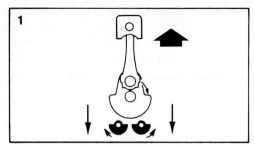

1. At top dead center (TDC) the forces will be going up at the piston.

   A. To offset the vibration — notice the crankshaft and balance shaft lobes, their force is down.

FIGURE 8-17 (Courtesy of Chrysler Canada Ltd.)

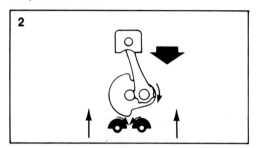

2. At 90 degrees of the crankshaft rotation, all pistons are level.

   A. On this downward stroke, a piston has travelled more than ½ of its travel distance in the cylinder.

   B. The engine forces are downward because the mass (downward pistons) have been accelerating since they have to travel a greater distance than the up-bound pistons to reach the point at which they are all level.

   C. The balance shafts, rotating at twice the speed of the crankshaft, exert an upward force to counter the forces in item B above.

FIGURE 8-18 (Courtesy of Chrysler Canada Ltd.)

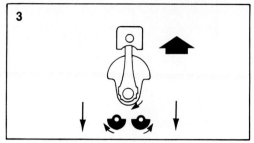

3. At 180 degrees, there is an upward force in the engine.

   A. This is due to the forces at the other pistons on the upward stroke.

   B. The dual counter-rotating balance shafts, travelling twice the speed of the crankshaft, have their lobes down, exerting a downward force to counteract the upward force.

FIGURE 8-19 (Courtesy of Chrysler Canada Ltd.)

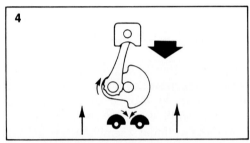

4. At 270 degrees of crankshaft rotation, there is a downward force again.

   A. This is not due to the piston going up (past 270 degrees of crankshaft rotation) but to the pistons coming down and accelerating in travel.

   B. Again, to offset this force, the dual counter-rotating balance shaft lobes move to the top position, minimizing the downward force.

FIGURE 8-20 (Courtesy of Chrysler Canada Ltd.)

CRANKSHAFT INSPECTION

The crankshaft should be inspected to determine whether it should be reground or scrapped. The following points should be considered.

1. Check the keyway for damage or wear. A keyway that is badly damaged or worn will allow movement of the key and sprocket.

2. Check the threads in the front of the crankshaft for damage. Chase the threads with a tap. If the threads are badly damaged, a thread repair insert can be used.

3. Check the flywheel flange for damage or distortion. A bent flange will cause flywheel or converter runout. Inspect the threaded holes in the flange. Chase the threads to clean them. Repair badly damaged threads with a thread repair insert.

4. Inspect the pilot bearing bore for wear or damage.

5. Check the rear main oil seal journal for wear or damage. It is located just ahead of the flange on some crankshafts or on the flange OD on others. Repair sleeves or oversized seals are available for some engines. The seal

journal surface must be smooth and completely free of nicks and burrs.

6. Visually inspect the main and rod bearing journals for damage. Rebuild damaged journals by metal spraying or submerged arc welding if cost-effective.

7. Check the journals with a micrometer to determine whether regrinding is feasible. If the journals have previously been ground to maximum undersize, the crankshaft is scrapped. Measure the width of all journals and thrust surfaces to determine if regrinding is feasible. If too little stock remains, the crankshaft may have to be scrapped, the thrust surfaces rebuilt by welding or by using oversize thrust bearings.

8. Place the crankshaft in V blocks and check for bends (with a dial indicator) in the snout area, in the area of main bearing journal alignment, and in the flywheel mounting flange. Straighten the crankshaft as necessary in a press fixture designed for the purpose. (*CAUTION:* Do not bend nitrided crankshafts since cracks may be produced. Check for nitriding with a file on a noncritical area of the crankshaft. If the metal is easily filed, it has not been nitrided.)

9. Inspect the crankshaft for cracks with magnetic crack detection equipment. Areas where cracks are most commonly found are shown in Figure 8-21.

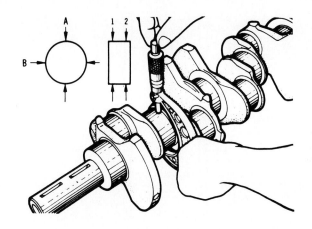

A VS B = VERTICAL TAPER
C VS D = HORIZONTAL TAPER
A VS C AND B VS D = OUT OF ROUND
CHECK FOR OUT-OF-ROUND AT EACH END OF JOURNAL

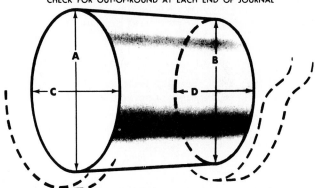

FIGURE 8-22 Measuring bearing journals. (Courtesy of Federal-Mogul Corporation, top, and Ford of Canada, bottom.)

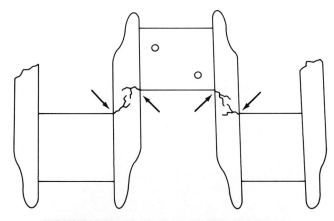

FIGURE 8-21 Critical load areas of crankshaft are most subject to cracking as shown here. (Courtesy of FT Enterprises.)

MEASURING THE CRANKSHAFT

Crankshaft main and connecting rod journals must be measured with an outside micrometer and a dial indicator to determine the following (Figure 8-22).

Journal size and Whether Standard or Undersized. If the crankshaft has been reground, the journals will be undersized. Some engine crankshafts may have one or more main or rod journals that are one or two thousandths of an inch undersized from the factory.

Journal Out-of-Round. The difference between vertical and horizontal diameters of each journal taken at the front and rear determine their roundness. A maximum out-of-round limit of 0.001 in. is usually acceptable.

Journal Taper. The difference between the vertical or horizontal measurements at the front and rear of each journal is the amount of taper present. To be acceptable, this should not exceed 0.001 in. Out-of-round and taper in excess of .001 in. indicate that the crankshaft must be reground.

Barrel-Shaped Wear (Figure 8-23). This is checked vertically and horizontally at the front, center,

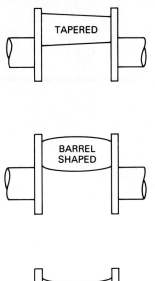

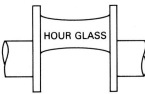

FIGURE 8-23 Bearing journals wear and may become tapered, or hourglass- or barrel-shaped. (Courtesy of FT Enterprises.)

and rear of each journal. A maximum of 0.001 in. is acceptable.

Ridging. Ridging is the result of journal wear occurring on both sides of a grooved bearing. If ridging exceeds 0.0003 in. the crankshaft should be reground (Figure 8-24).

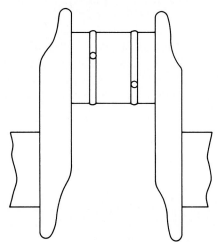

FIGURE 8-24 Typical journal ridging. (Courtesy of FT Enterprises.)

Crankshaft Runout. This must be checked with a dial indicator and a set of V blocks. The crankshaft is supported by V blocks at the two end main journals. (The V blocks must be smooth and well oiled to prevent the journals from being scored.) A dial indicator is mounted with the plunger travel perpendicular to the journal surface being measured. All unsupported main journals and the crank snout should be checked for runout as the crankshaft is turned (Figure 8-25). The total indicator reading for one revolution is the amount of runout for each position measured. Max-

FIGURE 8-25 Checking crankshaft runout with a dial indicator. (Courtesy of Federal-Mogul Corporation.)

imum runout should not exceed 0.001 or 0.002 in. Check against crankshaft specifications in the service manual. If runout is excessive, the crankshaft must be straightened in a press fixture designed for the purpose.

Thrust Surface Wear. This can be measured with an inside micrometer or a telescoping gauge and outside micrometer. Measure between the thrust surfaces parallel to the journal (Figure 8-26). Compare measurements with specifications. Excessive thrust surface wear requires crankshaft replacement, re-

FIGURE 8-26 Measuring thrust surface wear. (Courtesy of Federal-Mogul Corporation.)

conditioning by welding to replace lost metal and machining, or the use of oversized thrust bearings.

CRANKSHAFT STRAIGHTENING

Crankshafts that exceed maximum runout specifications are straightened in a fixture equipped with a hydraulic press prior to grinding (Figure 8-27). The V blocks are positioned to support the crankshaft to be straightened and the crankshaft positioned on the press. The hold-down clamps are positioned on the main journals on each side of the section to be straightened. The hydraulic ram is positioned under the rod journal affected. The dial indicator is positioned over the main journal that is out of alignment and set at zero (Figures 8-28 and 8-29). Hydraulic pressure is applied until the desired reading on the dial indicator is obtained. A round-edged chisel and hammer are used to "set" the correction. A runout check is made after straightening to ensure accuracy.

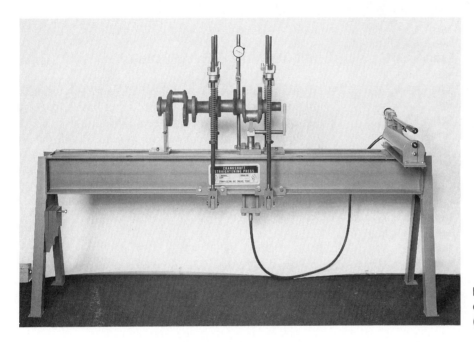

FIGURE 8-27 Hydraulic crankshaft straightening press. (Courtesy of Storm Vulcan Co.)

FIGURE 8-28 Crankshaft in position for straightening. Hydraulic ram pushes up against crankshaft journal. (Courtesy of Storm Vulcan Co.)

FIGURE 8-29 Straightening a large diesel engine crankshaft. (Courtesy of Western Engine Ltd.)

FACTORY-UNDERSIZED CRANKSHAFTS

During the manufacturing process, minor damage may occur from handling or machining errors which would require discarding the crankshaft if nothing but standard journal diameters were to be produced. Rather than discarding such crankshafts, they are salvaged simply by machining the slightly damaged journals to an undersize of 0.001 in. and up to 0.010 in.

Engine manufacturers use different methods to indicate that crankshaft journals are undersized. For example, Chrysler, on their 318- and 360-CID engines, stamps the number 8 counterweight with an R-1, R-2,

R-3, or R-4 to indicate that crankpin 1, 2, 3, or 4 is 0.001 in. undersized. A stamp on the 1 and 8 counterweights of M-1, M-2, M-3, M-4, or M-5 indicates that main journal 1, 2, 3, 4, or 5 is 0.001 in. undersized, whereas R-10 or M-10 indicates 0.010 in. undersized (Figure 8-30). Other designations are used on other Chrysler engines. On their 231-, 350-, and 400-CID engines, General Motors uses a spot of orange paint on the counterweights to indicate a rod journal undersize of 0.010 in. Obviously, the most positive method of determining crankshaft journal sizes is by measurement with an outside micrometer. This method assures accuracy.

NITRIDED CRANKSHAFT CHECK

Nitriding is a process of crankshaft hardening using an ammonia gas furnace or a special salt bath. To determine if a crankshaft has been nitrided, use a 10% aqueous solution of copper ammonia chloride, available from a pharmacy. Using an eyedropper, apply 1 drop of the solution to the bearing journal. If the drop turns to a copper color within about 10 seconds, the crankshaft has not been nitrided.

REBUILDING JOURNALS

Crankshaft bearing journals are rebuilt only if the required repair is cost-effective. If a number of journals or all the journals require rebuilding, the cost would

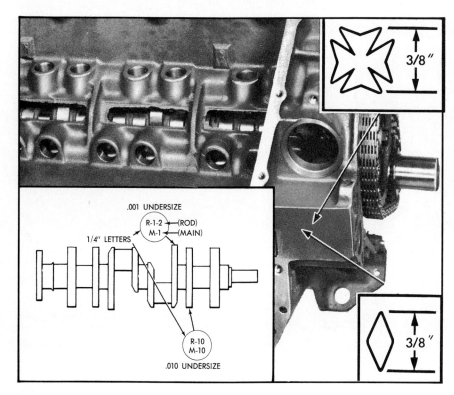

FIGURE 8-30 A Maltese cross stamped on the engine ID pad indicates that rod or main bearing journals are 0.001 in. undersized. If an X is added, 0.010 in. undersize is indicated. A diamond stamped on the pad indicates that lifter bores are 0.008 in. oversized. Stampings on the crankshaft indicate rod and main journal undersizes as shown. These markings are typical for some Chrysler V-8 engines. (Courtesy of Chrysler Canada Ltd.)

FIGURE 8-31 Submerged arc welding equipment for rebuilding crankshaft journals. (Courtesy of Peterson Machine Tool, Inc.)

1. Welding Head vertical adj.
2. Welding tip
3. Ammeter
4. Wire speed knob
5. Volt-ammeter
6. Rotation speed control knob
7. Inch wire
8. Travel Speed
9. Polish & Weld switch
10. Weld switch
11. Direction switch
12. Hand knob and clutch
13. Flux valve
14. Flux nozzle
15. Stroke adjustment
16. Chuck lock
17. Adjustment for counterweight
18. Locking pin
19. Throwhead locks
20. Clutch lever
21. Ammeter
22. Volt-ammeter
23. Coarse voltage knob
24. Fine voltage knob
25. Power switch
26. Coarse voltage readout
27. Flux vacuum switch

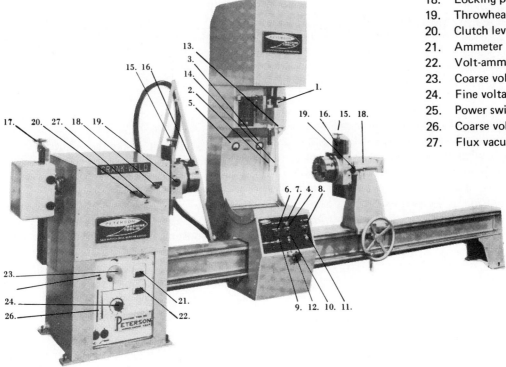

be prohibitive. However, many crankshafts are salvaged by journal rebuilding. Several methods are used to replace metal on badly worn journals: (1) metal spraying, (2) submerged arc welding, and (3) chrome plating.

Metal Spraying

In this procedure the crankshaft is mounted in the lathe to provide crankshaft rotation. While the journal is turning, molten steel is sprayed evenly over the bearing surface.

Submerged Arc Welding

This popular arc welding procedure deposits a steel alloy on the journal surface in a continuous bead as the crank is rotated. The metal is fed evenly from the welder in wire form, while the actual arc is submerged in a special flux to prevent porosity and contamination. The automatic welding head can be positioned over

any journal. Submerged arc welding is the preferred method of journal rebuilding (Figures 8-31 and 8-32).

FIGURE 8-32 Crankshaft journal rebuilt by submerged arc welding process. (Courtesy of Jasper Engine and Transmission Exchange.)

Chrome Plating

In this process the crankshaft is mounted and submerged in electrolyte in a chrome plating tank. Chrome plating material is in the electrolyte. As electrical current is passed through the electrolyte to the rotating crankshaft, the chromium is deposited on its surface. This method involves heating, magnetic, and chemical processes. This produces a very hard, nonporous, wear-resistant surface. Chrome plating is restricted to heavy-duty and some high-performance engines. After chrome plating the journals are machined to size and polished.

SHOT PEENING

Shot peening is a process whereby very fine steel shot (tiny pellets) is blasted under high velocity at the part being treated. Shot peening is done in a cabinet equipped with a turntable. The part to be peened is placed on the turntable and the cabinet closed. In operation the shot is blasted at the part as it turns. Shot peening improves the fatigue life of the part and imparts a slight hardening to the surface.

CROSS DRILLING

Cross drilling of crankshaft bearing journals is often done to provide additional lubrication for high-performance and competition engines. Cross drilling provides two oil holes 180° apart on all journals. On V engines with two rod journals on the same crank throw there would be four oil holes. All oil holes must be chamfered to blend smoothly into the journal surface, with the maximum chamfered area diameter to be no more than twice the diameter of the oil hole. Chamfering improves oil distribution.

CRANKSHAFT GRINDING

Crankshaft grinding requires a sophisticated, expensive machine and a highly skilled operator. The process requires close observation, constant checking and adjustments, and proper maintenance of machining equipment. Accuracy of precision measuring, machining, and calculations is of utmost importance. The objectives in crankshaft grinding are to achieve the following results on the main and connecting rod bearing journals (refer to Figure 8-33 and the Reference Data on page 135).

1. *A and D* (journal diameter). Journals are machined to the undersize required to clean up

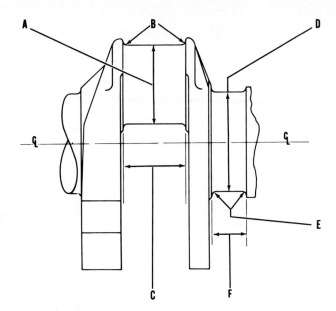

FIGURE 8-33 Critical crankshaft grinding dimensions are indicated by the letters A to F. (Courtesy of Federal-Mogul Corporation.)

all journal imperfections and for which undersize bearings are available. Journals must be machined to ensure that journal taper and out-of-round dimensions do not exceed allowable limits. The surface finish produced must meet the microinch specifications for the crankshaft being ground.

2. *B and E.* The radii must be machined to the dimensions and specifications for each crankshaft being reground. The radius must blend smoothly with the newly machined journal surface. The surface finish must meet the microinch specifications for the crankshaft being reground.

3. *C and F.* The width dimensions at C and F must meet the specifications for the crankshaft being reground.

Crankshaft Grinding Limits

Most automotive crankshafts are not reground beyond 0.030 in. undersize. Removing too much metal weakens the crankshaft and can result in breakage. Larger industrial and transport engine crankshafts can be reground to 0.060 in. or more undersize, depending on the make, model, and size of the engine. However, some crankshafts are limited to a maximum undersize of 0.010 in., as is the case with the Ford 3.8-liter V6. This is also true of many other modern engines. Follow the grinding limit specifications of the crankshaft manufacturer to avoid possible crankshaft breakage.

Dressing the Grinding Wheel

The crankshaft grinding wheel should be dressed frequently to ensure accuracy and quality of work. Dress the face, corners, and sides of the grinding wheel as required (Figure 8-34). Follow the procedures recommended by the equipment manufacturer. The quality and accuracy of the grinding job are only as good as the condition of the grinding wheel. A grinding wheel that is out-of-round or in static unbalance will result in uneven grinding and an uneven bearing journal surface. Grinding wheel balance is checked on a balancing stand (Figure 8-35).

FIGURE 8-35 Balancing stand used to check balance of crankshaft grinding wheel. (Courtesy of Storm Vulcan Co.)

Grinding Direction

For best results, turn the crankshaft in the direction of engine rotation while grinding against this direction (Figure 8-36). Do the opposite when polishing the journals after grinding. This ensures that the minute projections of a 10- to 15-μin. finish will slide easily across the bearing surface when installed (Figure 8-37).

FIGURE 8-34 Dressing the face, sides, and corners of the grinding wheel to ensure accurate grinding of the journal surface and radii. (Courtesy of Peterson Machine Tool, Inc.)

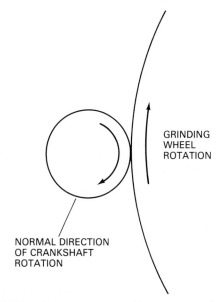

FIGURE 8-36 Proper rotation of crankshaft and grinding wheel. (Courtesy of FT Enterprises.)

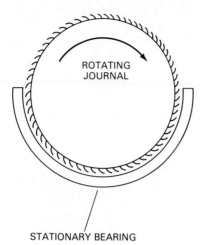

ROTATING
JOURNAL

STATIONARY BEARING

FIGURE 8-37 Minute projections (exaggerated here) slide easily over bearing surface if journals are ground and polishes in the proper direction. (Courtesy of FT Enterprises.)

Fillet Radii

Check the radius on each side of each journal with the radius gauge specified for the crankshaft being ground. Some touchup grinding may be required after journal grinding. Some crankshafts have a deep fillet between the bearing surface and the radius, which ensures that the radius will not be ground away while regrinding the journals.

The fillet radius on both sides of the bearing journal must be a continuous quarter round that blends smoothly into the bearing journal and crank cheek without increasing rod-to-cheek clearance. Fillet radii should not be ground below minimum specified tolerance since this weakens the crankshaft. The radius must not be above the maximum specified since this would cause interference with the bearing. Fillet radii should be ground to the surface finish specified (Figures 8-38 to 8-44). Some heavy-duty crankshafts have cold-rolled fillets which must be shot peened after grinding but before final polishing.

Oil Hole Chamfer

Check the chamfer at each oil hole to make sure that it blends smoothly into the bearing surface. Correct with a fine rotary file if necessary. The maximum diameter of the chamfer should not exceed twice the diameter of the oil hole.

Reference Data: Main and Crankpin Journals

The following reference data are reproduced courtesy of Federal Mogul Corporation. Items marked by asterisks apply to heavy-duty and/or highly loaded engines.

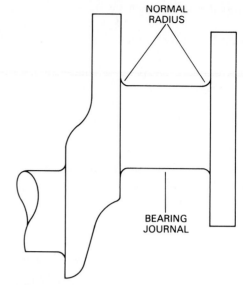

NORMAL
RADIUS

BEARING
JOURNAL

FIGURE 8-38 Typical journal radii. (Courtesy of FT Enterprises.)

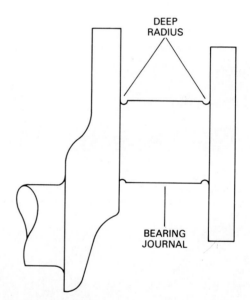

DEEP
RADIUS

BEARING
JOURNAL

FIGURE 8-39 Deep fillet radii. (Courtesy of FT Enterprises.)

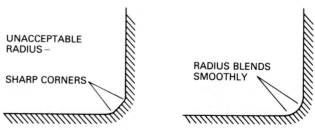

UNACCEPTABLE
RADIUS —

SHARP CORNERS

RADIUS BLENDS
SMOOTHLY

FIGURE 8-40 Unacceptable (top) and acceptable (bottom) fillet radii. (Courtesy of FT Enterprises.)

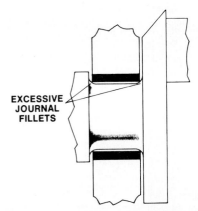

FIGURE 8-41 Too large a fillet radius results in fillet ride and rapid wear. (Courtesy of Federal-Mogul Corporation.)

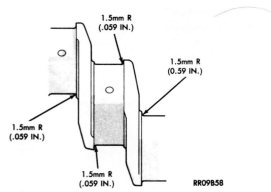

FIGURE 8-42 Typical radius dimensions. (Courtesy of Chrysler Canada Ltd.)

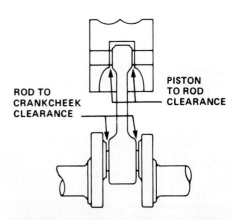

IF ROD TO CRANKSHAFT CLEARANCE EXCEEDS PISTON TO ROD CLEARANCE ON EITHER SIDE, THE ROD MAY BOSS IN PISTON RATHER THAN ON CRANKSHAFT AS IT SHOULD. THIS MAY CAUSE COCKING OF PISTON AND EXCESSIVE OIL CONSUMPTION.

FIGURE 8-43 Rod-to-crank check clearance must not be increased when grinding crankshafts. (Courtesy of Federal-Mogul Corporation.)

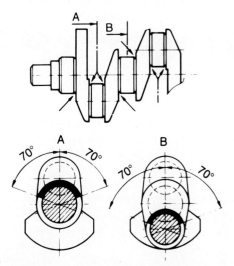

FIGURE 8-44 Roll-hardened area on rod and main journals of this Chrysler 2-liter turbo diesel engine extends over 140° as shown. This hardened area must remain intact after grinding. (Courtesy of Chrysler Canada Ltd.)

Diameter Tolerance

Shaft Diameter [in. (mm)]	Tolerance [in. (mm)]
Up to 1½ (38)	0.0005 (0.013)
1½–10 (38–250)	0.001 (0.025)
10 (250) and over	0.002 (0.050)

Crankshaft Taper (Diametral Taper Tolerance). Taper across the bearing in the most heavily loaded area is more critical than in other areas.

Journal Length [in. (mm)]	Tolerance [in. (mm)]	Tolerance* [in. (mm)]
Up to 1 (25)	0.0002 (0.005)	0.0001 (0.003)
1–2 (25–50)	0.0004 (0.010)	0.0002 (0.005)
2 (50) and over	0.0005 (0.013)	0.0003 (0.007)

Hourglass or Barrel-Shape Condition. The standards are the same as for taper. Following are the out-of-round (OOR) conditions.

Shaft Diameter [in. (mm)]	Maximum Allowed OOR [in. (mm)]
Up to 3 (75)	0.0005 (0.013); 0.0002 (0.005)*
3–5 (75–125)	0.0005 (0.013); 0.0003 (0.007)*
5 (125) and over	0.001 (0.025); 0.0004 (0.010)*

FIGURE 8-45 Typical crankshaft grinder with crankshaft in position. (Courtesy of Storm Vulcan Co.)

Axial and Circumferential Surface Variations. It is suggested that journal traces (on precision surface and roundness instruments) be made periodically to monitor the axial and circumferential geometry of the journals. Standards for *all* engines:

1. *Waviness:* a widely spaced axial variation in surface linearity. Waviness should be held within 0.0001 in. (0.0025 mm) T.I.R.
2. *Lobing:* a circumferential journal irregularity consisting of a gradually undulating surface contour with several peaks and corresponding valleys which are spaced approximately equally around the entire journal. The presence of definite three to seven lobe patterns in the circumference is highly undesirable. Lobing should be held within 0.0001 in. (0.0025 mm) T.I.R.
3. *Chatter:* a circumferential journal surface irregularity marked by a large number of peaks and corresponding valleys which are approximately evenly spaced, but are not necessarily in evidence around the entire journal. Maximum chatter should not exceed 0.000050 in. (0.0013 mm).

Crankshaft Machining and Finishing. For *all* engines, the best machining practice is as follows: Turn in the direction of engine shaft rotation, then grind against this direction toward the high limit, and then lightly lap (polish) in the direction of engine shaft rotation only to remove grinding fuzz. A maximum of

FIGURE 8-46 Grinding a crankshaft journal. (Courtesy of Jasper Engine and Transmission Exchange.)

136   *Crankshafts, Flywheels, and Balancers*

FIGURE 8-47 Polishing a crankshaft journal. (Courtesy of Western Engine Ltd.)

FIGURE 8-48 Close-up of journal being polished. (Courtesy of Jasper Engine and Transmission Exchange.)

FIGURE 8-49 Centering rod attachment with dial indicator for centering crankshaft journals. (Courtesy of Peterson Machine tool, Inc.)

240-grit-size polish paper should be used with maximum stock removal of 0.0001 in. (0.003 mm). The final surface finish should be no greater than 15 μin. (10 μin.)*. To prevent disruption of the critical oil film and resultant bearing damage, oil holes should be well blended into the journal surface. The maximum diameter at the runout of the blend should not exceed twice the hole diameter.

See Figures 8-45 to 8-51 for typical crankshaft grinding equipment and procedures.

FIGURE 8-50 Continuous measurement attachment allows continuous monitoring of journal size. (Courtesy of Peterson Machine Tool, Inc.)

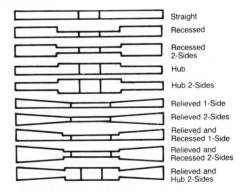

Straight
Recessed
Recessed 2-Sides
Hub
Hub 2-Sides
Relieved 1-Side
Relieved 2-Sides
Relieved and Recessed 1-Side
Relieved and Recessed 2-Sides
Relieved and Hub 2-Sides

FIGURE 8-51 A variety of crankshaft grinding wheels is available to suit any requirement. (Courtesy of Silver Seal Products Company, Inc.)

VIBRATION DAMPER INSPECTION

The vibration damper (harmonic balancer) hub must be inspected for grooved wear caused by the timing cover seal. Grooved wear can be repaired by installing a repair sleeve on the hub. This should only be done if the damper is otherwise in good condition (Figures 8-52 and 8-53). Check the resilience of the rubber ring by pressing a screwdriver tip against it. It should feel lively. If it is badly cracked, is loose, or there are pieces missing, the damper must be replaced. If the pulley is part of the balancer, inspect the pulley groove for stepped wear on the sides. A badly worn pulley should be replaced (Figure 8-54).

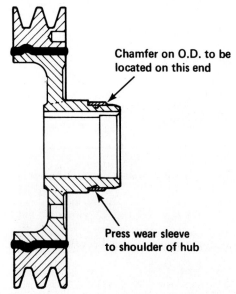

Chamfer on O.D. to be located on this end

Press wear sleeve to shoulder of hub

FIGURE 8-52 Repair sleeve can be used to correct grooved wear on vibration damper hub. (Courtesy of Ford of Canada.)

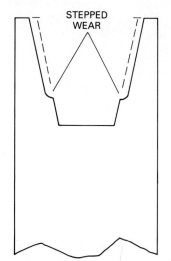

STEPPED WEAR

FIGURE 8-53 Stepped wear on sides of pulley groove require pulley replacement. (Courtesy of FT Enterprises.)

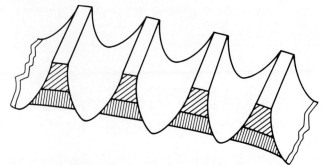

FIGURE 8-54 Damaged teeth on starter ring gear. (Courtesy of FT Enterprises.)

FLYWHEEL INSPECTION

The flywheel clutch friction surface must be flat and true and must be perpendicular to the crankshaft centerline to avoid face runout. The friction surface must be flat, free from heat discoloration, heat checks (small very shallow surface cracks), cracks, or grooved wear. The ring gear must be tight on the flywheel and the teeth must be in good condition. The following flywheel problems may be encountered.

Ring Gear Damage. The ring gear teeth should be inspected for wear and chipping. Worn ring gear teeth cause poor starter engagement. The metal loss also affects flywheel balance (Figure 8-54).

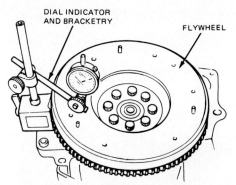

DIAL INDICATOR AND BRACKETRY

FLYWHEEL

FIGURE 8-55 Checking flywheel runout with a dial indicator. (Courtesy of Ford of Canada.)

Overheating. Overheating due to clutch slippage creates discoloration, hard spots, heat checks, and cracks on the friction surface of the flywheel. Discoloration, hard spots, and minor heat checks can be removed by resurfacing. In most cases hard spots are best removed by grinding since they are usually too hard for milling. Resurfacing must maintain the parallelism of the mounting flange and the friction surface. A badly cracked flywheel must be replaced.

Runout. Runout must be checked with a dial indicator while the flywheel is mounted on the crank-

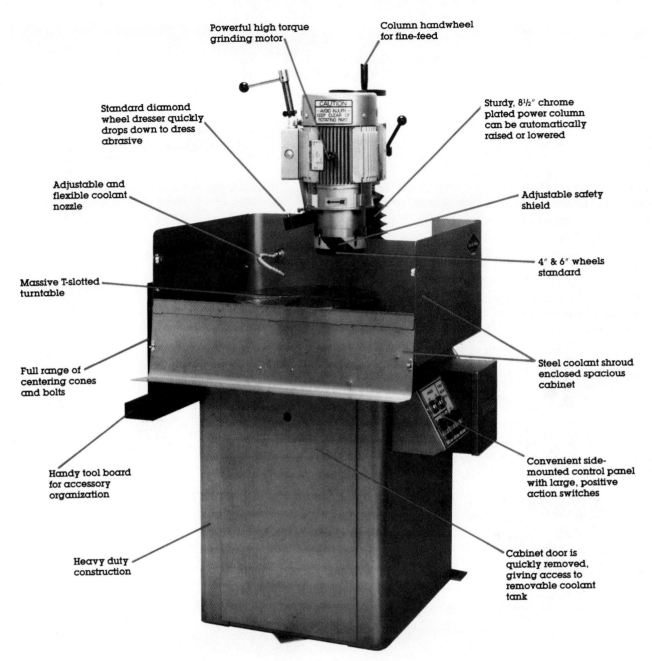

Figure 8-56 Typical flywheel surface grinder with features identified. (Courtesy of Kwik-Way Manufacturing Company.)

Labels in figure:

- Powerful high torque grinding motor
- Column handwheel for fine-feed
- Standard diamond wheel dresser quickly drops down to dress abrasive
- Sturdy, 8½″ chrome plated power column can be automatically raised or lowered
- Adjustable and flexible coolant nozzle
- Adjustable safety shield
- Massive T-slotted turntable
- 4″ & 6″ wheels standard
- Full range of centering cones and bolts
- Steel coolant shroud enclosed spacious cabinet
- Handy tool board for accessory organization
- Convenient side-mounted control panel with large, positive action switches
- Heavy duty construction
- Cabinet door is quickly removed, giving access to removable coolant tank

CAUTION — AVOID INJURY — KEEP CLEAR OF ROTATING PART

shaft. While turning the flywheel, make sure that crankshaft end play is not a factor in the measuring procedure. Flywheel face runout should not exceed 0.0005 in. per inch of diameter. Maximum runout for a 10-in. clutch would be 0.005 in. (Figure 8-55). Flywheel runout may be caused by crankshaft runout, a bent mounting flange on the flywheel or on the crankshaft, by foreign material between the mounting flanges, by raised metal on either of the mounting surfaces, or by improper resurfacing procedures. Flywheel resurfacing must be referenced from the mounting flange surface.

Dished Surface. The friction surface of the flywheel may become dished, especially on lighter flywheels. Check with a straightedge and feeler gauge. A dished flywheel must be resurfaced or replaced.

Grooved Wear. Grooved wear results from hard spots on the clutch friction disc. These hard spots may develop as a result of contamination and embedded foreign material. Grooved wear can also result from a severely worn clutch disc. A grooved flywheel must be resurfaced.

Flywheel Resurfacing

Resurfacing a flywheel restores flatness, removes minor heat checks, and provides a good surface finish (Figures 8-57 to 8-59). Prior to resurfacing the flywheel, it should be inspected for cracks by magnafluxing. A cracked flywheel must be replaced.

The flywheel friction surface must be machined parallel to the mounting flange surface on the side that bolts to the crankshaft. The amount of material that can safely be removed is often not specified. The best practice is to remove only enough material to clean up the surface. Some shops remove no more than 0.030 in. To machine a stepped flywheel requires two machining procedures if the pressure plate attaches to the outside surface. To maintain proper clutch spring pressure, the same amount of material is removed from the pressure plate mounting surface as was removed from the friction surface.

FIGURE 8-59 Heating a flywheel ring gear expands it for easy removal. (Courtesy of Ford of Canada.)

RING GEAR REPLACEMENT

The starter ring gear is a shrink fit on the flywheel. It can be removed (if not welded) by heating to expand the ring gear and then driving if off the flywheel with a hammer and drift punch (Figure 8-59).

To install a new ring gear, make sure that it is installed with the proper side toward the engine. If the gear teeth are beveled on one side, that side should face the starter drive pinion. If the inner edge is beveled, make sure that it faces the stepped corner of the flywheel. First heat the ring gear uniformly until about 400°F (never over 500°F). Overheating the ring gear will cause it to lose temper and become soft. When heated, place the ring gear on the flywheel and make sure that it bottoms against the flange as it cools and shrinks.

FIGURE 8-57 Flywheel in position ready for grinding. (Courtesy of Peterson Machine Tool, Inc.)

FIGURE 8-58 Radius cutting attachment for recessed flywheels. (Courtesy of Kwik-Way Manufacturing Company.)

FIGURE 8-60 Measuring pilot bushing wear. (Courtesy of Ford of Canada.)

PILOT BUSHING REPLACEMENT

The pilot bushing is located in a machined hole in the center of the rear end of the crankshaft. It supports the front end of the transmission input shaft or clutch

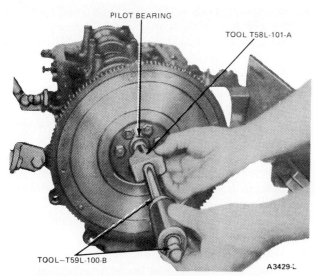

PILOT BEARING

TOOL T58L-101-A

TOOL-T59L-100-B

A3429-L

FIGURE 8-61 Removing a pilot bushing with a slide hammer puller. (Courtesy of Ford of Canada.)

TOOL-T71P-7137-C

TOOL-T71P-7137-H

A3430-C

FIGURE 8-62 Installing a pilot bushing. (Courtesy of Ford of Canada.)

shaft on vehicles equipped with manual transmissions or transaxles. The bushing is usually of the porous bronze type and is oil soaked for permanent lubrication.

The pilot bushing should be inspected for looseness and measured for wear with an inside micrometer or telescopic gauge and outside micrometer (Figure 8-60). If it is loose or worn, it should be replaced. An expanding jaw-type slide hammer puller can be used

to remove the bushing (Figure 8-61). The new porous bronze bushing should be soaked in engine oil before installation. This can easily be done by holding the bushing closed at one end and filling the bushing with oil. Pressure is then applied to the oil at both ends of the bushing until the oil is squeezed through to the outside and it looks like the bushing is sweating oil. Install the bushing with the appropriate-size bushing driver and a hammer. The bushing must be installed to the proper depth as specified in the service manual (Figure 8-62).

REVIEW QUESTIONS

1. Automotive crankshafts are:
   (a) cast, filleted, or forged
   (b) die cast, billeted, or forged
   (c) filleted, die cast, or forged
   (d) billeted, cast, or forged

2. How is the crankshaft held in place?

3. Compare the crank throw dimension with the stroke dimension.

4. What does an oil slinger do?

5. How is crankshaft end play controlled?

6. Technician A says that crankshaft counterweights are used to offset the weight of the crank throws and connecting rods. Technician B says that balance shafts are used on some engines to improve engine balance. Who is right?
   (a) technician A
   (b) technician B
   (c) both are right
   (d) both are wrong

7. Explain how crank throws are indexed on an in-line six-cylinder engine.

8. Why is a flywheel needed in an engine?

9. How is the starter ring gear attached to the flywheel?

10. Technician A says that a harmonic balancer dampens crankshaft torsional vibrations. Technician B says that a vibration damper is used for this purpose. Who is right?
    (a) technician A
    (b) technician B
    (c) both are right
    (d) both are wrong

11. There are two forces at work in an engine that tend to create an unbalanced condition. What are they?
    (a) rotating and centrifugal
    (b) centrifugal and inertial
    (c) inertial and rotating
    (d) none of the above

12. List the areas on a crankshaft that should be inspected visually.

13. How many measurements are required on a bearing journal to determine size, taper, and out-of-round?
    (a) two   (c) four
    (b) three   (d) six

14. What equipment is needed to check crankshaft runout?

15. What is crankshaft runout?

16. How should a crankshaft be checked for cracks?

17. Crankshafts are straightened with a _____ and the correction set with a hammer and _____ .

18. True or false: All crankshafts that have not been reground have standard-size bearing journals.

19. What is "nitriding"?

20. How can a crankshaft be checked for nitriding?

21. What three methods can be used to replace lost metal on bearing journals?

22. Shot peening a crankshaft improves its _____ life.

23. Why are oil holes in bearing journals chamfered?

24. Why is crankshaft grinding usually limited to no more than 0.030 in. undersize.

25. When grinding a crankshaft the dimensions the machinist must be concerned with are:
    (a) journal diameter, width, and radii
    (b) journal diameter, radii, and length
    (c) journal diameter and radii
    (d) journal diameter and width

26. What effect do grinding wheel runout and static imbalance have on the bearing journal surface?

27. What direction should the grinding wheel turn when grinding a crankshaft?

28. What is crankshaft "lobing"?

29. What microinch finish is required on crankshaft journals?

30. Maximum stock removal while polishing journals should not exceed _____ in.

31. How can a vibration damper with grooved hub wear be repaired?

32. Describe what condition the rubber should be in on a vibration damper in good condition.

33. List the points that must be checked during flywheel inspection.

34. List four causes of flywheel runout.

35. What causes the friction surface on a flywheel to crack?

36. How can grooved wear on a flywheel be corrected?

37. How should a flywheel be checked to determine if it is dished?

38. Describe how to remove a damaged starter ring gear.

CHAPTER 9

Connecting Rods and Piston Pins

OBJECTIVES

1. To develop an understanding of the function and design of connecting rods and piston pins.
2. To develop the ability to recondition connecting rods to meet the manufacturer's specifications.
3. To recognize the different types of piston pin mounting methods.
4. To develop the ability to precision fit piston pins to connecting rods and pistons.

CONNECTING ROD FUNCTION

Connecting rods are designed to connect the reciprocating piston to the rotating crankshaft. Connecting rods transmit all the energy produced in the cylinder from the piston to the crankshaft during the power stroke. They also move the pistons up and down through the exhaust, intake, and compression strokes.

During the power stroke the connecting rod is subjected to a force of as much as 4 to 5 tons. The rotating crankshaft causes the connecting rod to reverse its direction thousands of times a minute. At the other end of the rod, the piston is started in one direction, stopped, and its direction reversed. At 3500 rpm the piston is stopped and started 7000 times every minute. In addition, the connecting rods are subjected to heating and cooling cycles, which adds to stress. Over time the constant pounding and temperature changes result in the elongation of the big-end bore of the connecting rod. Connecting rods must either be replaced or reconditioned when overhauling an engine.

CONNECTING ROD MATERIALS AND DESIGN

Connecting rods are designed to be as light as possible and still provide the strength required. They are manufactured either by casting or forging processes from iron or steel alloys. Most of today's engines use forged steel connecting rods. Some high-speed engines used in racing use forged aluminum connecting rods to reduce the rotating and reciprocating masses (Figures 9-1 and 9-2).

The hole in the upper end of the rod allows the rod to be connected to the piston by means of a steel piston pin. The lower or big end of the rod is a split ring, which provides the means to connect it to the crankshaft. A split-type precision bearing is fitted to the rod and bearing journal. Precision-dimensioned bolts and matching holes in the rod yoke and cap ensure cap alignment with the rod. The cap must not be reversed since this would result in cap misalignment and an offset bearing bore. Bolt heads fit into machined recesses in the rod. Notches in the cap and yoke are provided to accommodate locating tangs on the bearing inserts (Figure 9-3).

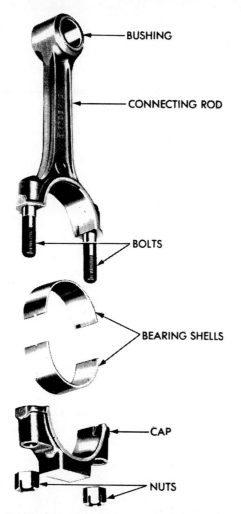

FIGURE 9-1 Typical connecting rod assembly. (Courtesy of Chrysler Canada Ltd.)

FIGURE 9-2 Forged and shot-peened connecting rod (top). Lightweight aluminum rod for high-rpm drag racing (bottom). (Courtesy of Chrysler Canada Ltd.)

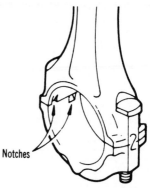

FIGURE 9-3 Lugs on bearing inserts fit into notches in rod cap and yoke to keep bearings in position. (Courtesy of Chrysler Canada Ltd.)

ROD BALANCING

Connecting rods are designed with balance pads or bosses. Balancing bosses are located at the bottom of the cap or on the sides of the beam near the bottom. Some rods also have a balancing boss at the top of the rod eye (Figure 9-4) to allow balancing the rods so they are all of equal weight at both ends.

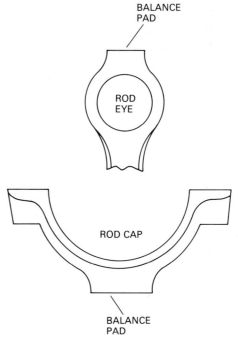

FIGURE 9-4 Balance pad locations on connecting rod. (Courtesy of FT Enterprises.)

CONNECTING ROD OFFSET

Some connecting rods are designed with the rod beam offset from the center of the big-end bore. This is done to provide clearance at the crankshaft cheek. This re-

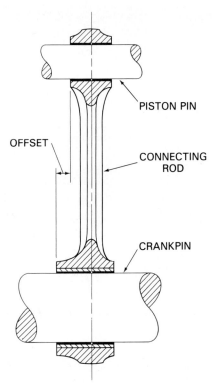

FIGURE 9-5 Connecting rod offset. (Courtesy of FT Enterprises.)

sults from an engine design where the cylinder bore centerline and the center of the crankpin journal are offset or because of crankshaft design (Figure 9-5).

OIL HOLES

Many engines are designed to provide piston pin and cylinder wall lubrication from oil spurt holes in the connecting rods. The spurt hole is located in the yoke of the connecting rod and is positioned to aim the oil at the pin and cylinder wall. Some connecting rods have an oil bleed hole at the parting surface of the rod. Bleed holes and spurt holes aid in proper oil flow through the rod bearing (Figure 9-6). Spurt holes and bleed holes must not be restricted by foreign material.

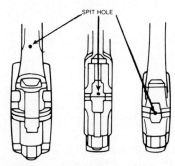

FIGURE 9-6 Oil spit hole locations. (Courtesy of Chrysler Canada Ltd.)

During piston and rod assembly, careful attention must be paid to ensure that the oil spurt hole faces in the right direction, toward the major thrust side of the cylinder.

CONNECTING ROD FAILURE ANALYSIS

For a description of connecting rod failures and failure analysis, see the section, ''Connecting Rod Failure'' in Chapter 5.

CHECKING AND MEASURING CONNECTING RODS

Connecting rods that are to be used over again must be checked and measured for the following: (1) bend or twist (alignment; Figure 9-7); (2) big-end bore out-of-round (also known as stretch, eccentricity, or elongation; Figure 9-8); and (3) small-end bore for stretch or wear.

Rod Alignment

If the small-end bore and the big-end bore are not vertically parallel, the rod is bent. If the two bores are not parallel laterally, the rod is twisted. To check for the bend or twist, use the rod alignment checking equipment shown in Figures 9-9 and 9-10. Either the piston pin or the piston must be in place to check alignment. The big-end bore is used as the reference point from which to check.

To check for bend, place the rod on the mandrel, making sure that full bore contact is maintained with the mandrel contacts. While holding the piston in the vertical position, slide the V block around the piston to contact the surface plate of the alignment fixture. If the V block does not make full-length contact with the surface plate, the rod is bent and must be straightened or replaced. Bend must not exceed 0.001 in. over 6 in.

To check for twist, hold the V block against the piston (as above) and tilt the piston on the rod as far as possible. If the V block does not make full contact with the surface plate, the rod is twisted and must be straightened or replaced. Twist must not exceed 0.001 in. over 6 in. Maximum bend (parallelism) and twist limits are shown in Figure 9-7.

Bore Stretch

The big-end bore can be measured with a special gauge, a dial gauge, an inside micrometer, or a telescoping gauge and outside micrometer. Proceed as follows.

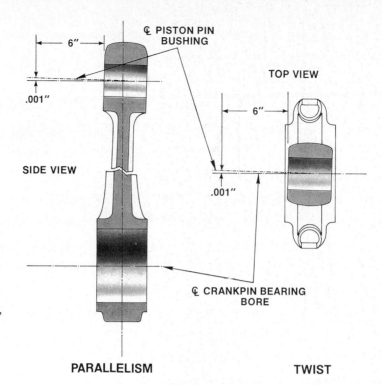

CL PISTON PIN BUSHING

SIDE VIEW

6"

.001"

TOP VIEW

6"

.001"

CL CRANKPIN BEARING BORE

PARALLELISM

TWIST

FIGURE 9-7 If the big-end and small-end bores are not parallel along lines through their centers, the rod is twisted or bent. Maximum limits for bend and twist are shown. (Courtesy of Federal-Mogul Corporation.)

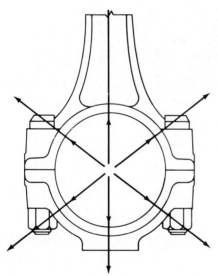

FIGURE 9-8 Big-end bore of connecting rod must not be out-of-round. (Courtesy of Sunnen Products Company.)

1. Make sure that the parting surfaces of the rod and cap are clean. Place the rod in a holding fixture (to prevent twist; Figure 9-11), install the cap and tighten the connecting rod bolts or nuts to the torque specified. Remove the rod from the holding fixture.

2. Measure the big-end bore (Figure 9-12) across the three positions shown in Figure 9-8. The maximum allowable out-of-round tol-

FIGURE 9-9 Checking connecting rod alignment with piston attached. (Courtesy of Red River Community College.)

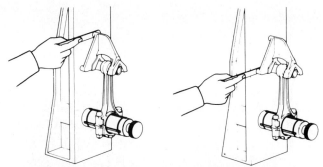

FIGURE 9-10 Checking connecting rod alignment with piston removed. Checking for bend (*left*) and twist (*right*). (Courtesy of Ford of Canada.)

FIGURE 9-11 Connecting rod holding fixture prevents rod from twisting during tightening or loosening of bolts. (Courtesy of Red River Community College.)

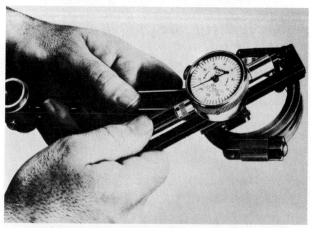

FIGURE 9-12 Measuring connecting rod big-end bore stretch or out-of-round. (Courtesy of Federal-Mogul Corporation.)

erance on new engines is 0.0003 in. The maximum tolerance for size is 0.001 in. If either out-of-round or size tolerances are exceeded, the rod should be reconditioned or replaced. The maximum big-end out-of-round usually occurs vertically or about 30° off of vertical. The difference between the two measurements taken across from each other is the amount of out-of-round at that point.

RECONDITIONING THE CONNECTING ROD

Reconditioning the Big-End Bore

To recondition the big-end bore the objectives are to achieve the following (Figures 9-13 and 9-14).

- A bore size that meets specifications
- A bore diameter that is round within 0.0003 in.
- A bore that has the proper rms finish (usually 30 to 90 μin.)
- A bore with parallel inside surfaces

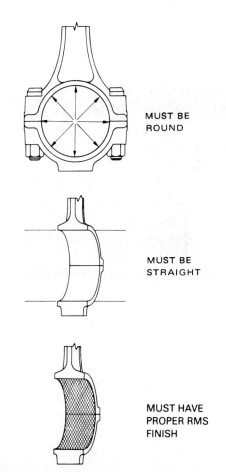

MUST BE ROUND

MUST BE STRAIGHT

MUST HAVE PROPER RMS FINISH

FIGURE 9-13 A reconditioned big-end bore must meet the above requirements. (Courtesy of Sunnen Products Company.)

Reconditioning the Connecting Rod **147**

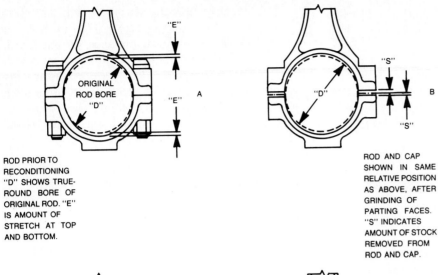

ROD PRIOR TO RECONDITIONING "D" SHOWS TRUE-ROUND BORE OF ORIGINAL ROD. "E" IS AMOUNT OF STRETCH AT TOP AND BOTTOM.

ROD AND CAP SHOWN IN SAME RELATIVE POSITION AS ABOVE, AFTER GRINDING OF PARTING FACES. "S" INDICATES AMOUNT OF STOCK REMOVED FROM ROD AND CAP.

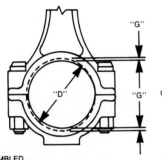
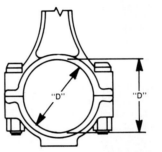

RE-ASSEMBLED ROD NOW HAS A SMALLER VERTICAL DIMENSION (BY AMOUNTS "G") THAN DIAMETER "D" OF ORIGINAL TRUE-ROUND FACTORY ROD.

COMPLETELY RECONDITIONED ROD AFTER HONING. ROD BORE IS AGAIN TRUE-ROUND, AND OF THE SAME DIAMETER "D" AS THE ORIGINAL FACTORY ROD.

FIGURE 9-14 Procedure for reconditioning the big-end bore. (Courtesy of Sunnen Products Company.)

The procedure involves three steps, as follows.

1. Remove 0.002 in. of stock from both the cap and the rod parting faces. This requires removing the bolts from the rod. To remove the bolts, tap them out with a brass hammer while supporting the connecting rod. If the rod is not supported while removing (or installing) the bolts, the rod yoke can be distorted. To remove stock from the parting surfaces, special equipment is required, as shown in Figure 9-15. This is a precision operation and must be done with precision equipment. First dress the stone with the diamond dresser. Clamp the rod or cap in place on the grinder so that grinding will take place equally across both parting surfaces. Use the micrometer feed to remove the precise amount desired (no more than 0.002 in.). Parting surfaces must be kept flat and parallel.

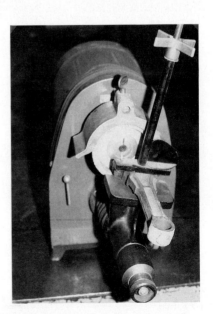

FIGURE 9-15 Grinding stock from the connecting rod parting line. Note micrometer depth of feed. (Courtesy of Red River Community College.)

2. After stock removal from the rod and cap, reinstall the rod bolts with the rod properly supported as shown in Figure 9-16. Make sure that the bolts are aligned to allow the heads to seat properly. Assemble the cap to the rod and tighten the bolts to specifications with the rod clamped in the holding fixture.

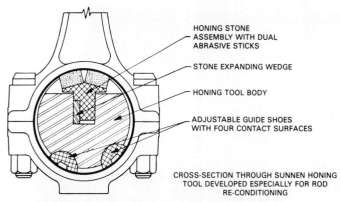

FIGURE 9-17A Courtesy of Sunnen Products Company.)

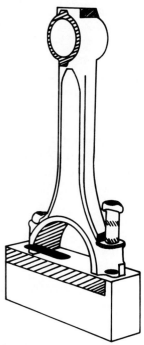

FIGURE 9-16 Connecting rod must be properly supported while installing rod bolts to avoid rod distortion. (Courtesy of Silver Seal Products Company, Inc.)

3. Select and insert the appropriate size mandrel and hone for the rod to be honed. Use the rod support fixture to ensure that the rod is honed straight. Carefully hone the bore to bring it to original bore size and finish (Figures 9-17 to 9-20). Make certain that an adequate supply of honing oil is directed at the rod at all times during honing. Measurements will not be consistent if the rod is allowed to overheat, and honing quality will be affected if an adequate supply of honing oil is not present. Check the bore size frequently to ensure that it is not increased beyond size limits. In some cases a small area at the parting lines may not clean up. This can be ignored since it has no effect on the life of the finished product.

Small-End Bore

Measure the small-end bore of the rod with an inside micrometer or precision gauge that measures accurately to 0.0001-in. increments. Compare measure-

FIGURE 9-17 Honing the big-end bore of a connecting rod. (Courtesy of Western Engine Ltd.)

FIGURE 9-18 Checking the big-end bore size on a precision gauge during the honing process. (Courtesy of Western Engine Ltd.)

FIGURE 9-19 Close-up of precision connecting rod and pin hole measuring gauge, showing bore contact points. (Courtesy of Sunnen Products Company.)

ments to specifications. Small-end bores are honed to fit oversize piston pins. For full-floating pins the small-end bushing is honed to provide from 0.0003 to 0.0005 in. of clearance while the piston pin boss is honed to provide from 0.0001 to 0.0003 in. of clearance. If the pin is a press-fit in the rod, the rod eye is honed to −0.0008 to −0.0012 in. smaller than the pin diameter to provide the interference fit needed to keep the pin tight in the rod. To make pin installation easier, a slight chamfer must be provided on each side of the rod eye.

Bushing Replacement. To remove a worn bushing from the small end of the rod, use a bushing driver of the proper size and drive or press the bushing out. In some cases a very thin walled bushing may have to be split with a cape chisel. Be sure not to damage the small-end bore when splitting the bushing. Press the new bushing into place; expand it to ensure that there will be no high spots (Figure 9-23). Expanding the bushing also assures good heat transfer. After ex-

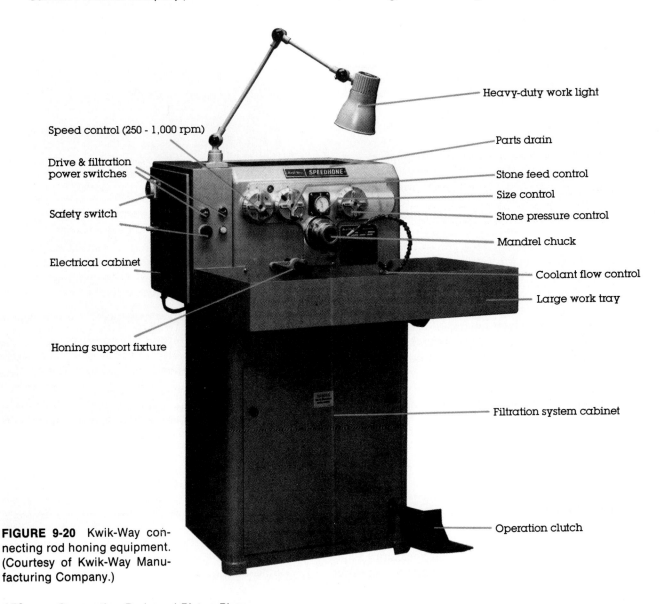

Heavy-duty work light

Speed control (250 - 1,000 rpm)

Drive & filtration power switches

Safety switch

Electrical cabinet

Honing support fixture

Parts drain

Stone feed control

Size control

Stone pressure control

Mandrel chuck

Coolant flow control

Large work tray

Filtration system cabinet

Operation clutch

FIGURE 9-20 Kwik-Way connecting rod honing equipment. (Courtesy of Kwik-Way Manufacturing Company.)

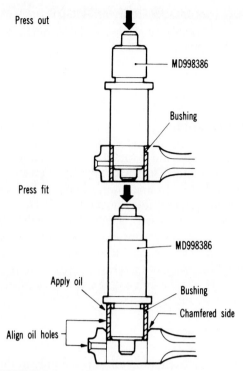

FIGURE 9-21 Replacing a small-end bushing in a connecting rod using a press. (Courtesy of Chrysler Canada Ltd.)

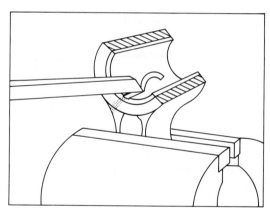

FIGURE 9-22 Using a cape chisel to split a bushing prior to removal from the connecting rod small end. (Courtesy of FT Enterprises.)

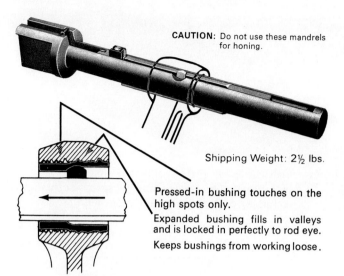

CAUTION: Do not use these mandrels for honing.

Shipping Weight: 2½ lbs.

Pressed-in bushing touches on the high spots only.

Expanded bushing fills in valleys and is locked in perfectly to rod eye.

Keeps bushings from working loose.

FIGURE 9-23 After the bushing is installed, it should be expanded to lock it in place. This bushing expander by Sunnen is used with the Sunnen honing machine. Various sizes are available. (Courtesy of Sunnen Products Company.)

FIGURE 9-24 Chamfering cone for small-end bushing of connecting rod. Cone is used with Sunnen honing machine. (Courtesy of Sunnen Products Company.)

panding, hone or diamond bore the bushing to the required size and finish. A slight chamfer at each end of the bushing will remove any burrs and aid in pin installation. (Figures 9-21 to 9-23). Figure 9-24 shows a chamfering cone for use in a Sunnen honing machine.

PISTON PIN FUNCTION

The pistons are attached to the small end of the connecting rods by means of piston pins. All the force applied to the piston during combustion is transmitted to the connecting rod through the piston pin. This can be as much as 2 to 4 tons of load under certain conditions. The piston pushes down at each end of the pin, while the connecting rod opposes this force due to vehicle load. In addition, the pin is subjected to wide temperature variations, which cause it to expand and contract as it heats and cools.

PISTON PIN MATERIALS AND DESIGN

To provide strength, piston pins are made from very high quality steel alloys. The surface is a very smooth, mirror-like finish, to minimize friction (Figure 9-25). Most piston pins are hollow to keep the weight of the reciprocating piston assembly at a minimum. In some

FIGURE 9-25 Piston pins are made from high-grade steel and appear mirror smooth. (Courtesy of Sealed Power Corporation.)

SOLID

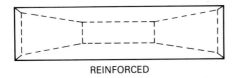

REINFORCED

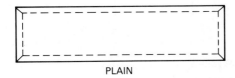

PLAIN

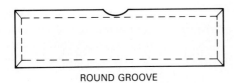

ROUND GROOVE

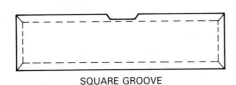

SQUARE GROOVE

FIGURE 9-26 Typical piston pin designs. (Courtesy of FT Enterprises.)

PISTON PIN MOUNTING METHODS

The piston pin must not be allowed to make contact with the cylinder wall. This requires limiting endwise movement of the pin. Several methods can be used to keep the pin centered in the piston.

Full-Floating Pin

In this design the pin is retained in the piston by a lock ring at each end. The lock rings fit in grooves in the piston pin holes. In operation the pin is free to rotate in the piston and in the connecting rod. The small end of the rod is fitted with a bushing to reduce friction and allow replacement when worn. Normal clearance between the pin and piston pin bosses is from 0.0001 to 0.0003 in. and from 0.0003 to 0.0005 in. in the connecting rod. Oscillation can take place in both the piston and rod (Figure 9-27).

Press-Fit in Rod

In this design oscillation takes place only in the piston (Figure 9-28). The pin is held tightly in the rod as a

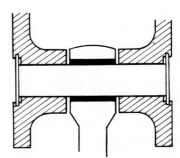

FIGURE 9-27 Full-floating pin oscillates in both piston and rod. Pin is held in place by lock rings in piston. (Courtesy of Sunnen Products Company.)

ALUMINUM PISTON
NO BUSHING

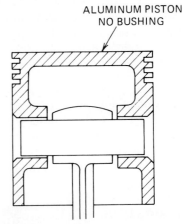

FIGURE 9-28 In this design the bushing is a press-fit in the rod. The pin oscillates only in the piston. (Courtesy of Sunnen Products Company.)

pins the hole is tapered (smaller in the center section) to increase strength. Other design features include a notch, or flat, to lock the pin in place (used with the clamp and bolt rod) and a flat to provide lubrication. Several pin designs are shown in Figure 9-26.

result of a press-fit. The pin hole in the connecting rod is smaller than the pin diameter. An interference fit from −0.0008 to 0.00012 in. is required for pin retention. Clearance between the pin and piston pin bosses is from 0.0003 to 0.0005 in.

Pin Clamped in Rod

In this design the rod eye is split and forms a clamp. A cap screw is used to tighten the rod eye around the pin to control endwise movement. Clearance between the pin and the pin bosses is from 0.0003 to 0.0005 in. This design is limited mostly to older engines (Figure 9-29).

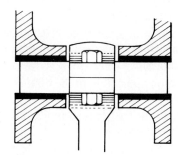

FIGURE 9-29 The piston pin is locked to the connecting rod by the clamping action of the split small end of the rod as the bolt is tightened. The pin oscillates only in the piston. (Courtesy of Sunnen Products Company.)

Pin Locked by Setscrew

In this design the rod oscillates on the pin while the pin is held in place in the piston by a setscrew (Figure 9-30). A setscrew located in a threaded hole in one of the piston bosses is tightened against the pin to lock it in place. Clearance between the pin and connecting rod bushing is from 0.0007 to 0.0009 in. clearance. The

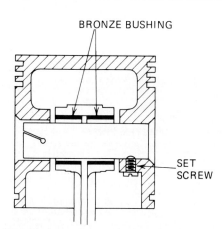

FIGURE 9-30 In this design the pin is locked to the piston by a setscrew. Oscillation occurs only in the rod. (Courtesy of Sunnen Products Company.)

screw side fit in the piston should be a press-fit from −0.0002 to −0.0003 in. interference fit. On the free side the fit should be from 0 to 0.0001 in. clearance.

PISTON PIN OFFSET

Many engine manufacturers offset the piston pin toward the major thrust side. The pin is offset by as much as 0.060 or 0.090 in. (Figure 9-31). Piston pin offset provides a more gradual change in thrust pressure of the piston against the cylinder wall as the piston moves through TDC on the compression stroke to the power stroke. This reduces piston slap (Figure 9-32). When the pin is offset to the minor thrust side, as with some racing engines, a slight increase in mechanical advantage and engine torque is gained.

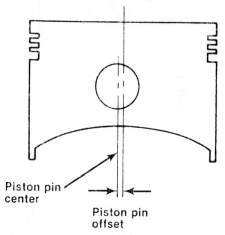

FIGURE 9-31 Piston pin is offset from center by from 0.060 to 0.090 in. to the minor thrust side to reduce piston slap. (Courtesy of Chrysler Canada Ltd.)

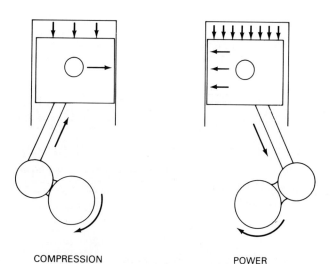

FIGURE 9-32 Piston slap occurs as the connecting rod angle changes from the compression stroke to the power stroke. (Courtesy of FT Enterprises.)

PIN FITTING

Pin fitting is a precision operation performed on precision equipment shown in Figures 9-17 to 9-20. Precise measurements and close tolerances are required. To achieve a pin fit that will provide the maximum bearing surface while ensuring free oscillation requires precision equipment and an experienced operator. If the pin fit is too loose, the bearing area is reduced and pin knock will result. If the pin fit is too tight, the pin will seize and cause piston and rod failure. For pin fit failure analysis, see Chapter 5.

The following results must be achieved when fitting piston pins.

1. A 360° bearing surface must be achieved; no taper, no bell mouth, no high spots.
2. The pin holes in the piston and rod must be parallel to the big-end bore centerline. This means that the pin holes in the piston and in the rod eye must also be in perfect alignment.
3. The proper rms surface finish must be obtained.
4. The precise amount of clearance as specified for each type of pin mounting must be provided (Figure 9-33).

Piston pins are generally available in standard sizes and 0.0015, 0.003, and 0.005 in. oversized. Piston pin holes in the piston and the rod small end are resized by honing or boring to achieve the desired clearances and surface finish. After resizing the piston bosses and rod eye, all grit must be completely removed to prevent damage after engine assembly.

ASSEMBLING THE PISTON AND THE ROD

The correct assembly of the pistons and rods is essential to proper engine operation. Reversing the position of pistons or connecting rods can cause serious engine damage. Typical reference points on pistons and rods are shown in Figures 9-34 and 9-35. When assembled and installed, the following criteria must be met.

1. The front of the piston must face forward in the engine because of piston design, such as pin offset, differences in major and minor thrust sides, and valve relief.
2. The oil spurt hole in the connecting rod must face the major thrust side (right side) of the engine to ensure piston pin and cylinder wall lubrication.
3. The front side of connecting rods designed with offset must face forward in the engine.
4. Numbered pistons and rods must match (i.e., number 1 piston and number 1 rod must go together, etc.).

Identifying marks on pistons indicating front are: the word "front," the letter F, an arrow, or a notch.

Pin Type	Description	Aluminum Piston	Cast Iron Piston	Connecting Rods
A	Full Floating	.0001″ to .0003″ clearance	.0003″ to .0005″ clearance	.0003″ to .0005″ clearance (all pressure feed, .0005″ to .0007″ clearance.)
B	Oscillating in bushed piston		.0003″ to .0005″ clearance	clamped in Rod
C	Oscillating in piston (no bushing)	.0003″ to .0005″ clearance	.0006″ to .0008″ clearance	clamped in Rod
D	Oscillating in piston—press fit in Rod	.0003″ to .0005″ clearance		−.0008″ to −.0012″ press fit
E	Set Screw Type Piston	Screw Side −.0002″ to −.0003″ press fit / Free Side 0 to .0001″ clearance	Screw Side −.0001″ to −.0002″ press fit / Free Side 0 to .0001″ clearance	When locked in piston—and all pressure feed, .0007″ to .0009″ clearance.

PRECISION PIN FITS ON ENGINES WITH 1¼″ AND 1½″ DIAMETER PINS*

Pin Type	Description	Aluminum Piston	Cast Iron Piston	Connecting Rods
A	Full Floating 1¼″ dia. pin holes	.0003″ to .0005″ clearance		.0007″ to .0009″ clearance (all pressure feed, .0009″ to .0011″ clearance.)
A	Full Floating 1½″ dia. pin holes	.0005″ to .0007″ clearance		.0010″ to .0012″ clearance (all pressure feed, .0013″ to .0015″ clearance.)

*On large diameter pins check Engine Manufacturers Manual for recommended clearances.

FIGURE 9-33 Very precise pin fitting tolerances are required to assure trouble-free operation. (Courtesy of Sunnen Products Company.)

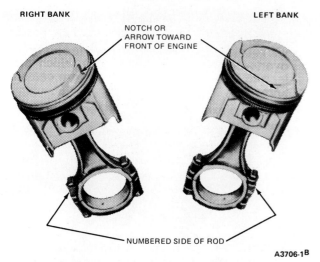

RIGHT BANK   LEFT BANK

NOTCH OR ARROW TOWARD FRONT OF ENGINE

NUMBERED SIDE OF ROD

A3706-1B

FIGURE 9-34 Piston and rod assembly reference points for 302- and 351-CID Ford engines. (Courtesy of Ford of Canada.)

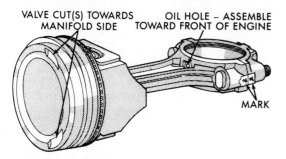

VALVE CUT(S) TOWARDS MANIFOLD SIDE   OIL HOLE – ASSEMBLE TOWARD FRONT OF ENGINE

MARK

FIGURE 9-35 Piston and rod assembly reference points typical for a Chrysler four-cylinder engine. (Courtesy of Chrysler Canada Ltd.)

The piston must be installed with this mark to the front of the engine. Rod positioning is usually determined with respect to the oil spurt hole or bleed hole, a mark on the rod beam, rod beam offset, bearing locating tangs, number marking, or a chamfer on one side of the big-end bore.

FULL-FLOATING PIN INSTALLATION

In this design, pin retaining rings limit the endwise movement of the pin. Two types of retaining rings or circlips are used: the round steel wire type and the Tru-Arc type. Always use new lock rings. Both must be installed with the open end down. The wire type with a tang must be installed with the tang pointing away from the pin. To assemble, first install one retaining ring. Start the pin in the other side of the piston. Position the rod in the piston and push the pin through the rod and into the piston far enough to clear the lock ring groove. Install the remaining lock ring and check to ensure that both rings are fully seated in their groove

FIGURE 9-36 Installing a piston pin lock ring. (Courtesy of Sealed Power Corporation.)

(Figure 9-36). When Tru-Arc lock rings are used they must be installed with the rounded edges facing the piston pin to prevent them from popping out of the groove.

PRESS-FIT PIN INSTALLATION

Two different methods may be used to install press-fit piston pins: the hydraulic press or the rod heater. A hydraulic press equipped with either a hand-operated or electric-motor-driven pump may be used. A set of pin installing adapters and a connecting rod support

FIGURE 9-37 Benchtop hydraulic press for replacing press-fit piston pins. (Courtesy of Red River Community College.)

FIGURE 9-38 Press-fit pin lubricant prevents galling when installing piston pins with a press. (Courtesy of Sunnen Products Company.)

FIGURE 9-39 Spraying pin lubricant on ID of rod eye before pin installation. (Courtesy of Sealed Power Corporation.)

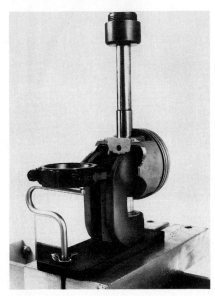

FIGURE 9-40 Installing a press-fit pin using a hydraulic press. (Courtesy of OTC Division, Sealed Power Corporation.)

FIGURE 9-41 Piston pin removing/installing kit. (Courtesy of OTC Division, Sealed Power Corporation.)

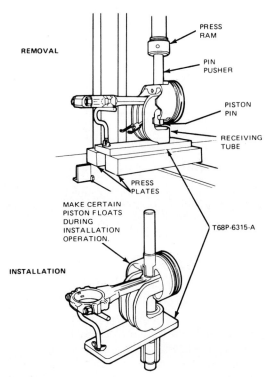

FIGURE 9-42 Piston pin removal (top) and installation (bottom). (Courtesy of Ford of Canada.)

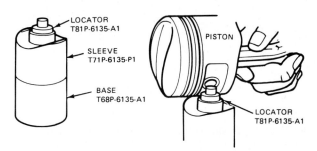

FIGURE 9-43 How to set up the locator and piston for pin installation. (Courtesy of Ford of Canada.)

are required. These include pin press adapters and pin guides. The correct-size adapters and pin guides appropriate for the piston and pin design must be used to ensure that the piston floats free during pin installation. Pressure against the piston must be avoided to prevent piston damage. The rod and the pin must be centered in the piston. This can be done with special locators designed for the purpose (Figures 9-37 to 9-43).

Many rebuilding shops use either electric or gas-fired rod heating equipment to aid in pin installation. In this method the rod eye is expanded by heating to about 400°F. The pin is then pushed into place using a centering device. The device must be adjusted for the particular piston and pin design being assembled (Figures 9-44 to 9-46).

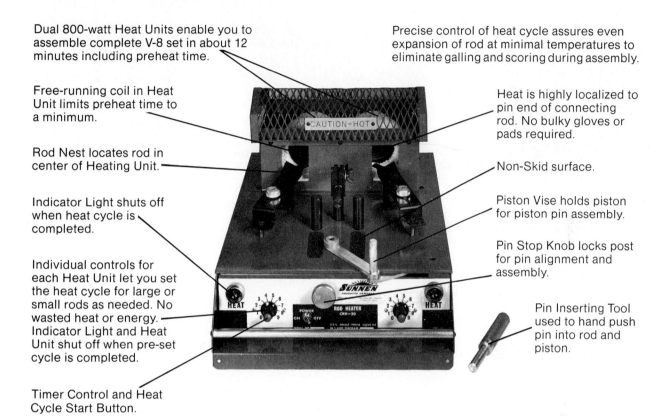

Dual 800-watt Heat Units enable you to assemble complete V-8 set in about 12 minutes including preheat time.

Free-running coil in Heat Unit limits preheat time to a minimum.

Rod Nest locates rod in center of Heating Unit.

Indicator Light shuts off when heat cycle is completed.

Individual controls for each Heat Unit let you set the heat cycle for large or small rods as needed. No wasted heat or energy. Indicator Light and Heat Unit shut off when pre-set cycle is completed.

Timer Control and Heat Cycle Start Button.

Precise control of heat cycle assures even expansion of rod at minimal temperatures to eliminate galling and scoring during assembly.

Heat is highly localized to pin end of connecting rod. No bulky gloves or pads required.

Non-Skid surface.

Piston Vise holds piston for piston pin assembly.

Pin Stop Knob locks post for pin alignment and assembly.

Pin Inserting Tool used to hand push pin into rod and piston.

FIGURE 9-44 Electric rod heater. (Courtesy of Sunnen Products Company.)

FIGURE 9-45 Gas-fired rod heater. (Courtesy of Western Engine Ltd.)

FIGURE 9-46 Assembling the rod, piston and pin after preheating the rod. (Courtesy of Western Engine Ltd.)

REVIEW QUESTIONS

1. What function does the connecting rod perform?
2. During the power stroke the connecting rod is subjected to a force of as much as:
   (a) 1000 to 2000 lb
   (b) 10,000 to 20,000 lb
   (c) 4 to 5 tons
   (d) 40 to 50 tons
3. At 3500 rpm the piston is started and stopped:
   (a) 3500 times a minute
   (b) 7000 times a minute
   (c) 350 times a minute
   (d) 700 times a minute
4. Connecting rods are designed to be:
   (a) heavy and strong
   (b) short and light
   (c) strong and light
   (d) short and heavy
5. What causes elongation of the connecting rod big-end bore?
6. Why is the split design of the big-end bore of a connecting rod necessary?
7. Connecting rods are manufactured by the _____ or _____ process.
8. What provision is made on connecting rods to balance their weight?
9. What is the reason that some connecting rods are offset vertically?
10. What is the reason for oil spurt holes in connecting rods?
11. "Connecting rod alignment" refers to:
    (a) bend
    (b) twist
    (c) parallelism
    (d) all of the above
12. Connecting rod big-end bore stretch is measured with _____.
13. Why is it important to clamp a connecting rod yoke securely when tightening connecting rod bolts?
14. The maximum allowable big-end bore out-of-round on new engines is:
    (a) 0.03 in.
    (b) 0.003 in.
    (c) 0.0003 in.
    (d) none of the above
15. To recondition the rod big-end bore, just remove ___ in. of stock from the parting lines of the cap and rod.
16. What four criteria must a properly reconditioned connecting rod, big-end bore meet?
17. After replacing a bushing in the small end of a connecting rod, why should the bushing be expanded?
18. What is the function of the piston pin?
19. List four methods used to retain piston pins.
20. What is the reason for piston pin offset toward the major thrust side?
21. What is a "negative" pin fit?
22. What four criteria must be met to achieve a proper pin fit?
23. Piston pins are generally available in which oversizes?
24. What four guidelines are used to ensure that the piston and connecting rod are properly assembled?
25. What two methods are used to install press-fit piston pins?

CHAPTER 10

Pistons and Piston Rings

OBJECTIVES

1. To develop an understanding of the function, operation, and design of pistons and piston rings.
2. To develop the ability to measure accurately pistons, piston-to-cylinder wall clearance, and ring groove wear.
3. To develop the ability to select and install piston rings and piston and rod assemblies.

PISTON FUNCTION

Automotive engine pistons are designed to transmit the force of combustion to the connecting rod and crankshaft. They provide the pumping action required during the intake and exhaust strokes and compress the air-fuel mixture during the compression stroke. They contain and support the piston rings, which form the seal between the piston and cylinder wall. Pistons must be as light as possible to minimize reciprocating forces, yet strong enough to withstand the repeated loads of combustion. They must provide the means for controlling thermal expansion in order to maintain proper piston-to-cylinder wall clearance. The pistons must maintain their ability to function under wide temperature variations.

PISTON TERMINOLOGY

It is important to be able to understand and use the proper terms when discussing piston operation and design. The terms and descriptions listed here are commonly applied to piston components.

1. *Piston head:* top surface area of the piston against which combustion pressures act. Head designs include the flat, stepped, angled step, crowned, wedge or beveled, cupped or concave, dished or recessed, and domed shapes (Figure 10-1). In some cases piston heads are chamfered to reduce the formation of unburned hydrocarbons between the piston and cylinder above the top ring (Figure 10-2).
2. *Ring area:* that section of the piston which carries the piston rings. It consists of the ring grooves and lands (Figure 10-3).
3. *Compression ring grooves:* grooves cut into the piston circumference between the piston head and oil ring groove. These grooves carry the compression rings. There are two compression ring grooves on most pistons.
4. *Oil ring groove:* groove cut into the circumference of the piston just below the compression ring grooves. This groove is designed to carry the oil control ring. The

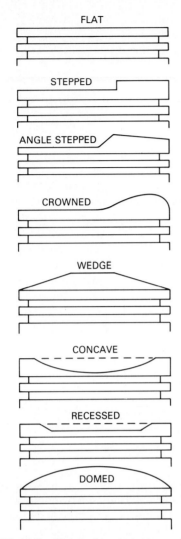

FIGURE 10-1 Piston head designs. (Courtesy of FT Enterprises.)

FIGURE 10-2 Forged aluminum piston with chamfered head. (Courtesy of Sealed Power Corporation.)

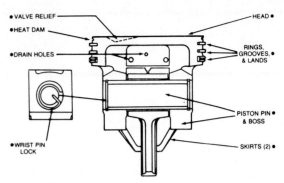

FIGURE 10-3 Piston nomenclature. (Courtesy of Chrysler Canada Ltd.)

bottom of the groove is usually equipped with oil drain back holes or slots.

5. *Heat dam:* narrow groove cut into the circumference of the piston just above the top compression ring. This groove is cut deeper than the top ring groove. This creates a path for heat transfer that bypasses the rings to the lower, heavier part of the piston.

6. *Skirt:* that area of the piston from the ring groove just above the pin hole down to the bottom of the piston. Some engines are designed with full skirt or trunk-type pistons, while others use slipper pistons that have two skirt pads, one on the minor thrust side and one on the major thrust side.

7. *Tapered skirt:* difference in piston diameter when measured at the bottom of the skirt and at the top of the skirt, the larger diameter being at the bottom. This helps to maintain proper piston-to-cylinder wall clearance since piston temperature is lower at the bottom than at the top of the skirt, where greater expansion takes place.

8. *Cam ground or elliptical pistons:* oval or elliptical shape of the piston skirt (Figure 10-4). Piston skirt diameter across the thrust surfaces is larger than the measurement taken 45° away from the pin axis. This creates a curved skirt, which results in only a small area of skirt contact with the cylinder wall when the piston is cold and almost full contact when at operating temperature (Figure 10-5). This is called cam action. During temperature increase the pin bosses actually slide outward on the piston pin. As the piston cools the pin bosses slide back again. If the pin fit is too tight, the pin bosses can seize on the pin and cause piston breakage.

9. *Pin bosses:* heavily reinforced section around each piston pin hole.

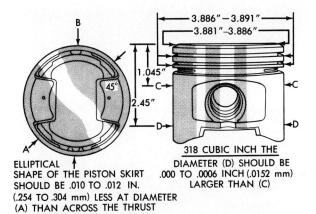

FIGURE 10-4 Example of elliptical shape of piston. (Courtesy of Chrysler Canada Ltd.)

ELLIPTICAL SHAPE OF THE PISTON SKIRT SHOULD BE .010 TO .012 IN. (.254 TO .304 mm) LESS AT DIAMETER (A) THAN ACROSS THE THRUST FACES AT DIAMETER (B)

318 CUBIC INCH THE DIAMETER (D) SHOULD BE .000 TO .0006 INCH (.0152 mm) LARGER THAN (C)

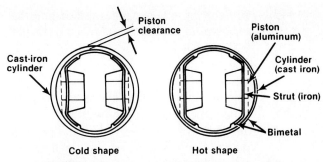

FIGURE 10-6 Action of steel strut method of piston expansion control. (Courtesy of Chrysler Canada Ltd.)

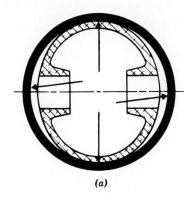

(a)

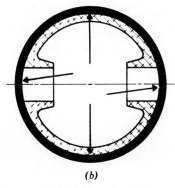

(b)

FIGURE 10-5 Cam action of elliptical piston (cam ground). Piston in cold engine (a) and at operating temperature (b). (Courtesy of Sunnen Products Company.)

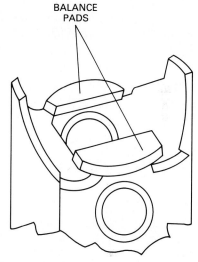

FIGURE 10-7 Balance pads on piston. (Courtesy of FT Enterprises.)

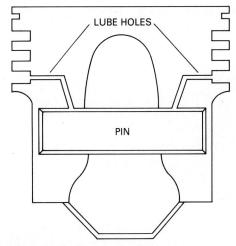

FIGURE 10-8 Lubrication for the piston pin is provided by passages from the oil ring groove to the pin boss on some pistons. (Courtesy of FT Enterprises.)

10. *Steel strut:* used in some aluminum pistons to control expansion across the pin bosses (Figure 10-6).

11. *Balance pads:* pads located just below the pin bosses. The extra piston material in these pads is provided to allow balancing the piston weight with other pistons. Material is removed from the heavier pistons until the piston weights are equal (Figure 10-7).

12. *Pin lubrication holes:* holes drilled into the pin bosses to provide pin lubrication (Figure 10-8).

13. *Ring groove insert:* steel insert in the top compression ring groove is used in some diesel engines pistons. The insert reduces ring groove wear (Figure 10-9).

14. *Ring groove spacer:* when worn top compression ring grooves are machined to restore squareness, a steel spacer is installed to restore groove width to the original dimensions (Figure 10-10).

15. *Pin offset:* used by some engine manufacturers to offset the piston pin from the piston centerline toward the major thrust side. This reduces piston slap as the piston passes through TDC from the compression stroke to the power stroke (Figures 10-11 and 10-12).

16. *Control buttons:* another method used in some racing engines to control piston slap.

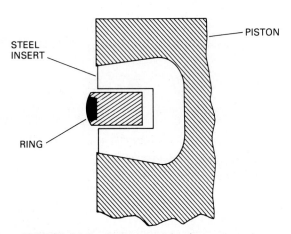

FIGURE 10-9 Steel insert in ring groove reduces wear. (Courtesy of FT Enterprises.)

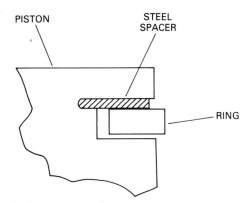

FIGURE 10-10 A steel insert is used to restore a reconditioned ring groove to proper width. (Courtesy of FT Enterprises.)

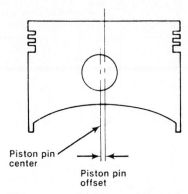

FIGURE 10-11 Piston pin offset from center of piston. (Courtesy of Chrysler Canada Ltd.)

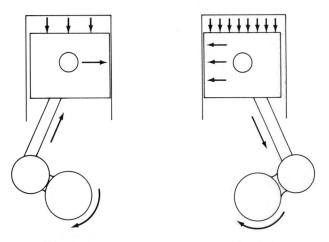

COMPRESSION   POWER

FIGURE 10-12 As the piston moves through TDC between the compression and power strokes it is slammed from the minor thrust side to the major thrust side. (Courtesy of FT Enterprises.)

FIGURE 10-13 Cast aluminum piston with recessed areas in head for valve clearance. (Courtesy of Sealed Power Corporation.)

Holes are machined in the piston skirt and Teflon buttons installed. An interference fit is obtained by heating the piston and cooling the buttons before installation.

17. *Oil drainback holes or slots:* located in the bottom of the oil ring groove and designed to allow oil scraped from the cylinder walls to drain back to the crankcase.

18. *Valve relief:* recessed areas on top of the piston designed to prevent contact between piston and valves (Figure 10-13)

PISTON MATERIAL

Modern automotive engine pistons are made from aluminum. They are either cast or forged. Cast aluminum pistons are made by pouring a molten aluminum alloy into a piston mold. After cooling, the piston is removed and machined to exact dimensions. Figure 10-13 shows a cast aluminum piston.

Forged pistons are stronger and more durable than cast pistons. Forging increases metal density and creates a directional grain in the metal, particularly in critical areas such as the head and pin bosses. Forged pistons are made from an aluminum alloy slug. These slugs are formed under extreme forging pressures. Forged pistons are the same weight as cast pistons but operate at a lower temperature than cast pistons under the same conditions. Figure 10-2 shows a forged aluminum piston.

Piston-to-Cylinder Wall Clearance

Proper piston-to-cylinder clearance is critical to good engine performance. Excessive clearance results in noisy operation, oil consumption, and rapid wear. Too little clearance causes scoring, rapid wear, and possible seizure.

The clearance between the piston and cylinder is the difference between the piston skirt diameter and the cylinder bore diameter. Clearance requirements vary with engine type, piston design, piston material, bore diameter, type of engine service (normal or severe), and whether the engine is equipped with a turbocharger. Clearance specifications for production passenger car engines range from approximately 0.0005 to 0.003 in.

To maintain proper piston clearance, good piston cam action is required. Factors that affect clearance during engine operation include overheating, overcooling, incorrect ignition timing, overloading, overspeeding, and the like. All of these (with the exception of overcooling) will increase piston temperatures and reduce piston clearance. Engines used for high-performance, heavy-duty use or severe operating conditions may require piston clearance that is greater than average.

Piston clearance should be measured with the pistons and cylinders at room temperature (around 70°F). Pistons must be measured across the thrust sides at the height point specified by the piston manufacturer. The height at which measurement is made varies with different manufacturers from a point just below the oil ring groove, at the pin centerline, or midway between the bottom of the skirt and the balance pad.

PISTON SKIRT FINISH

The surface finish of piston skirts is an important factor in determining its ability to carry oil for lubrication and its friction characteristics. Machining tool marks leave a series of very fine horizontal grooves and ridges across the skirt surface. The smoothness of this finish varies with different piston manufacturers. The depth of these grooves ranges around 0.0005 in. Some pistons are plated with a very thin tin plating of about 0.0005 in. to help reduce scuffing and scoring during periods of minimal lubrication, such as startup. A curvilinear surface is used on some pistons to improve oil retention and resist seizure (Figure 10-14).

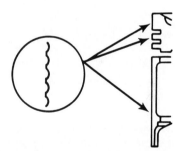

FIGURE 10-14 A tin-plated curvilinear surface is used on this piston to improve oil retention and resist seizure. (Courtesy of Chrysler Canada Ltd.)

PISTON TEMPERATURES

Piston head temperatures can range from well below zero (in cold winter climates) to as hot as 550°F for cast-aluminum pistons. The head is the hottest part of the piston. The piston head diameter (and the ring area) is 0.020 to 0.030 in. smaller than the skirt diameter. This is necessary since the heavy concentration of metal in this area is subjected to the highest temperatures and therefore the most expansion.

Temperatures in other parts of the piston are progressively lower toward the bottom of the piston. Figure 10-15 shows typical piston operating temperatures

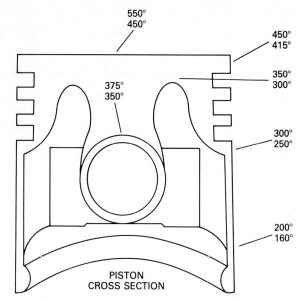

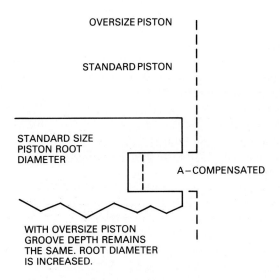

FIGURE 10-15 Typical piston operating temperatures in degrees Fahrenheit. Upper figures are for cast aluminum pistons and lower figures are for forged aluminum pistons. (Courtesy of FT Enterprises.)

for cast and forged pistons. The rate and direction of piston expansion is controlled by such features as cam-shaped piston skirts, heat dams, pin boss relief, and steel inserts.

PISTON SIZES

Pistons are available in standard. +0.030-, +0.040-, and +0.060-in. sizes. Standard pistons can vary in size by as much as 0.0006 in., due to production tolerances. Some manufacturers have as many as four or five different-size standard pistons to accommodate cylinder size variations during production. For example the Ford 2.3-liter four-cylinder engine has standard piston sizes available as follows:

Coded red	3.7764–3.7770 in.
Coded blue	3.7776–3.7782 in.
0.003 in. oversize	3.7788–3.7794 in.

The Ford 351 V8 has the following sizes available:

Coded red	3.9978–3.9984 in.
Coded blue	3.9990–3.9996 in.
Coded yellow	4.0014–4.0020 in.
0.003 in. oversize	4.0002–4.0008 in.

These figures indicate a production tolerance range of 0.0006 in. With a piston-to-bore size of 0.0018 to 0.0026 in., it is easy to see that piston selection and finished bore size after honing are critical to maintaining proper piston-to-bore clearance. Most piston man-

ufacturers build the operating clearance into the piston. A piston marked 0.030 in. oversize is in fact slightly smaller for this reason. This kind of piston sizing allows cylinders to be rebored and finish honed to the size stated on the piston.

Some piston manufacturers finish pistons to the exact size stated on the piston. In this case cylinder boring and honing must take into consideration the required operating clearance. Since piston sizes vary to such an extent, it is important that cylinders be bored and honed to fit the pistons to be used.

Oversized Piston Groove Root Diameter

Oversized pistons supplied by original-equipment manufacturers are normally of the compensated type. This means that the ring groove depth of oversize pistons is the same as for standard-sized pistons. The groove root diameter is increased (compensated) by the amount of piston oversize in order to maintain

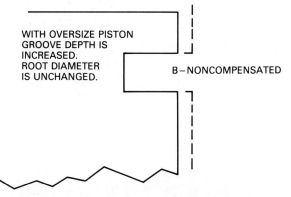

FIGURE 10-16 Effect of root diameter on ring groove depth on compensated and noncompensated oversize pistons. (Courtesy of FT Enterprises.)

groove depth. Some aftermarket piston manufacturers' oversized pistons are of the noncompensated type. In this piston design the groove root diameter is noncompensated in order to maintain the same root diameter as for standard-sized pistons and the groove depth is increased. Whether compensated or noncompensated pistons should be used must be determined by the engine rebuilder based on the type and dimension of piston rings to be used (Figure 10-16).

PISTON SELECTION

Pistons must be purchased in the size to which the cylinders are to be rebored. Factory-finished oversized pistons will fit and provide the required clearance in cylinders bored to their oversize. Although standard bore cylinders are sometimes several thousandths larger than standard, replacement standard-sized pistons are large enough to fit the larger cylinders, while some honing of the smaller cylinder may be required.

USING USED PISTONS

While new pistons are normally used by engine rebuilders and remanufacturers to fit rebored cylinders, used pistons are often used by smaller independent shops in less extensive engine overhaul procedures. Using pistons over again requires careful piston inspection and measuring to determine their serviceability. If the piston is not cracked and the skirt is not collapsed (Figure 10-17), it can be fitted with an over-

sized piston pin, and if worn, the top ring groove can be reconditioned and the skirts expanded by knurling.

Machining Worn Top Ring Groove

If pistons are not to be replaced during an engine overhaul, the top ring groove may require reconditioning (Figure 10-18). A worn top ring groove allows excessive blowby even when new rings are installed. A new ring does not seal properly against the sides of a worn ring groove. Measure the ring groove with a ring groove wear gauge and compare to specifications (Figure 10-19). If worn, the groove must be machined and a groove insert installed to restore proper sealing of the ring. Figure 10-20 shows power regrooving equipment.

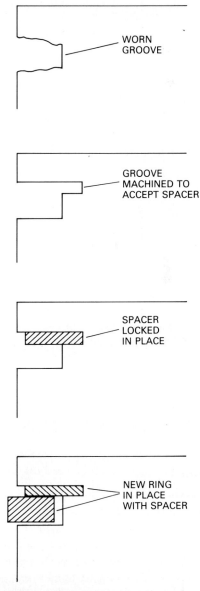

FIGURE 10-18 Worn top ring groove machined and spacer installed. (Courtesy of FT Enterprises.)

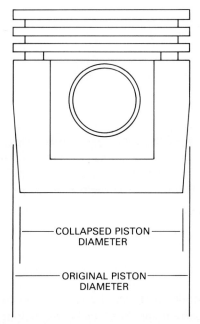

FIGURE 10-17 Effect of collapsed skirt on piston diameter. (Courtesy of FT Enterprises.)

FIGURE 10-19 Measuring top ring groove for wear. (Courtesy of Federal-Mogul Corporation.)

FIGURE 10-21 Knurling a piston. (Courtesy of Hastings Manufacturing Company.)

FIGURE 10-20 Power ring regrooving equipment. (Courtesy of Hastings Manufacturing Company.)

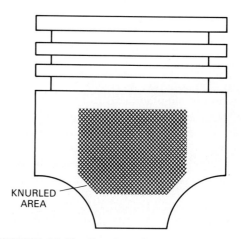

KNURLED AREA

FIGURE 10-22 Knurled piston. (Courtesy of FT Enterprises.)

PISTON FAILURE ANALYSIS

Piston failures result from such causes as preignition, detonation, lack of lubrication, too tight a pin fit, misaligned connecting rod, a galled piston pin, a broken valve, loss of pin lock ring, coolant in cylinder, and deck height too low. For more detail on piston failures and failure analysis, see Chapter 5.

PISTON RING FUNCTION

Piston rings are designed to act as a dynamic seal between the piston and the cylinder wall. They must keep the escape of combustion gases or blowby past the rings to a minimum. They must also help to maintain an oil film on the cylinder wall and prevent oil from entering the combustion chamber.

Piston Resizing

In situations where pistons are to be used over again, piston-to-cylinder clearance can be restored by resizing the piston skirts on the thrust sides. Piston skirts can be expanded by a process of metal displacement called knurling. Metal is forced up between the teeth of the knurling tool (Figures 10-21 and 10-22). Increasing the tool apply pressure increases the amount of metal displacement and therefore the skirt diameter. Knurled pistons are fitted slightly tighter than new pistons. The knurled skirt surface carries lubricating oil to maintain an oil film between the tighter-fitting piston and the cylinder wall.

PISTON RING TYPES

Most modern passenger car engines are fitted with pistons that carry three rings: two compression rings and an oil control ring or scraper (Figure 10-23). The primary function of the top compression ring is to seal against combustion gases. The ring is held against the cylinder wall by static ring tension. On the power stroke combustion pressure forces the ring down against its lower land. Gas pressure gets in behind the ring, increasing ring pressure against the cylinder wall. The second compression ring seals against any gases that get past the top ring. The second compression ring also aids in oil control. The oil ring is designed to control the amount of oil distributed over the cylinder wall. It must scrape off the excess oil and return it to the crankcase.

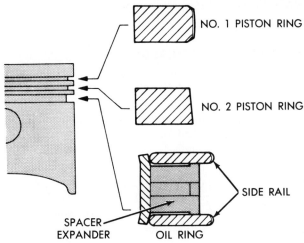

FIGURE 10-23 Cross section of typical piston rings for three-ring piston. (Courtesy of Chrysler Canada Ltd.)

PISTON RING ACTION AND DESIGN

Static tension keeps the piston rings tightly in contact with the cylinder wall. In their relaxed state, piston rings are larger than the cylinder bore and are slightly ovate. When compressed to fit the cylinder they become round and develop the required static tension. Originally, piston rings were of simple rectangular cross-section design. The use of tapers, chamfers, and counterbores have improved ring action.

A tapered faced compression ring scrapes oil from the cylinder wall when the piston moves down, and slides over the oil film when the piston moves up in the cylinder. A line contact is achieved between the lower edge of the ring and the cylinder wall.

A compression ring with a counterbore or taper on the upper inner edge of the ring causes the ring to twist slightly when installed. This results in ring action

similar to that of the taper faced ring. A counterbore on the lower outer edge of the ring creates similar action.

A ring with a lower inside bevel or counterbore provides a reverse twist when installed but has a barrel or tapered face to provide sliding action when moving up in the cylinder. All beveled and counterbored rings provide line contact sealing between the side of the ring and the ring groove. The reverse torsional twist ring reverses the line contact seal between the sides of the ring and the ring groove to prevent oil from entering the ring groove and causing ring float. This reverse torsional twist ring design is widely used as the second compression ring to reduce oil consumption caused by high intake manifold vacuum (Figure 10-24). High manifold vacuum draws oil around the back of the rings and into the cylinder if the rings do not seal well against the ring lands. A plain rectangular cross-section ring provides no torsional twist and therefore no line contact sealing.

The chrome-plated ring is designed to provide cylinder contact with the lower edge of the ring. The ring face maintains good cylinder wall contact despite slight piston rock or groove misalignment. Chrome-plated rings are widely used as the top compression ring (Figure 10-25).

A unique top compression ring design is the head land ring used by many racing engines. The ring and

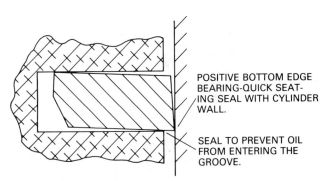

FIGURE 10-24 Taper-faced reverse twist compression ring for second groove. (Courtesy of Muskegon Piston Ring Company.)

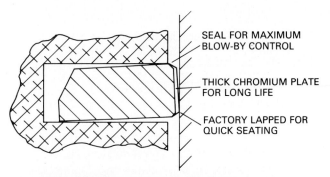

FIGURE 10-25 Chrome-plated top compression ring action. (Courtesy of Muskegon Piston Ring Company.)

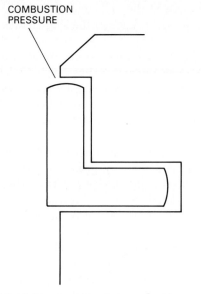

FIGURE 10-26 Headland ring used in some racing engines. (Courtesy of FT Enterprises.)

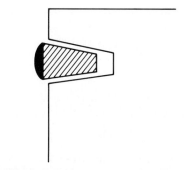

FIGURE 10-27 Keystone ring used in some diesel engines. (Courtesy of FT Enterprises.)

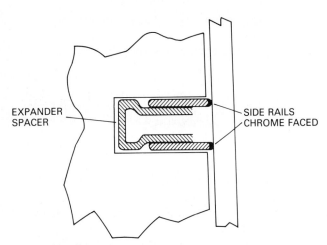

FIGURE 10-28 Cross section of three-piece oil ring with chrome-faced side rails. (Courtesy of FT Enterprises.)

the ring groove are L-shaped in cross section. This allows combustion pressures to enter into the area behind the ring to force the ring tightly against the cylinder wall (Figure 10-26). Keystone compression rings are used in heavy-duty diesel engine applications (Figure 10-27).

Oil ring designs include the simple slotted one-piece cast-iron ring and the segmented multipiece steel oil ring. The oil ring must scrape excess oil from the cylinder wall and return it to the inside of the piston, where it can drain back to the pan. Oil rings are designed to direct oil through the ring to the drain back holes or slots in the piston. A popular oil ring design consists of two chrome-plated steel side rails or scrapers with an expander-spacer. The expander-spacer keeps the side rails in place and provides mechanical pressure to keep the scrapers against the cylinder wall (Figure 10-28). A variety of ring designs are shown in Figure 10-29.

PISTON RING MATERIALS

Compression rings for passenger cars are usually made of a cast-iron alloy. Nodular iron compression rings are used in heavy-duty and diesel applications. Nodular iron is stronger and more durable. Nodular or flake graphite may be included in the ring material.

The popular three-piece oil control ring includes two circumferential chrome-faced steel side rails and a steel expander-spacer. This design is lightweight with low inertia and is very open to provide free oil flow through the ring. A two-piece oil ring design combines the expander-spacer with a side rail on one side. A separate side rail is used on the other side.

PISTON RING COATINGS

Most ring manufacturers use chrome or molybdenum on the face of the top compression ring (Figure 10-30). An electroplating process applies approximately 0.005 in. of chrome to the ring face. Moly coating of rings is done by first grooving the ring face, then applying the moly by a plasma-spraying process. Chrome is a very dense, hard, wear-resistant material with excellent wear resistance and good frictional characteristics. Chrome is also extremely heat resistant with a melting point of 3212°F. These qualities result in high resistance to embedding of foreign material and excellent resistance to scuffing and scoring.

Moly coated rings are designed to survive under severe engine operating conditions. Molybdenum has a lower coefficient of friction than chrome, is harder, and is more heat resistant. The porosity of the moly coating provides oil retention, which reduces friction.

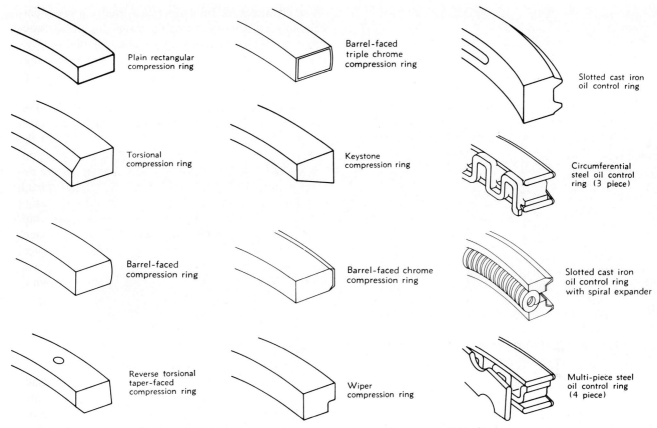

Figure 10-29 Typical piston ring designs. (Courtesy of Hastings Manufacturing Company.)

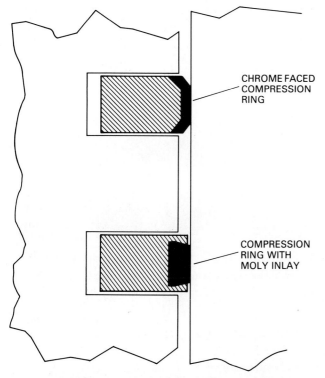

FIGURE 10-30 Chrome-faced compression ring (top) and compression ring with molybdenum (moly) inlay (bottom). (Courtesy of FT Enterprises.)

When both moly-coated and chrome-faced compression rings are used together, the moly ring is used in the top groove and the chrome ring in the second groove.

New ring facing materials are constantly being developed. Among these is the plasma ceramic composition of aluminum oxide and titanium oxide over a nickel aluminide bond. This material is applied with a plasma arc torch at between 30,000 and 40,000°F. This provides an exceptionally durable porous facing.

PISTON RING GAPS

The amount of piston ring gap is an important factor in ring function. Too much ring gap allows excessive blowby and oil consumption. Too little ring gap allows piston ring ends to butt when the engine reaches operating temperature. Butting increases ring pressure against the cylinder walls and causes scuffing, scoring, and rapid wear to take place. Automotive piston rings are designed with straight or butt ends. Some heavy-duty, low-speed engines use a stepped or angled seal cut gap. Figure 10-31 shows various ring gap designs.

The minimum ring gap required is normally 0.0035 per inch of cylinder bore diameter. A 4-in. cyl-

45° ANGLE JOINT

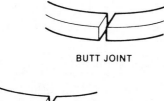

BUTT JOINT

STEP JOINT

FIGURE 10-31 Butt-type ring end design is used with automotive engines. (Courtesy of Ford of Canada.)

inder bore would therefore require a minimum of 0.014-in. ring gap. The following figures represent the specifications for compression ring gap as outlined by the Society of Automotive Engineers (SAE) as standards.

Cylinder Diameter (in.)	Ring Gap Tolerance (in.)
$1-1\frac{31}{32}$	0.005–0.013
$2-2\frac{31}{32}$	0.007–0.017
$3-3\frac{31}{32}$	0.010–0.020
$4-4\frac{31}{32}$	0.013–0.025
$5-6\frac{31}{32}$	0.017–0.032
$7-8$	0.023–0.040

FIGURE 10-32 To position the ring squarely in the cylinder for gap measurement, push the ring down with an inverted piston. (Courtesy of Ford of Canada.)

To measure ring gap, place the ring squarely in an unworn area (below ring travel) in the cylinder. Square the ring by using a piston to push the ring down (Figure 10-32). Measure the gap with a feeler gauge (Figure 10-33). To increase the ring gap, use a thin disc grinder as shown in Figure 10-34. This will keep the butt ends of the ring square.

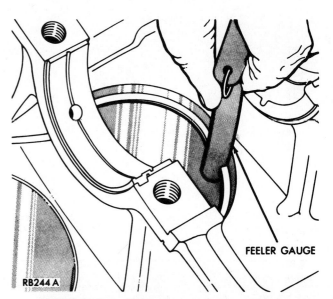

FEELER GAUGE

RB244 A

FIGURE 10-33 Measuring piston ring gap. (Courtesy of Chrysler Canada Ltd.)

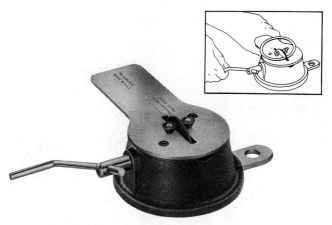

FIGURE 10-34 Thin disc piston ring end gap grinder. (Courtesy of Easco/KD Tools, Inc.)

PISTON RING CLEARANCE

Side Clearance

Ring side clearance is an important factor in ring performance and service life. Ring side clearance is the clearance between the piston ring and the ring land.

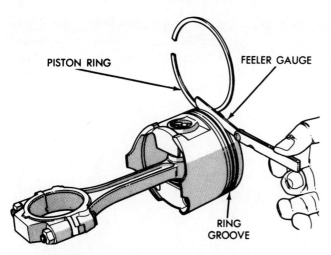

FIGURE 10-35 Measuring ring-to-land (ring side) clearance. (Courtesy of Chrysler Canada Ltd.)

This clearance is measured with a new piston ring and a feeler gauge (Figure 10-35). Excessive side clearance increases inertial forces and causes rapid ring groove wear. Insufficient side clearance causes ring sticking and seizure. To check ring side clearance, roll a new ring around the piston in the ring groove. If it does not bind or drag, there is enough side clearance. Side clearance should not exceed 0.006 in. for compression rings when measured with a feeler gauge and new ring.

Back Clearance

Ring back clearance is a product of ring groove depth and piston ring width. Too little back clearance increases ring pressure against the cylinder wall and causes scuffing, scoring, and rapid wear. To check back clearance of compression rings, place a small straightedge across the piston lands and check the clearance between the rings and straightedge with the

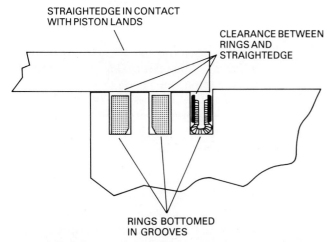

FIGURE 10-36 Checking ring back clearance. (Courtesy of FT Enterprises.)

rings bottomed in their grooves. Minimum clearance should be 0.005 in. (Figure 10-36).

To check oil ring back clearance, place the straightedge across the piston skirt and measure the clearance between the face of the side rails and the straightedge with the side rails and expander-spacer bottomed in the groove. Back clearance can also be checked with a depth micrometer to measure groove depth and comparing to ring width (Figure 10-37). A minimum of 0.006 in. is required. If back clearance is not adequate, the ring grooves will have to be machined to increase their depth.

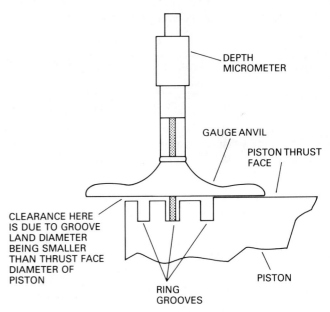

FIGURE 10-37 Measuring ring groove depth. (Courtesy of FT Enterprises.)

PISTON RING SELECTION

The type of piston rings to be used in any particular engine is based on engine design factors such as the following:

- Bore-to-stroke ratio
- Compression ratio
- Cylinder bore material
- Method of cylinder lubrication
- Peak intake manifold vacuum
- Piston design
- Piston displacement
- Piston speed
- Cooling capacity of engine

Soft coated and porous metal faced piston rings are designed to wear in quickly to form a good seal against the cylinder wall. This type of ring is usually recommended for use in cylinders that are slightly worn and have not been rebored or rigid honed. Piston rings with hard coatings such as chrome or molybdenum are normally recommended for use in new or resized cylinders that are perfectly round and not worn. A very shallow ribbed face on the ring helps to hold oil and allows the ring to wear in quickly and good sealing to take place.

The original equipment manufacturer's ring recommendations, the piston ring manufacturer's recommendations, and the experience of the engine rebuilder combine to make the most appropriate ring selection. Piston ring size requirements are related directly to the cylinder bore size, in which they are to be installed as follows:

Bore Size (in. OS)	Ring Size (in. OS)
Standard and 0.010	Standard
0.020	0.020
0.030	0.030
0.040 and 0.050	0.040
0.060 and 0.070	0.060

PISTON RING INSTALLATION

When installing piston rings, the following guidelines should be used.

1. Pistons must be new or in excellent condition.
2. Pistons must be absolutely clean with no carbon residue in ring grooves.
3. Make sure that the correct ring size and type are used.
4. Check to make sure that compression ring end gaps are correct.
5. Check to ensure that compression ring side clearance is correct.
6. Check to ensure that ring back clearance is adequate.
7. Install the segmented oil ring as follows. First install the expander-spacer in the oil ring groove. Follow the ring set instructions as to placement of the expander ends. Make sure that the expander ends do not overlap at any time. Install the top side rail first by spiraling it into place with the end gaps about 1 in. away from the ends of the expander-spacer. Hold the expander-spacer ends in place while installing the upper side rail (Figure 10-38). Install the lower side rail in the same manner, positioning the end

FIGURE 10-38 Installing the side rail of a three-piece oil ring with the expander-spacer already in place. (Courtesy of Chrysler Canada Ltd.)

gaps to the other side of the expander-spacer ends. Segmented oil ring installation is done entirely by hand.

8. Examine the instructions that come with the ring set to distinguish between the top and second compression rings. Examine the rings to establish which side is up by noting the markings on the rings (Figure 10-39).

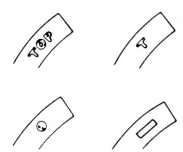

FIGURE 10-39 Typical ring markings indicating which side is up. (Courtesy of Chrysler Canada Ltd.)

9. Using a good ring installing tool, install the number 2 compression ring with the top facing the top of the piston (Figure 10-40). Install the top compression ring in the same manner. Never allow rings to become distorted in any way since a distorted ring results in poor sealing and excessive oil consumption. Rings installed upside down or placed in the wrong groove will cause oil to be pushed into the combustion chamber.

10. Just before installing the piston and rod assembly in the engine, dip the piston in clean engine oil past the piston pin to ensure ring and pin lubrication. Squirting oil at the rings

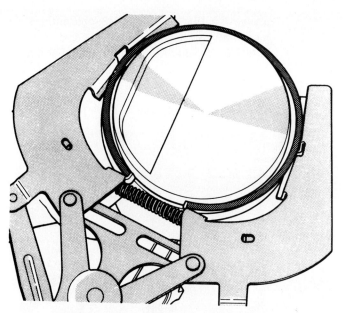

FIGURE 10-40 Using a ring expander to install a compression ring. (Courtesy of Chrysler Canada Ltd.)

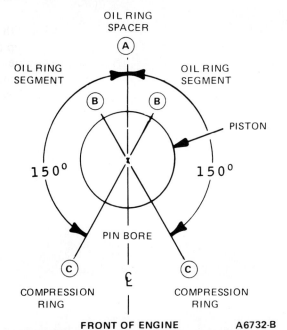

OIL RING SPACER

(A)

OIL RING SEGMENT

OIL RING SEGMENT

(B) (B)

PISTON

150° 150°

PIN BORE

(C) (C)

COMPRESSION RING

COMPRESSION RING

FRONT OF ENGINE   A6732-B

FIGURE 10-41 Ring gaps should be staggered as shown before installing the piston assembly. (Courtesy of Ford of Canada.)

with an oil can often results in leaving the pin or some ring area dry.

11. Stagger the ring gaps as shown in Figure 10-41. Never position ring gaps in line with the piston thrust faces or the pin hole. Install the bearing inserts into the rod yoke and cap.

12. Position the rod journal to which the rod is to be attached at the bottom center. Use a good ring compressor to squeeze the rings

FIGURE 10-42 Piston ring compressor. (Courtesy of Easco/KD Tools, Inc.)

FIGURE 10-43 Use rod bolt covers to avoid nicking crankshaft. Pieces of rubber hose will do. (Courtesy of Chrysler Canada Ltd.)

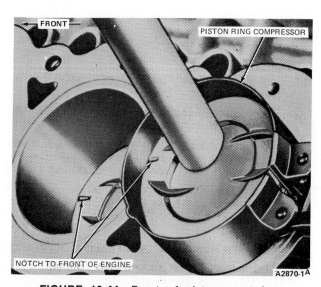

FRONT

PISTON RING COMPRESSOR

NOTCH TO FRONT OF ENGINE

A2870-1A

FIGURE 10-44 Front of piston must face front of engine. Other piston markings are also used. (Courtesy of Ford of Canada.)

into their grooves (Figure 10-42). Use rod bolt covers to prevent nicking the crank pin during assembly (Figure 10-43). Place the assembly in the appropriate cylinder, making sure that the front of the piston faces forward (Figure 10-44), and the numbered side of the rod faces in the right direction. Carefully bump the piston into the cylinder while guiding the rod over the bearing journal (Figures 10-45 and 10-46). Thoroughly lubricate the rod journal with engine oil, install the rod cap (do not reverse the cap), and tighten the nuts alternately using new nuts, to specified torque. Check the rod side clearance to make sure that it is adequate (Figure 10-47).

13. It is good practice to check the effort required to turn the crankshaft after each piston assembly is installed (Figure 10-48). If excessive turning effort is apparent after any one piston is installed, the last installation is at fault. Remove the last piston installed, correct the problem, and reinstall the assembly.

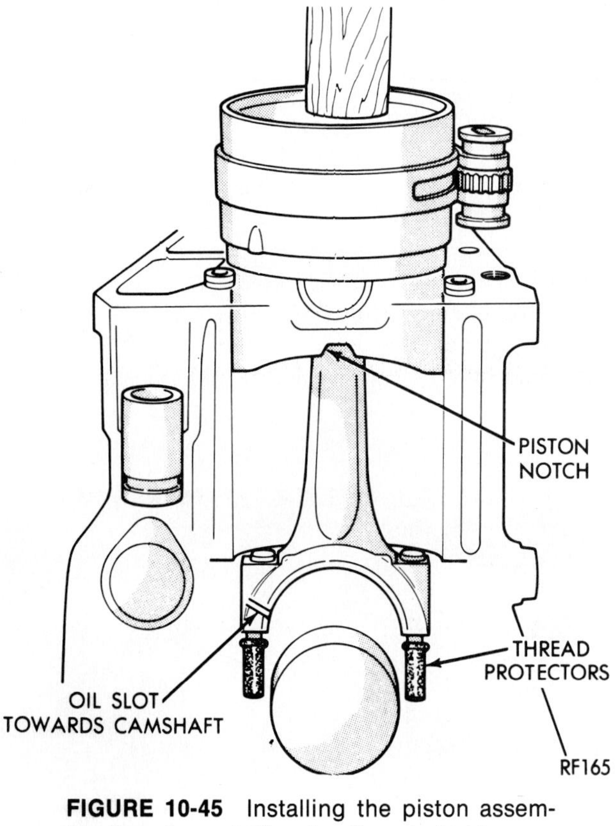

FIGURE 10-46 Installing piston assemblies in V-8 engine. (Courtesy of Federal-Mogul Corporation.)

FIGURE 10-45 Installing the piston assembly (Courtesy of Chrysler Canada Ltd.)

PISTON NOTCH

THREAD PROTECTORS

OIL SLOT TOWARDS CAMSHAFT

RF165

FIGURE 10-47 Checking connecting rod side clearance. (Courtesy of Ford of Canada.)

REVIEW QUESTIONS

1. What function do pistons perform in an engine?
2. Pistons should be:
   (a) as light as possible but still strong enough
   (b) able to maintain proper piston to cylinder clearance
   (c) both (a) and (b)
   (d) neither (a) nor (b)
3. List six different piston head designs.
4. Why are some piston heads chamfered?

FIGURE 10-48 Check the turning effort required after each piston is installed. If excessive effort is required, check the last installation and correct the problem. (Courtesy of Federal-Mogul Corporation.)

5. Why are oil drain back holes needed in a piston?

6. Technician A says that a heat dam keeps the engine at the proper operating temperature. Technician B says that a heat dam is a metal ridge at the top of the cylinder. Who is right?
   (a) technician A
   (b) technician B
   (c) both are right
   (d) both are wrong

7. The piston skirt is tapered:
   (a) to create a wedging action
   (b) to fit a tapered cylinder
   (c) to be smaller at the bottom
   (d) to maintain proper piston to cylinder clearance

8. Why are pistons cam ground or elliptical in shape?

9. Balance pads on pistons are:
   (a) attached to the piston head
   (b) extra piston material used to balance pistons
   (c) bolted to the piston skirt
   (d) removed to balance the piston

10. What is the difference between a top ring groove insert and a ring groove spacer?

11. True or false: Piston pins are sometimes offset to the major thrust side.

12. Modern automotive pistons are made from _____ and are produced by either a _____ or a _____ process.

13. Why are some pistons tin plated?

14. Are all pistons exactly the same size in an engine? Explain your answer.

15. What is the difference between a compensated piston and a noncompensated piston?

16. Used pistons that are not going to be replaced are sometimes knurled to:
    (a) improve their friction characteristics
    (b) increase their taper
    (c) expand them
    (d) provide a tighter pin fit

17. What are the three basic functions of piston rings?

18. The piston rings are kept in contact with the cylinder wall by:
    (a) static tension
    (b) dynamic pressure
    (c) both (a) or (b)
    (d) neither (a) nor (b)

19. A counter bore on the lower inside edge of the ring provides:
    (a) a reverse twist
    (b) sliding action when moving up in the cylinder
    (c) scraping action when moving down in the cylinder
    (d) all of the above
    (e) none of the above

20. Two popular piston ring face coatings are:
    (a) steel and chrome

(b) chrome and molybdenum
(c) molybdenum and tin
(d) tin and steel

21. Why is a piston ring gap needed?
22. How much ring gap should there be for each inch of bore diameter?
    (a) 0.35 in.
    (b) 0.035 in.
    (c) 0.0035 in.
    (d) 0.00035 in.
23. Piston ring side clearance should not exceed:
    (a) 0.6 in.
    (b) 0.06 in.
    (c) 0.006 in.
    (d) 0.0006 in.
24. Too little ring back clearance will cause:
    (a) too much ring pressure against the cylinder wall
    (b) too little ring pressure against the cylinder wall

(c) excessive side clearance
(d) insufficient side clearance

25. When installing a three-piece oil ring, install the:
    (a) upper rail first
    (b) lower rail first
    (c) expander-spacer first
    (d) none of the above
26. Installing a piston ring upside down will cause:
    (a) oil to be pushed into the combustion chamber
    (b) oil to be pushed down the cylinder wall
    (c) poor ring lubrication
    (d) none of the above
27. Before installing the piston and rod assembly:
    (a) install rod bolt covers
    (b) position the crankpin at BDC
    (c) compress the piston rings
    (d) all of the above

CHAPTER 11

Camshafts, Lifters, and Camshaft Drives

OBJECTIVES

1. To develop an understanding of the function, construction, design and operation of camshafts, valve lifters, and cam-shaft drive mechanisms.
2. To develop the ability to inspect and determine the service ability of camshafts, valve lifters, and camshaft drive mechanisms.
3. To develop the ability to install camshaft bearings, camshafts, valve lifters, and camshaft drive mechanisms to meet factory specifications.

CAMSHAFT FUNCTION

The camshaft is designed to control the opening and closing of the intake and exhaust valves. It controls when the valves start to open, how far they will open, how long they remain open, and when they will close. Cam lobe design also determines how fast the valves will open and close. In many cases the camshaft also drives the distributor, oil pump, fuel pump, and accessory shaft.

CAMSHAFT LOCATION

Some engines are designed with the camshaft located in the engine block. This design requires pushrods to transfer cam action to the rocker arms and valves in overhead valve engines (Figures 11-1 and 11-2). On the older L-head engines with the valves located in the block, pushrods were not required. Both in-line and V engines require only one camshaft to operate all the valves.

Overhead cam engines have one or more camshafts located at the top of the cylinder head. In-line OHC engines may have one camshaft (SOHC, single overhead cam) to operate all the valves or two camshafts (DOHC, double overhead cam), one to operate the intake valves and one to operate the exhaust valves

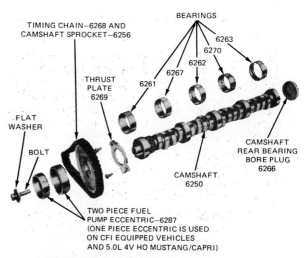

FIGURE 11-1 Chain driven in-block camshaft. (Courtesy of Ford of Canada.)

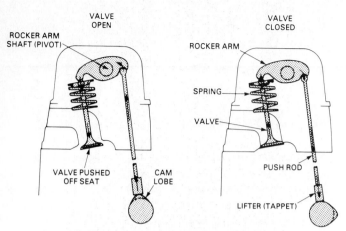

FIGURE 11-2 In-block camshaft operation. (Courtesy of Ford of Canada.)

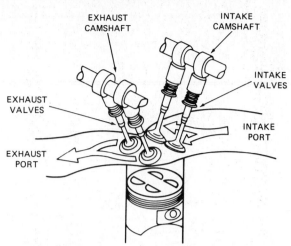

FIGURE 11-4 Direct-acting double overhead cam (DOHC) engine has no pushrods or rocker arms. (Courtesy of Prentice-Hall, Inc., from *Automotive Principles and Service* by Thiessen and Dales.)

FIGURE 11-3 Indirect-acting single overhead cam (SOHC) operates valves through rocker arms. (Courtesy of Ford of Canada.)

(Figures 11-3 and 11-4). V-type OHC engines have either two or four camshafts. Overhead camshaft engine designs require more extensive camshaft drive mechanisms since the camshafts are farther away from the crankshaft.

CAMSHAFT MATERIALS

Most camshafts that use spherical-based valve lifters use camshafts made from hardenable cast-iron alloy. This material has the required strength and wear resistance needed for spherical-based lifters. The molten cast iron is poured into a mold, allowed to cool, and then removed for machining and hardening. A forged steel camshaft is used in many race engines.

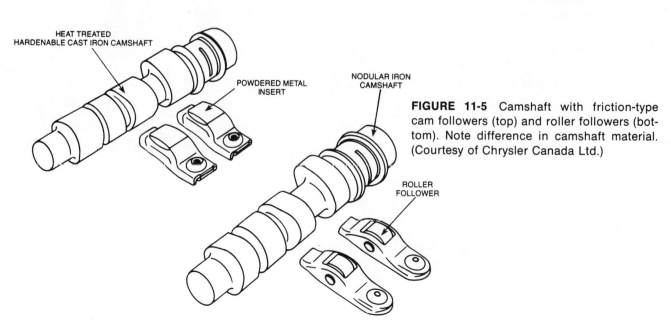

FIGURE 11-5 Camshaft with friction-type cam followers (top) and roller followers (bottom). Note difference in camshaft material. (Courtesy of Chrysler Canada Ltd.)

FIGURE 11-6 Typical camshaft for friction valve lifters. (Courtesy of Sealed Power Corporation.)

FIGURE 11-7 Roller lifter camshaft. Note the difference in the shape of the cam lobes compared to those in Figure 11-6. (Courtesy of Chrysler Canada Ltd.)

Most engines equipped with roller-based valve lifters use a high-carbon-steel or hardened cast iron camshaft. This is necessary due to the smaller contact area of the roller with the camshaft. The smaller contact area results in much greater unit pressure and increased Brinell stress. The camshaft is machined and induction hardened to achieve the desired durability (Figures 11-5 to 11-7).

CAMSHAFT DESIGN

Camshafts are of one-piece design with cam lobes, bearing journals, and a drive flange. Many camshafts also have a distributor and oil pump drive gear and a fuel pump eccentric. Gear blanks are cut into shape by a gear cutter while the cam lobes, bearing journals, and fuel pump eccentric are ground to the desired shape. The rest of the camshaft remains unfinished. Some camshafts have the bearing journals drilled to meter oil to the rocker shaft each time they index with the oil hole in the bearing. Another design has a drilled oil passage the full length of the camshaft with metered oil delivery holes through the cam lobe to lubricate the cam followers. For diagrams of camshaft lubrication, see Figures 14-1 to 14-4.

On engines where the cam bearing bores are full circle, the bearing journals must be larger than the cam lobes to allow camshaft installation through the bearings. Some overhead camshaft bearings are split to allow the camshaft to be removed or installed when the bearing caps are removed. To make camshaft removal and installation easier some camshafts are designed with progressively larger bearing journals, with the smallest journal at the rear. Other camshafts have the same-size bearing journals throughout.

CAMSHAFT END THRUST

End thrust of the camshaft is controlled by several methods. One way is by means of a thrust plate between the cam drive gear or sprocket and a flange on the camshaft (Figures 11-8 and 11-9). The thrust plate is attached to the engine block with cap screws. On overhead camshafts the thrust plate is attached to the cylinder head and fits into a groove in the camshaft (Figure 11-10). Another method is to use the thrust resulting from the distributor and oil pump drive gear load. The angle of the drive gear teeth holds the camshaft in the block. A thrust surface on the back of the drive gear or sprocket limits rearward movement of the camshaft and absorbs the thrust load. A spring-loaded thrust button at the front of the camshaft (Figure 11-11) contacts a thrust surface on the inside of the timing cover to limit forward movement of the cam-

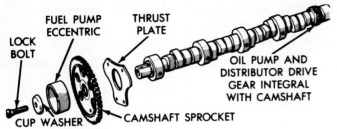

FIGURE 11-8 End play in this in-block camshaft is controlled by a thrust plate located between the front bearing journal and the sprocket. The plate is bolted to the engine block. (Courtesy of Chrysler Canada Ltd.)

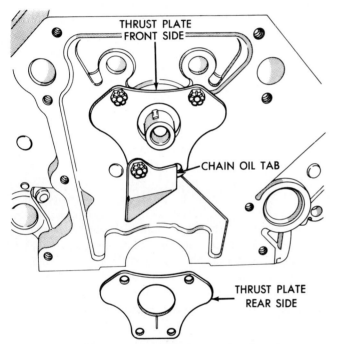

FIGURE 11-9 Camshaft thrust plate location for arrangement shown in Figure 11-8. (Courtesy of Chrysler Canada Ltd.)

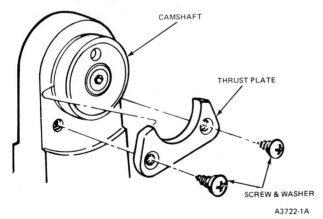

FIGURE 11-10 Camshaft thrust plate on 2.3-liter Ford OHC engine. (Courtesy of Ford of Canada.)

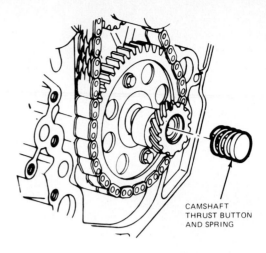

FIGURE 11-11 Camshaft thrust button bears against thrust surface of the timing cover on this engine. (Courtesy of Ford of Canada.)

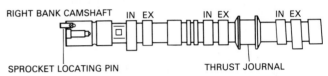

FIGURE 11-12 Thrust flanges on each side of the bearing journal control endwise movement of this camshaft. (Courtesy of Chrysler Canada Ltd.)

shaft. Some overhead camshafts are designed with a flanged thrust journal which bears against machined surfaces on the bearing pedestal and cap (Figure 11-12).

CAM LOBE DESIGN

Cam lobes have an eccentric shape (Figure 11-13) which when turning converts rotary motion to reciprocating motion of the valve train. The shape of the cam lobe is a major factor in engine performance. Altering the cam lobe contour changes the performance of an engine. Cam contours for production passenger

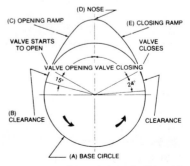

FIGURE 11-13 Cam lobe design terminology. (Courtesy of Chrysler Canada Ltd.)

car engines are designed to provide good performance at engine speeds from idle to cruising at highway speeds. The design and position of the cam lobe determines the following:

- When the valves begin to open (valve-open timing)
- How fast the valves will open (acceleration)
- How far the valves will open (lift)
- How long the valves remain open (duration)
- How fast the valves will close (deceleration)
- When the valves will close (valve-closed timing)

The cam lobe design for any specific engine is based on the following factors:

- Engine speed range
- Displacement
- Intake and exhaust port size
- Valve head diameter
- Combustion chamber design
- Intake manifold design
- Exhaust manifold design
- Compression ratio
- Vehicle weight
- Type of service

CAM ACTION

The cam profile consists of the base circle, opening ramp, opening flank, nose, closing flank, and closing ramp. The opening ramp controls the speed (relative to engine speed) at which the valve is lifted off its seat. The opening ramp blends into the opening flank to control the speed at which the valve opens. The width of the nose area determines how long the valve will remain at the fully open position. The closing flank regulates the rate of deceleration and blends into the closing ramp, which controls the speed at which the valve is seated. The heel is the area between the closing and opening ramps. The valve is seated when the heel of the cam is in contact with the cam follower.

Camshafts designed for use with mechanical lifters have a clearance ramp designed to take up valve lash at a controlled rate just before valve opening. On pushrod engines the cam acts on the lifter, which moves up and down to operate the pushrod, rocker arm, and valve. On overhead cam engines the cam may act directly on the valve stem through a bucket-type cam follower, or it may act near the center of a rocker arm that pivots at one end and operates the valve at the other (Figures 11-2 to 11-5).

Valve Lift and Lobe Lift

Lobe lift is the difference between measurements taken across the nose and the heel of the cam and across the base circle (Figure 11-14). On direct-acting camshafts, lobe lift and valve lift are the same since no lever action by rocker arms is involved. In engines with rocker arms, valve lift is a product of lobe lift and rocker arm ratio. Rocker arm ratio is the difference in dimension from the center of the rocker arm pivot (fulcrum) to the valve stem and the pushrod. A lobe lift of 0.265 in. and a rocker arm ratio of 1.5 yields a valve lift of 0.397 in. (Figure 11-15).

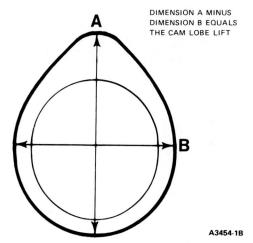

DIMENSION A MINUS DIMENSION B EQUALS THE CAM LOBE LIFT

A3454-1B

FIGURE 11-14 Camshaft lobe life. (Courtesy of Ford of Canada.)

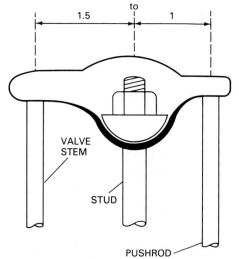

FIGURE 11-15 Rocker arm ratio. (Courtesy of FT Enterprises.)

Solid Lifter versus Hydraulic Lifter Cams

Camshafts designed for use with hydraulic lifters should not be used with solid lifters. The reverse is also true. Using solid lifters with a hydraulic lifter cam results in a noisy valve train and rapid wear. Using hydraulic lifters with a solid lifter cam would result in quiet operation but would change the valve timing and duration.

Roller Lifter versus Conventional Lifter Cam

Roller lifters provide rolling friction versus the sliding friction of conventional lifters. Rolling friction reduces wear and results in a cooler camshaft. Roller lifters allow a much greater rate of lift, which improves engine breathing. The reduced friction also allows operation at higher engine speeds. These advantages add up to longer-lasting camshafts and lifters. As can be seen in Figure 11-16, the roller lifter cam has a much broader profile than a conventional lifter cam. This results from the fact that lifter base design affects when the valve will begin to open in relation to the position of the cam lobe.

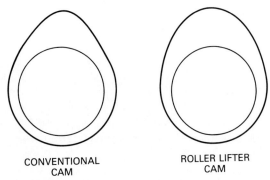

FIGURE 11-16 Comparison of cam lobe profiles. (Courtesy of FT Enterprises.)

VALVE TIMING

The time at which valves open and close (valve timing) and the duration of valve opening is stated in degrees of crankshaft rotation. For example, the intake valve normally begins to open just before the piston has reached top dead center. The valve remains open as the piston travels down to BDC and even past BDC. This is intake valve duration. An example of this could be stated as follows (Figure 11-17): IO at 20° BTDC, IC at 50° ABDC (or, intake opens 20° before top dead center, intake closes 50° after bottom dead center). Intake valve duration in this case is 250° of crankshaft rotation. This leaves 130° duration for the compression stroke since compression ends when the piston reaches TDC. At this point the power stroke begins. The power stroke ends when the exhaust valve begins to open approximately at 50° before bottom dead center. The duration of the power stroke in this case is also 130°

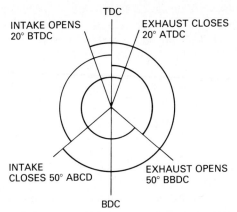

FIGURE 11-17 Valve timing diagram representing two revolutions of the crankshaft. (Courtesy of FT Enterprises.)

Since the exhaust valve is opening at 50° BBDC, this begins the exhaust stroke. The exhaust stroke continues as the piston passes BDC and moves upward to and past TDC. With the exhaust valve closed at 20° ATDC, the duration of the exhaust stroke is 250°.

It is apparent from this description that the exhaust valve stays open for a short period of time during which the intake valve is also open. In other words, the end of the exhaust stroke and the beginning of the intake stroke overlap for a short period of time. This is called valve overlap (Figure 11-18). Valve timing and valve overlap vary on different engines.

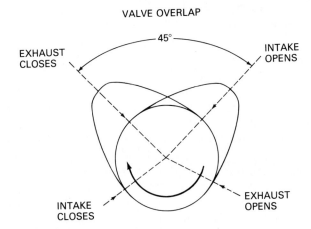

FIGURE 11-18 Valve overlap. (Courtesy of FT Enterprises.)

Opening the intake valve before TDC and closing it after BDC increases the fill of air–fuel mixture at the beginning of the intake stroke, while leaving the intake valve open after BDC takes advantage of the kinetic inertia of the moving air–fuel mixture. This increases volumetric efficiency. As the piston moves down on the power stroke past the 90° ATDC position, pressure in the cylinder has dropped, and the leverage to the crankshaft has decreased due to connecting rod angle and crankshaft position. This ends the effective length of the power stroke, and the exhaust valve can now be opened to begin expelling the burned gases. The exhaust valve remains open until the piston has moved as much of the burned gases as is possible and increases volumetric efficiency.

Valve timing in a production-line engine is designed to provide the best overall performance and economy for normal city and highway driving. However, for high-performance driving or racing, valve timing and cam lobes are designed to increase intake and exhaust duration and provide more power. An offset camshaft drive key is sometimes used to improve valve timing and make it more precise.

Effect of Valve Timing on Exhaust Emissions

Valve timing has been used to reduce exhaust emissions. At its peak during the mid-1970s, valve overlap could be as high as 120° on some engines. The large valve overlap allows some of the exhaust gases to remain in the cylinder with the fresh air–fuel mixture. This results in lower peak combustion temperatures and therefore lower nitrogen oxide emissions. Improvements in engine design, particularly the combustion chamber, intake and exhaust ports, and the development of three-way catalysts and exhaust gas recirculation have reduced the need for large overlap. Large valve overlap was a major cause of rough engine idle operation during the 1970s.

VALVE LIFTER FUNCTION

Valve lifters are also known as valve tappets or cam followers. They are designed to follow the contour of the cam lobe and transmit the resulting reciprocating motion to the valve train and valve. Valve lifters are either solid or hydraulic. Solid lifters are also known as mechanical lifters, while hydraulic lifters may also be called automatic lash adjusters.

Solid lifters are rigid mechanical devices used to transmit motion from the cam lobe. Hydraulic lifters act like solid lifters when the valve is opened but automatically maintain zero clearance (or lash) in the valve train as well. The solid lifter valve train must have some clearance to prevent valves from being held open when they should be closed. Valve lash or clearance allows parts to expand from heat without affecting valve closing. Hydraulic lifters eliminate the need for periodic valve lash adjustment, eliminate tappet noise, and provide more precise control of valve timing, which results in smoother engine operation and longer life of valve train parts. A variety of valve lifters is shown in Figure 11-19.

FIGURE 11-19 Valve lifter designs vary considerably. (Courtesy of Sealed Power Corporation.)

HYDRAULIC LIFTERS

Lifter Operation

Hydraulic lifter operation relies on engine oil pressure. When the intake or exhaust valve is closed, and the lifter is on the base circle or heel of the cam lobe, engine oil pressure is fed into the lifter body and the lifter plunger. Oil flows through the check valve and fills the area below the lifter plunger. This takes up all valve lash. As the camshaft rotates, the cam lobe begins to push against the bottom of the lifter body. Valve spring pressure through the rocker arm and pushrod attempts to keep the lifter plunger down. This causes pressure in the lifter below the plunger to increase and close the check valve. The oil is trapped below the plunger, and the lifter, in effect becomes solid since oil is not compressible.

A small amount of oil leakage past the plunger allows the lifter to leak down should the lifters pump up. Lifter pump-up is caused by anything that causes momentary clearance anywhere in the valve operating train. Valves that are sticky in the valve guides, weak valve springs, and overspeeding of an engine can cause lifter pump-up. Another reason for this slight leakage past the plunger is to allow oil to escape as parts expand as their temperatures increase. If oil could not escape, the valves would not be able to seal. On some models, a metering disc just below the pushrod seat meters the amount of oil delivered through the pushrod to the rocker arm. On others, the oil is fed down from the rocker arm through the pushrod to the lifter. Hydraulic lifters for pushrod engines and hydraulic lash adjusters for overhead cam engines operate in the same manner (Figures 11-20 to 11-26).

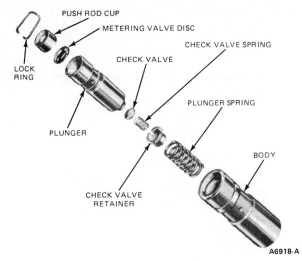

FIGURE 11-20 Exploded view of hydraulic lifter with disc-type check valve. (Courtesy of Ford of Canada.)

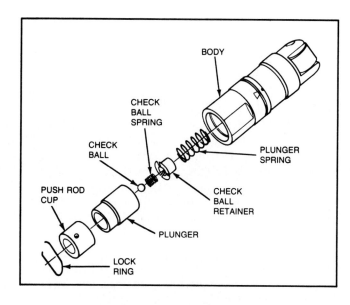

FIGURE 11-21 Exploded view of roller-type hydraulic lifter with ball check valve. (Courtesy of Ford of Canada.)

Automatic Lash Adjusters

Some engines are designed with automatic lash adjusters located in a bucket type cam follower (Figure 11-25) or in the rocker arm (Figure 11-26). They serve the same function as the hydraulic lifter described earlier and operate in a similar manner.

Lifter Design

Hydraulic lifter design varies considerably among manufacturers. In addition to the obvious differences in outside appearance, such as height, diameter, groov-

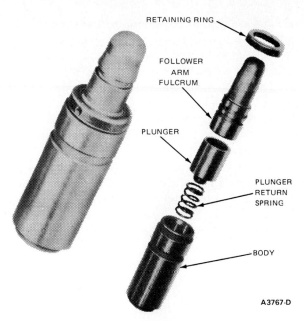

FIGURE 11-22 Hydraulic lash adjuster for overhead cam engine. (Courtesy of Ford of Canada.)

ing, pushrod seats, and retainers, there are internal differences. These include the following:

- Type of valve—ball or disc
- Type of plunger spring—straight or conical
- Shape of plunger
- Shape of pushrod seat

Replacement hydraulic lifters for the same engine may vary in design and appearance, depending on the lifter manufacturer's design choice.

Oversized Lifters

Engine manufacturers often do not scrap an engine block that has a minor lifter bore fault. The bore is machined to oversize to remove the fault. An oversized lifter is then installed in the refinished bore. Such engines may have one or more oversized lifter bores. Chrysler-produced V-8 engines with oversized lifters have a diamond-shaped mark on the engine ID pad. Some General Motors engines have a O stamped on the lifter bore OD to indicate 0.010 in. oversize. Lifter oversizes can range from 0.001 to 0.010 in. Replacement lifters must always match the lifter bore size.

Lifter Rotation

Lifter rotation is designed to promote even wear distribution. Lifter rotation is achieved by the precise relationship between the cam lobe and lifter base in the following ways. (Figure 11-27).

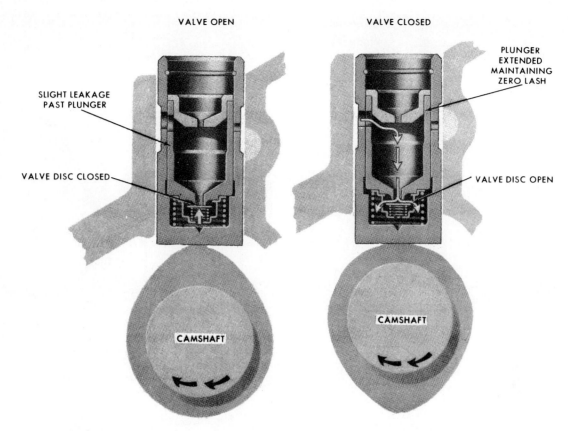

VALVE OPEN

VALVE CLOSED

PLUNGER EXTENDED MAINTAINING ZERO LASH

SLIGHT LEAKAGE PAST PLUNGER

VALVE DISC CLOSED

VALVE DISC OPEN

CAMSHAFT

CAMSHAFT

FIGURE 11-23 Hydraulic lifter operation. (Courtesy of Ford of Canada.)

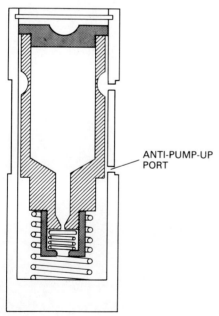

ANTI-PUMP-UP PORT

FIGURE 11-24 Anti-pump-up hydraulic lifter cross section. (Courtesy of FT Enterprises.)

1. The lifter centerline is offset from the lobe centerline. This results in offsetting the friction forces on the base of the lifter to one side, thereby inducing slight rotation every time the valve is opened.

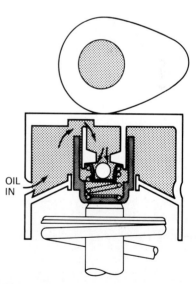

OIL IN

FIGURE 11-25 Bucket-type hydraulic lash adjuster for 16-valve DOHC engine. (Courtesy of FT Enterprises.)

2. The lifter base is ground to a convex or crowned shape and the surface of the cam lobe is ground to a slope. The amount of crown and slope is 0.0007–0.002 in. each. This arrangement offsets the area of contact on the lifter base, causing a slight rotation.

Hydraulic Lifters **185**

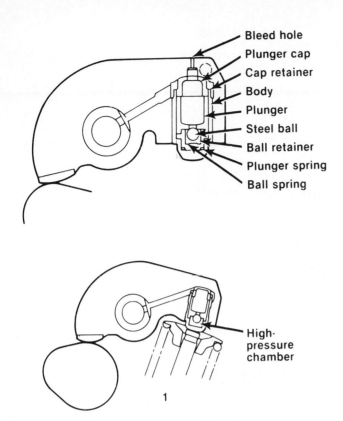

Bleed hole
Plunger cap
Cap retainer
Body
Plunger
Steel ball
Ball retainer
Plunger spring
Ball spring

High-pressure chamber

1

2

Reservoir chamber

3

FIGURE 11-26 Hydraulic lash adjuster for 3.0-liter V-6 Mitsubishi engine is incorporated into the rocker arm. (1) At start of lift check ball is seated. (2) During lift a small amount of oil passes between the plunger and body. (3) At the end of the lift the lash adjuster returns to its original mode. (Courtesy of Chrysler Canada Ltd.)

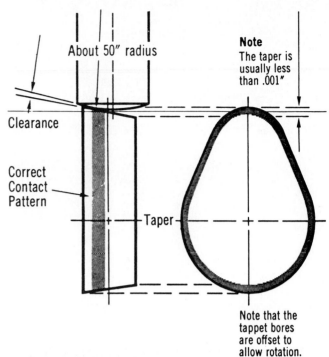

About 50" radius

Note
The taper is usually less than .001"

Clearance

Correct Contact Pattern

Taper

Note that the tappet bores are offset to allow rotation.

FIGURE 11-27 Lifter rotation is caused by offset contact between the cam lobe and lifter base and the shape of the lifter base and cam nose. (Courtesy of Sealed Power Corporation.)

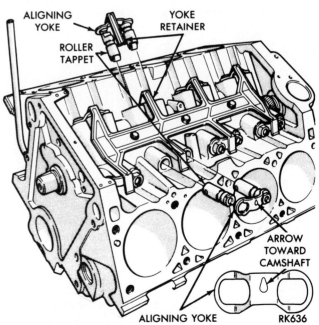

ALIGNING YOKE
YOKE RETAINER
ROLLER TAPPET
ARROW TOWARD CAMSHAFT
ALIGNING YOKE
RK636

FIGURE 11-28 Aligning yoke and retainer prevent roller lifter rotation. (Courtesy of Chrysler Canada Ltd.)

When lifters do not rotate, increased wear will take place on the cam lobe, lifter base, and pushrod ends. For this reason new lifters should always be used with a new or reground camshaft. Using worn lifters with a new cam or new lifters with a worn cam results in poor contact between the cam lobes and the lifters.

Roller Lifters

Roller lifters and cam followers are more durable and last longer than sliding types. Roller lifters have less friction but develop increased surface stresses on the camshaft. A camshaft with a harder surface is required for roller-type cam followers. Roller hydraulic lifters must be prevented from turning in the lifter bore since the rollers must remain aligned with the cam lobe surface. One method of preventing roller lifter rotation is by means of an aligning yoke and yoke retainer as shown in Figure 11-28.

HYDRAULIC LIFTER FAILURE ANALYSIS

For detailed information on hydraulic lifter noise and failure analysis, refer to Chapter 5.

CAMSHAFT BEARING WEAR

Camshaft bearings normally have a lower wear rate than crankshaft bearings since camshafts run at only half crankshaft speed. High pressure on one side of the camshaft causes bearings to wear out-of-round. This results in increased clearance and reduced engine oil pressure. Worn camshaft bearings must be replaced as outlined in Chapter 12.

CAMSHAFT INSPECTION AND MEASURING

Visually inspect the condition of the camshaft before measuring. A camshaft that has chipped, scored, pitted, worn, or overheated and discolored cam lobes or bearing journals must be replaced (Figures 11-29 to 11-32). Always use new lifters with a new or reground cam.

Measuring Lobe Lift and Journal Wear

Cam lobe wear results in reduced lobe lift and therefore less valve lift. Lobe lift can be measured on an OHV engine without removing the camshaft by using a dial indicator as shown in Figure 11-33. On OHC engines, cam lobe lift is easily measured with an outside micrometer or vernier caliper. The same method is used on camshafts removed from the engine. Lobe lift is the

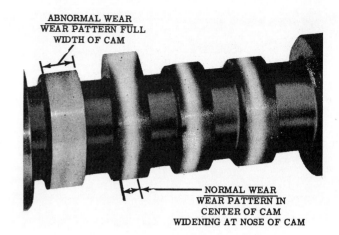

FIGURE 11-29 Normal and abnormal cam lobe wear patterns on in-block camshaft. (Courtesy of Chrysler Canada Ltd.)

FIGURE 11-30 Badly worn cam lobes. (Courtesy of McQuay Norris, Inc.)

FIGURE 11-31 Severely worn camshaft with accompanying lifter damage. (Courtesy of TRW, Inc., Automotive Aftermarket Division.)

difference between measurements taken at A and at B in Figure 11-34. Compare these measurements with specifications for a particular engine. Measure the bearing journal wear as shown in Figure 11-35.

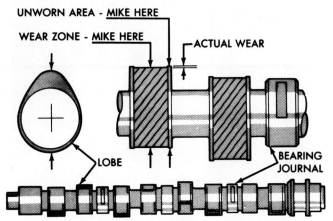

FIGURE 11-32 On some overhead camshafts wear does not reach the edges of the cam lobe. This must be taken into consideration when measuring. (Courtesy of Chrysler Canada Ltd.)

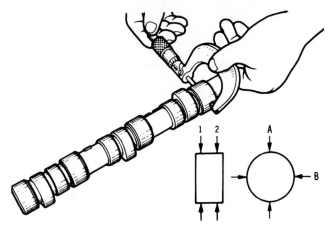

FIGURE 11-35 Measuring camshaft bearing journal wear. (Courtesy of Chrysler Canada Ltd.)

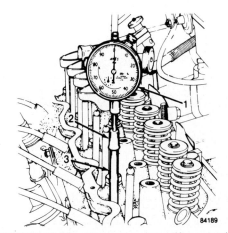

FIGURE 11-33 Measuring lobe lift on engine with in-block cam. (Courtesy of Chrysler Canada Ltd.)

Measuring Camshaft Runout

Place the camshaft in V blocks and mount a dial indicator as shown in Figure 11-36. Rotate the camshaft and observe the total indicator reading. Compare the reading with specifications. Runout should normally not exceed 0.001 to 0.002 in. Replace or straighten a bent camshaft.

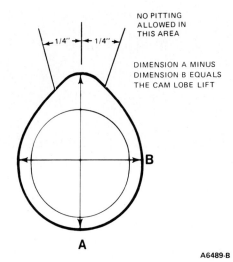

FIGURE 11-34 Cam lobe lift measurement is dimension A minus dimension B. (Courtesy of Ford of Canada.)

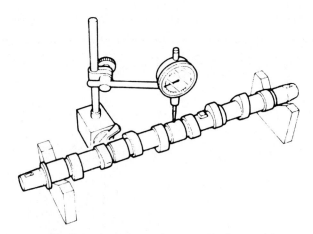

FIGURE 11-36 Measuring camshaft runout. (Courtesy of Ford of Canada.)

Measuring Base Circle Runout

With the camshaft mounted in V blocks, mount a dial indicator to contact the base circle of the cam lobe. Rotate the camshaft far enough to measure on the base circle only (Figure 11-37). Observe the dial indicator to obtain the total indicator reading. Maximum base circle runout should not exceed 0.001 in.

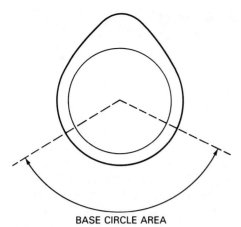

BASE CIRCLE AREA

FIGURE 11-37 Area of cam lobe base circle. (Courtesy of FT Enterprises.)

CAMSHAFT REGRINDING AND STRAIGHTENING

When the cam lobes are worn beyond the acceptable limit the camshaft must be replaced or reground. Regrinding a camshaft involves grinding a new cam profile on the worn lobes. The new profile retains the same lift and duration as the original cam lobe when it was new. Regrinding the cam lobe reduces its size and changes the lifter position slightly (Figure 11-38).

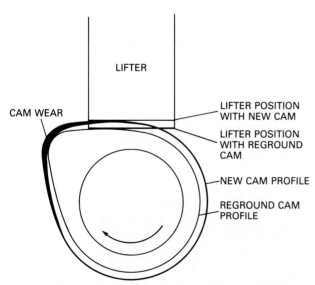

LIFTER

CAM WEAR

LIFTER POSITION WITH NEW CAM

LIFTER POSITION WITH REGROUND CAM

NEW CAM PROFILE

REGROUND CAM PROFILE

FIGURE 11-38 Change in cam lobe-to-lifter position with reground camshaft. (Courtesy of FT Enterprises.)

Camshafts are reground on a cam grinding machine. The machine operator selects the desired cam master and inserts it in the machine. The camshaft grinder then reproduces the cam lobe profile by fol-

lowing the cam master. Camshaft grinding is a very precise and exacting process and must be done by an experienced operator. Cam masters are designed by computer from mathematical calculations to produce the desired lift, duration, rate of lift acceleration, and closing deceleration. During grinding a large wheel follows the master cam profile. The large wheel is connected to the grinding wheel. As the large wheel follows the master profile, it causes the grinding wheel to move in the same manner (Figures 11-39 and 11-40).

FIGURE 11-39 Grinding a camshaft. Large wheel follows master cam lobe profile and carries grinding wheel in same profile. (Courtesy of Western Engine Ltd.)

FIGURE 11-40 Close-up of camshaft grinding process. (Courtesy Jasper Engine and Transmission Exchange.)

The camshaft is rough ground, straightened, then finish ground and surface treated. Rough grinding creates stress relief, which causes the camshaft to bend. For straightening, the camshaft is placed in the straightening fixture, where it is checked with a dial indicator. The low side is peened with a blunt chisel.

This moves the shaft ends toward the peened side. After straightening, the shaft is finish ground in the same equipment as used for rough grinding. Very little stock is removed in the finish grinding process. To ensure a smooth surface the camshaft is turned at a much slower speed than for rough grinding. After finish grinding the shaft is again checked for straightness and corrected if necessary. (Figures 11-41 and 11-42).

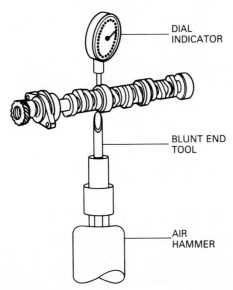

FIGURE 11-41 Peening a camshaft. (Courtesy of FT Enterprises.)

Surface Treatment

Reground camshafts are given a special surface treatment designed to hold lubricating oil to aid in the break-in process between the cam lobes and valve lifters. The treatment results in a dark, dull appearance that must not be removed from the cam lobes. The surface treatment can be either a moly disulfide coating applied by spraying, or the camshaft can be placed in a soak tank containing a manganese phosphate solution which etches the camshaft surface (Figure 11-43). This surface retains lubricant in order to reduce scuffing due to dry starts and high loads. The surface treatment is removed from the bearing journals since it is too rough for the soft-lined camshaft bearings. A belt-type sander-polisher is used for this procedure. Rebuilt camshafts must be stored properly to prevent damage (Figure 11-44).

INSTALLING THE CAMSHAFT IN THE BLOCK OR HEAD

Thoroughly lubricate the cam lobes and bearing journals with the special lubricant provided with the new camshaft kit (Figure 11-45). Proper lubrication is critical to good camshaft break-in. Poor lubrication can cause serious scoring of the camshaft and lifters during

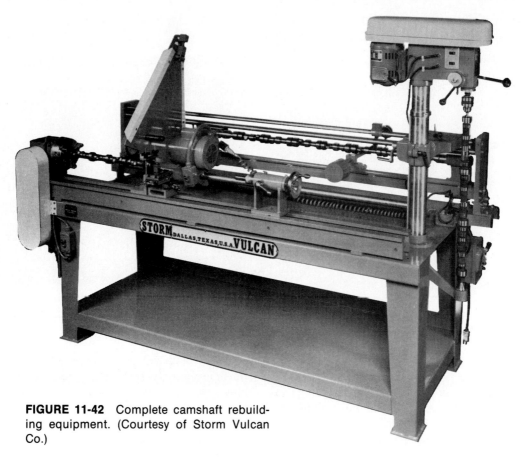

FIGURE 11-42 Complete camshaft rebuilding equipment. (Courtesy of Storm Vulcan Co.)

FIGURE 11-43 Surface treatment of reground camshafts. (Courtesy of Storm Vulcan Co.)

FIGURE 11-44 Rebuilt camshafts are stored by suspending them vertically. This prevents any change in straightness that could occur from improper storage. (Courtesy of Western Engine Ltd.)

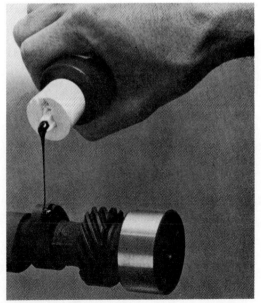

FIGURE 11-45 Special lubricant that will not drip off should be used on camshaft prior to installation. (Courtesy of Sealed Power Corporation.)

engine startup. Rapid wear will take place if scoring has occurred.

Carefully insert the camshaft, making sure that it is well supported and that the sharp edges of cam lobes and bearing journals do not scrape the bearing surfaces. A long bolt screwed into the drive end of the camshaft is helpful in supporting the camshaft during installation (Figures 11-46 to 11-48). The camshaft must be controlled during installation whether the block is in the upright position or if it is stood on end. Turn the camshaft slowly while it is inserted. If binding is encountered, do not force the camshaft. Remove the

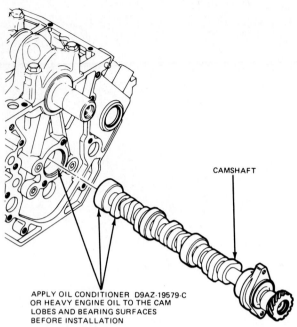

CAMSHAFT

APPLY OIL CONDITIONER D9AZ-19579-C
OR HEAVY ENGINE OIL TO THE CAM
LOBES AND BEARING SURFACES
BEFORE INSTALLATION

FIGURE 11-46 Camshaft lubrication instructions for Ford 3.8-liter V-6 engine. (Courtesy of Ford of Canada.)

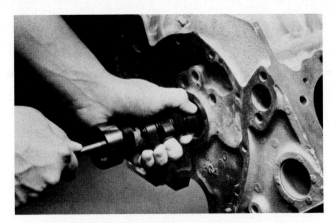

FIGURE 11-48 Camshaft must be properly supported during installation to prevent bearing damage. A long bolt screwed into the camshaft is helpful. (Courtesy of TRW, Inc., Automotive Aftermarket Division.)

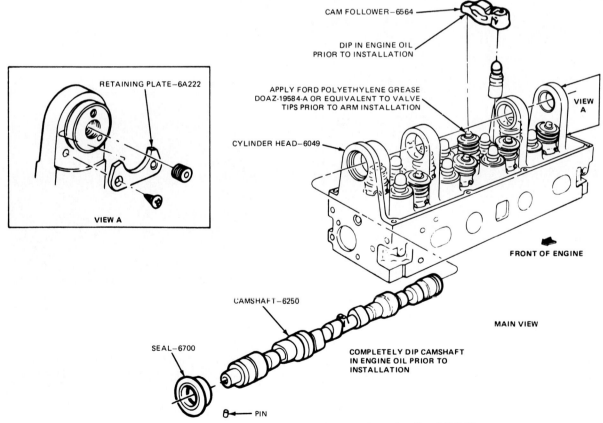

CAM FOLLOWER—6564

DIP IN ENGINE OIL
PRIOR TO INSTALLATION

RETAINING PLATE—6A222

APPLY FORD POLYETHYLENE GREASE
DOAZ-19584-A OR EQUIVALENT TO VALVE
TIPS PRIOR TO ARM INSTALLATION

CYLINDER HEAD—6049

VIEW A

VIEW A

FRONT OF ENGINE

CAMSHAFT—6250

MAIN VIEW

SEAL—6700

COMPLETELY DIP CAMSHAFT
IN ENGINE OIL PRIOR TO
INSTALLATION

PIN

FIGURE 11-47 Camshaft lubrication instructions for Ford 2.3-liter OHC engine. (Courtesy of Ford of Canada.)

camshaft, determine the cause of binding, and make the required correction before installing the camshaft.

Camshaft fit is acceptable if it can be turned by hand with moderate force. If binding is present, it will show up as a shiny area in the affected bearing. Very careful scraping of the area with a bearing scraper will correct the problem. An alternative method is to line hone the bearings lightly. All metal scrapings and honing abrasives must of course be removed prior to final assembly. On split bearing overhead camshafts bearing clearance can be checked with plastigage as shown in Figure 11-49. Check camshaft end play after installation as shown in Figure 11-50. Correct end play if necessary. For more detail on overhead camshaft installation see Chapter 18.

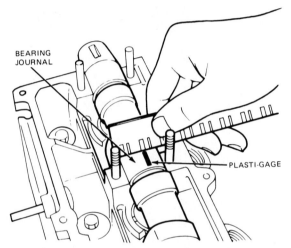

FIGURE 11-49 Measuring bearing clearance on OHC split bearing camshaft installation. (Courtesy of Ford of Canada.)

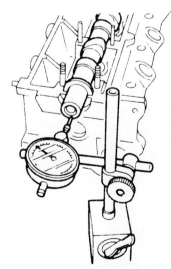

FIGURE 11-50 Measuring camshaft end play with a dial indicator. (Courtesy of Ford of Canada.)

CAMSHAFT DRIVE FUNCTION

The camshaft drive mechanism is designed to drive the camshaft from the crankshaft at one half crankshaft speed. It must turn the camshaft in the proper direction depending on cam lobe design. It must maintain an accurate timing relationship between the crankshaft and camshaft. It also maintains the timing relationship between the distributor and crankshaft on many engines.

CAMSHAFT DRIVE METHODS

Camshafts are driven by timing gears, a timing chain, and sprockets, or by a timing belt and sprockets. Each of these is described below.

Gear Drives

This system uses a small steel gear on the crankshaft which is in mesh with a larger cam drive gear (Figure 11-51). The crankshaft gear is indexed to the crankshaft by a straight key or a woodruff key to prevent the gear from turning on the shaft. The gear is normally a light press-fit on the crankshaft. The cam gear is usually fiber or aluminum to reduce gear noise and is a press-fit on the front end of the camshaft. The cam gear is indexed to the camshaft by a key, by a pin, or by unevenly spaced bolts.

FIGURE 11-51 Camshaft drive gears. Small crankshaft gear is steel. Camshaft gear is aluminum. (Courtesy of Sealed Power Corporation.)

Both gears are marked for timing purposes by one of the methods shown in Figure 11-52. If each gear has only one mark, the marks will align with the centerline across the two shafts. If one gear has two marks, they will straddle the centerline. Gear drives are used on some engines with an in-block camshaft.

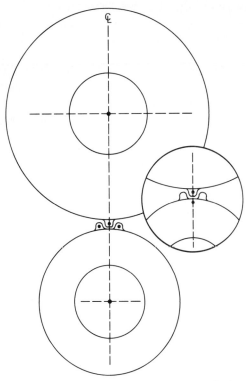

FIGURE 11-52 Typical cam gear timing marks. (Courtesy of FT Enterprises.)

Chain Drives

With a chain drive the crankshaft and camshaft are equipped with sprockets. These sprockets are not in contact with each other and are connected by a timing chain (Figures 11-53 to 11-56). The sprockets are indexed to the crankshaft and camshaft in a manner similar to the gears described above. The crankshaft sprocket is made of steel, while the camshaft sprocket is usually made of aluminum with a ring of plastic teeth to reduce noise.

FIGURE 11-53 Camshaft chain drive for in-block camshaft with steel crankshaft sprocket and plastic toothed aluminum camshaft sprocket. Chain is of the silent type. (Courtesy of Sealed Power Corporation.)

The chain can be of the silent or roller type. The silent chain is made of many flat links hinged together with pins. Special links either in the center of the chain or on each side of the chain keep the chain on the sprockets. Sprocket teeth for the center guide chain are grooved across the center. Timing marks are similar to those on gears for in-block camshaft engines. Specially colored links are used on some chains as timing marks. Although silent chains run quieter, roller chains last longer. Some engines use double roller chains for added strength. On some double overhead cam (DOHC) engines, one camshaft is driven at the front from the crankshaft, while the second cam is chain or gear driven off the rear of the first camshaft.

FIGURE 11-54 Double-roller-chain camshaft drive. (Courtesy of Chrysler Canada, Ltd.)

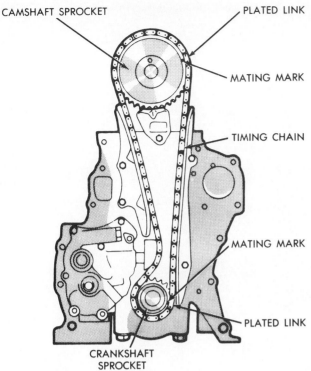

CAMSHAFT SPROCKET

PLATED LINK

MATING MARK

TIMING CHAIN

MATING MARK

PLATED LINK

CRANKSHAFT SPROCKET

FIGURE 11-55 Single-roller-chain camshaft drive for in line OHC engine. (Courtesy of Chrysler Canada Ltd.)

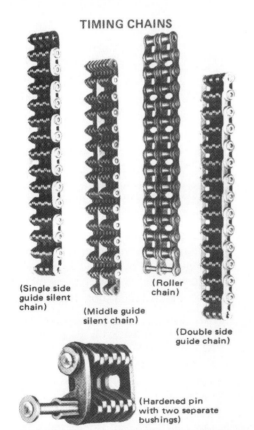

TIMING CHAINS

(Single side guide silent chain)

(Middle guide silent chain)

(Roller chain)

(Double side guide chain)

(Hardened pin with two separate bushings)

FIGURE 11-56 Camshaft drive chain types. (Courtesy of TRW, Inc., Automotive Aftermarket Division.)

Belt Drives

On many engines toothed belt drives are used to drive overhead camshafts, auxiliary shafts, and diesel fuel injection pumps. In all cases, a precise timing relationship must be maintained between the engine crankshaft and the driven components (Figures 11-57 and 11-58).

The camshaft and auxiliary shaft must be driven at exactly half the speed of the crankshaft. They must also be precisely timed to the engine crankshaft and piston position. This timing and speed ratio relationship must be maintained continuously during all phases of the operating life of the engine. This places more rigorous requirements on the cog belt drive than are required for V-belt drives. Timing marks on the sprockets and adjacent engine components are provided to ensure correct timing.

The cog belt must not stretch or lose its tension. Belt construction such as fiberglass reinforcement provides this characteristic. The cog belt must not slip. Teeth or cogs on the inner circumference of the belt, and corresponding teeth on the drive and driven sprockets, prevent slippage. The cog belt must not deteriorate over long periods of time from oil or water contamination. Synthetic rubber compounds assure long life under these conditions. The cog belt must not

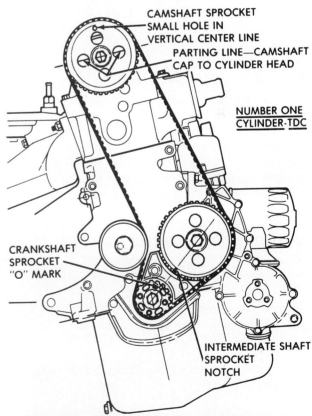

CAMSHAFT SPROCKET
SMALL HOLE IN VERTICAL CENTER LINE
PARTING LINE—CAMSHAFT CAP TO CYLINDER HEAD

NUMBER ONE CYLINDER-TDC

CRANKSHAFT SPROCKET "O" MARK

INTERMEDIATE SHAFT SPROCKET NOTCH

FIGURE 11-57 Timing belt camshaft drive for in-line OHC engine. (Courtesy of Chrysler Canada, Ltd.)

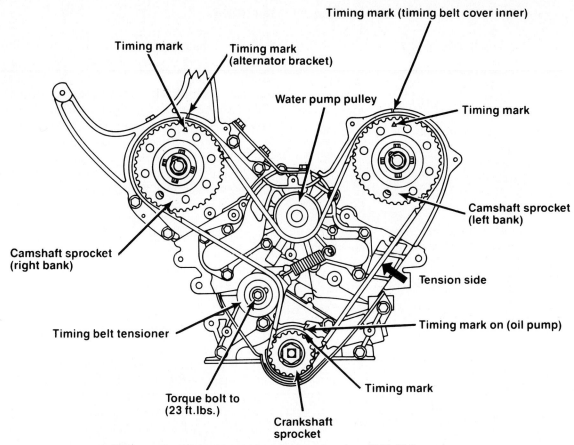

FIGURE 11-58 Timing belt camshaft drive for OHC V-6 engine. (Courtesy of Chrysler Canada Ltd.)

encounter foreign objects such as twigs, stones, ice, or snow during operation, which could cause the drive to fail. A shield almost completely enclosing the cog belt drive prevents entry of such foreign objects. Proper cog belt tension is provided by a belt tensioner adjustment. Proper cog belt operation (and engine operation) requires that precise belt tension specifications be followed when making adjustments.

CAMSHAFT DRIVE FAILURE ANALYSIS

For detailed information on camshaft drive mechanism failure analysis, see Chapter 5.

TIMING GEAR INSTALLATION

The following guidelines should be used to replace timing gears.

1. To install a press-fit camshaft gear, the camshaft must be removed from the engine. Attempting to do so with the camshaft in the engine can cause the cam bearing bore plug to be forced out as well as other damage.

2. Check the camshaft to ensure that all burrs are removed and that the shaft, keyway, and key are in good condition.

3. Check the gear and locate the timing mark. Gears are usually installed with the mark facing away from the shaft.

4. To install an all-metal (aluminum or steel) timing gear, heat the gear to about 200°F. The gear can be heated in a parts heater, boiling water, or hot cooking oil. The gear must be installed hot. Never heat fiber gears.

5. Turn the base plate on the press to position the appropriate opening below the arbor to fit the camshaft. Place the camshaft in the press.

6. If equipped with a thrust plate, place it on the camshaft. Make sure that the correct side of the plate faces forward.

7. Place the gear on the shaft (proper side forward) and align the keyways in the gear and

shaft. Make sure that the gear remains square on the shaft at all times during installation. A cocked gear will result in unacceptable gear runout.

8. Place the press adapter on the gear hub and carefully press the gear into place. Make sure that it is fully seated. Never press against any part of the gear except the hub to avoid gear damage or breakage.

9. Measure thrust plate clearance with a feeler gauge between the thrust plate and camshaft or gear thrust surfaces (Figure 11-59).

10. After camshaft installation, check gear backlash with a feeler gauge between the cam and crank gear teeth or use a dial indicator (Figures 11-60 and 11-61). Backlash is usually between 0.003 and 0.010 in. If new gears provide incorrect backlash, oversized or undersized gears must be considered.

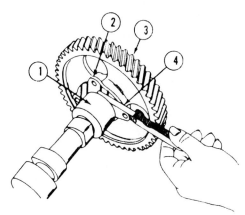

FIGURE 11-59 Measuring thrust plate clearance after gear installation: 1, bearing journal; 2, thrust plate; 3, gear; 4, feeler gauge. (Courtesy of Ford of Canada.)

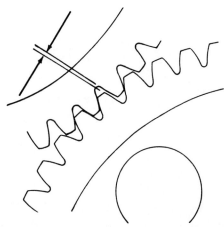

FIGURE 11-60 Timing gear backlash. (Courtesy of Ford of Canada.)

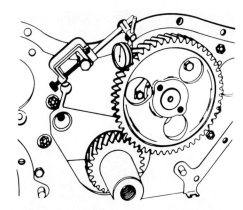

FIGURE 11-61 Measuring timing gear backlash. (Courtesy of Ford of Canada.)

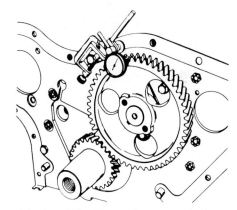

FIGURE 11-62 Measuring camshaft gear face runout. (Courtesy of Ford of Canada.)

11. Check timing gear face runout with a dial indicator (Figure 11-62). Face runout should not exceed 0.003 in.

TIMING CHAIN AND SPROCKET INSTALLATION

1. It is good practice to replace the chain and sprockets at the same time. Using a new chain with worn sprockets invites trouble.

2. Check the instructions that come with the replacement parts. In some cases a factory spacer is used in the original installation. Some replacement sprockets have the spacer built into the hub (Figure 11-63). Follow the instructions supplied with the sprocket to determine spacer use. The chain and sprockets must be installed as an assembly to avoid distorting the chain.

3. Position the crankshaft and camshaft to ensure alignment of keyways and timing marks

(Figure 11-64). Using an appropriate sleeve, gently tap both sprockets into place evenly if necessary to avoid chain distortion (Figure 11-65). Do not force one sprocket too far ahead of the other since this would stretch the chain. Do not apply any force against the chain.

4. Install and adjust any chain tensioner, damper, or guides and adjust them as specified in the appropriate service manual (Figure 11-66).

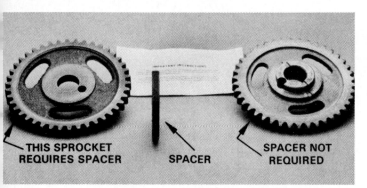

FIGURE 11-63 Camshaft sprocket with separate spacer (left) and built-in spacer (right). (Courtesy of TRW, Inc., Automotive Aftermarket Division.)

FIGURE 11-65 Tapping a keyed camshaft sprocket into place. (Courtesy of TRW, Inc., Automotive Aftermarket Division.)

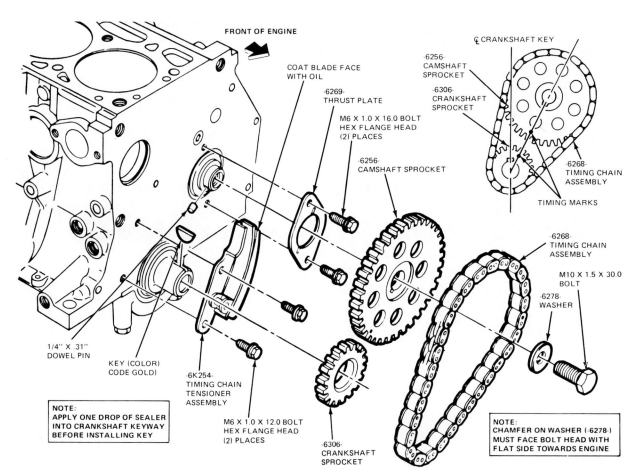

FIGURE 11-64 Typical camshaft timing gear and sprocket installation procedure. (Courtesy of Ford of Canada.)

198 *Camshafts, Lifters, and Camshaft Drives*

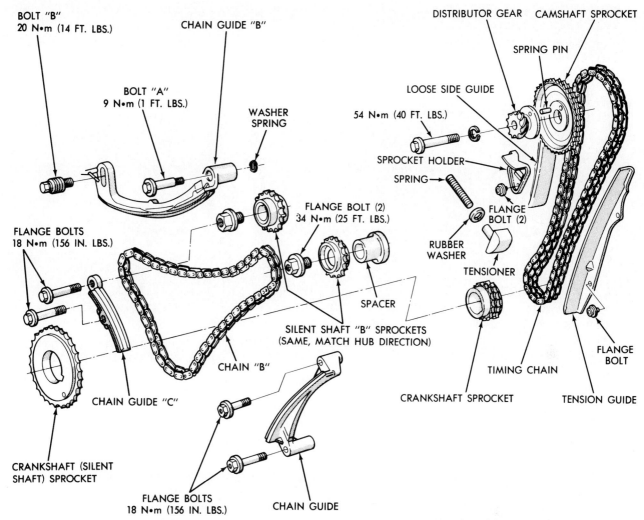

BOLT "B"
20 N•m (14 FT. LBS.)

CHAIN GUIDE "B"

BOLT "A"
9 N•m (1 FT. LBS.)

WASHER
SPRING

FLANGE BOLTS
18 N•m (156 IN. LBS.)

FLANGE BOLT (2)
34 N•m (25 FT. LBS.)

DISTRIBUTOR GEAR   CAMSHAFT SPROCKET

SPRING PIN

LOOSE SIDE GUIDE

54 N•m (40 FT. LBS.)

SPROCKET HOLDER

SPRING

RUBBER
WASHER

FLANGE
BOLT (2)

TENSIONER

SPACER

SILENT SHAFT "B" SPROCKETS
(SAME, MATCH HUB DIRECTION)

CHAIN "B"

CHAIN GUIDE "C"

CRANKSHAFT (SILENT
SHAFT) SPROCKET

FLANGE BOLTS
18 N•m (156 IN. LBS.)

CHAIN GUIDE

CRANKSHAFT SPROCKET

TIMING CHAIN

FLANGE
BOLT

TENSION GUIDE

FIGURE 11-66 Timing chain, balance shaft chain, and chain guides on 2.6-liter Mitsubishi engine. (Courtesy of Chrysler Canada Ltd.)

TIMING BELT AND SPROCKET INSTALLATION

1. Timing belt sprockets must be installed first. Since most sprockets are made of softer metal, careful handling is critical to avoid distortion or damage during installation. Apply force must always be directed only to the hub. Be sure to install sprockets with the proper side facing forward. Where spacers are required, be sure to install these with the proper side facing the sprocket (Figure 11-67).

2. Timing belts must be handled with care. They should not be twisted or kinked since this damages the internal belt structure and weakens the belt. Never pry on a timing belt.

3. Loosen the timing belt tensioner adjustment enough to allow easy belt installation.

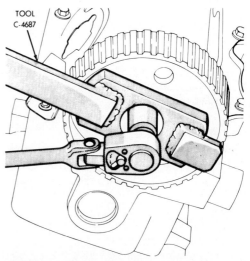

TOOL
C-4687

FIGURE 11-67 Installing belt drive cam sprocket on a 2.5-liter Chrysler engine. (Courtesy of Chrysler of Canada Ltd.)

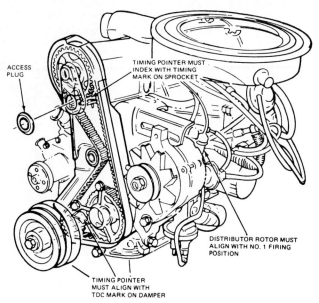

FIGURE 11-68 Typical timing mark alignment for timing belt installation. (Courtesy of Ford of Canada.)

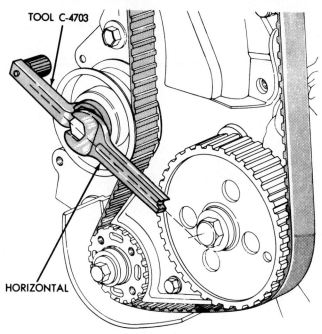

FIGURE 11-69 Adjusting timing drive belt tension. (Courtesy of Chrysler Canada Ltd.)

4. Position the crankshaft, camshaft, and auxiliary shafts to align the timing marks properly (Figure 11-68).

5. Place the belt in position, making sure that there is no slack on the nontensioned side of the belt and that timing marks remain in alignment.

6. Tighten the belt tensioner as outlined in the appropriate service manual. Belt tension is critical; the belt should not be too tight or too loose (Figure 11-69).

DEGREEING IN A CAM

Degreeing in a camshaft refers to the very precise timing of the camshaft to piston and crankshaft position. Some special equipment is required as well as careful attention to detail. See Chapter 4 for detailed information on degreeing a cam.

REVIEW QUESTIONS

1. Technician A says that the camshaft is designed to control the opening and closing of the valves. Technician B says that the camshaft is designed to control how far and how long the valves will be open. Who is right?
   (a) technician A
   (b) technician B
   (c) both are right
   (d) both are wrong

2. How many camshafts does an engine have?

3. True or false: Pushrods on overhead camshaft engines are shorter.

4. Overhead camshafts may operate:
   (a) only the intake valves
   (b) only the exhaust valves
   (c) both intake and exhaust valves
   (d) any of the above

5. Most camshafts are manufactured by the _____ or _____ process.

6. Why must roller lifter camshafts be harder than friction lifter camshafts?

7. The friction lifter cam lobe profile is more _____ than the roller lifter cam lobe.

8. True or false: Camshaft bearings may be of the full circle or split type.

9. What causes camshaft end thrust?

10. How is camshaft end thrust controlled?

11. List six valve operating factors that are controlled by the camshaft.

12. Name the four basic sections of a cam lobe profile.

13. Camshafts designed for use with mechanical lifters have a _____ ramp not found on hydraulic lifter cams.

14. What effect does rocker arm ratio have on the design of a cam lobe?

15. What is the advantage of roller lifters versus sliding lifters?

16. What is the duration in degrees of crankshaft rotation of each of the four strokes on an engine with the following valve timing: IO at 20° BTDC, IC at 50° ABDC, EO at 50° BBDC, EC at 20° ATDC.

17. Define "valve overlap."

18. How much valve overlap is there in the engine described in Question 16?

19. What effect does the amount of valve overlap have on exhaust emissions?

20. Technician A says that hydraulic lifters rely on engine oil pressure for proper operation. Technician B says that a small amount of oil leakage past the lifter plunger is necessary. Who is right?
    (a) technician A
    (b) technician B
    (c) both are right
    (d) both are wrong

21. True or false: Hydraulic lifters may be standard or oversized.

22. Technician A says that lifter rotation is necessary to ensure valve rotation. Technician B says that lifter rotation increases the wear rate. Who is right?
    (a) technician A
    (b) technician B
    (c) both are right
    (d) both are wrong

23. Camshaft inspection includes:
    (a) visual inspection
    (b) measuring cam lobe wear
    (c) measuring bearing journal wear
    (d) all of the above

24. Camshaft runout is measured with:
    (a) a dial indicator and straightedge
    (b) a straightedge and feeler gauge
    (c) V blocks and a straightedge
    (d) V blocks and a dial indicator

25. True or false: Regrinding a camshaft changes the lobe lift.

26. To regrind a camshaft requires a:
    (a) cam master
    (b) master lifter
    (c) lifter cam
    (d) laser beam

27. Why are reground camshafts surface treated?

28. In a two-gear camshaft drive:
    (a) the camshaft turns faster than the crankshaft
    (b) the camshaft and crankshaft turn at the same speed
    (c) the camshaft and crankshaft turn in opposite directions
    (d) none of the above

29. Camshaft chain drives use:
    (a) roller chains only
    (b) double roller chains only
    (c) gears on the crankshaft and camshaft
    (d) none of the above

30. When the cog belt becomes stretched and loose:
    (a) adjust the belt tightener
    (b) shorten the drive distance
    (c) lengthen the drive distance
    (d) replace the belt

31. Why must timing marks be aligned when installing timing gears?

32. True or false: Timing gears should be checked for backlash and face runout.

33. True or false: When a timing chain breaks replace the chain and the sprockets.

CHAPTER 12

Engine Bearings

OBJECTIVES

1. To develop an understanding of the design, construction, and operation of engine bearings.
2. To develop the ability to diagnose the causes of bearing failures.
3. To be able to select and install the correct bearings for any specific engine to meet manufacturer's specifications.

ENGINE BEARING FUNCTION

Engine bearings are designed to support rotating and oscillating engine components such as crankshafts, camshafts, intermediate shafts, balance shafts, and piston pins and to reduce friction. They are subjected to heavy loads, high temperatures, and severe operating conditions. The shafts they support often turn at very high speeds. They must not contact the metal of the shafts they support while the engine is in operation, or metal wiping and scoring will take place. Another function of engine bearings is that they serve as a replaceable wear surface for engine components.

BEARING OPERATION

When the engine is stopped, the shafts are at rest against the bearing surfaces. As soon as the shaft begins to turn it moves up slightly in the bearing, pulling some oil under the bearing journal. As the shaft continues turning at increasing speed, oil continues to be forced or wedged between the shaft and bearing. Oil under pressure is fed into the clearance between the bearing and bearing journal. Under ideal conditions the shaft actually floats on a film of oil inside the bearing (Figure 12-1). Under certain conditions the shaft may break through the oil wedge momentarily without causing any damage to the shaft or bearing since there is still enough oil present to provide lubrication. When the shaft floats on the oil film it is said to be in a hydrodynamic state. Bearing clearances, engine temperature control, and engine oils are designed to provide hydrodynamic lubrication as much as possible. When these factors are unable to maintain the oil film between the bearing and bearing journal, metal-to-metal contact takes place. This results in bearing material being "wiped" from the bearing surface and in scoring of the bearing journals. An oil clearance of from 0.001 to 0.003 in. is common for crankshaft bearings (Figure 12-2).

CRANKSHAFT BEARINGS

Crankshaft bearings are half-round split-type precision inserts. The two halves form a complete round bearing. The reason for the split design is to allow bearing in-

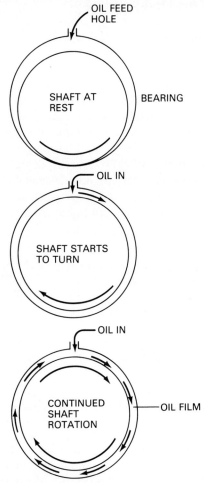

FIGURE 12-1 Action of shaft inside bearing as engine starts and runs. Under ideal conditions and shaft floats on film of oil. (Courtesy of FT Enterprises.)

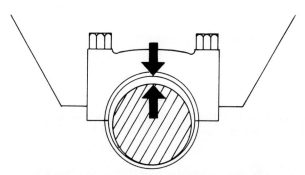

FIGURE 12-2 Bearing oil clearance (exaggerated). (Courtesy of Ford of Canada.)

stallation on bearing journals that are smaller in diameter than other parts of the shaft they support, such as the crankshaft and some overhead camshafts. Precision bearing inserts come in plain and flanged designs. The plain bearings support radial loads only, while the flanged bearing absorbs both radial and thrust (axial) loads (Figure 12-3).

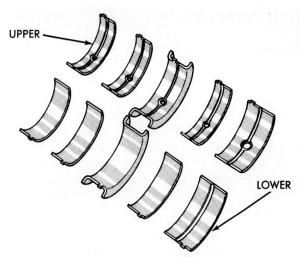

FIGURE 12-3 Set of crankshaft main bearings with one flanged thrust bearing. (Courtesy of Chrysler Canada Ltd.)

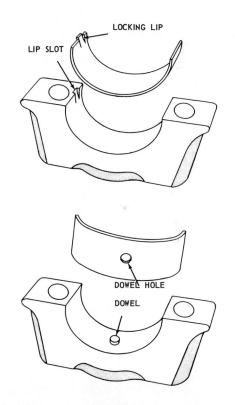

FIGURE 12-4 Bearing locating devices. (Courtesy of Federal-Mogul Corporation.)

There are several design features that are common to engine bearings. These include the locking lip (locating lug), bearing free spread, and crush height. The locking lip fits into a corresponding notch in the connecting rod bore or main bearing saddle bore. It keeps the bearing in place in the bore and prevents bearing rotation. Some bearing designs use a locating dowel with a dowel hole in the bearing and in the housing (Figure 12-4).

BEARING DESIGN

Bearing Spread

Free spread refers to the distance across the two parting edges of the bearing before it is installed in its bore. This dimension is slightly larger than the bore diameter into which it fits. There may be as much as 0.030 in. of spread in some bearings. Spread helps to ensure a tight fit against the housing bore and also keeps the bearing in place during assembly (Figure 12-5). It is not good practice to try to reshape a collapsed bearing to provide spread.

FIGURE 12-5 Bearing spread helps bearing fit snugly in bore. (Courtesy of Sunnen Products Company.)

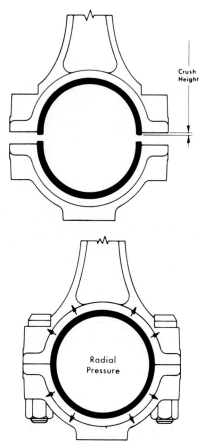

FIGURE 12-6 Bearing crush height results in good bearing-to-bore contact after bolts are tightened. (Courtesy of Sunnen Products Company.)

Bearing Crush Height

Crush height refers to the extra height of the bearing half when it is seated in the bearing housing (Figure 12-6). This extra height may be as little as 0.00025 in. When the two bearing halves are bolted together, this extra height is "crushed," setting up radial pressures that ensure good metal-to-metal seating of the bearing in its housing bore. This ensures that the bearing conforms to the roundness of the housing bore and provides a good path for heat transfer from the bearing to the housing. Most bearings are designed with some parting relief to prevent the inner parting edges from curling up when the bolts are tightened and the crush applied (Figure 12-7).

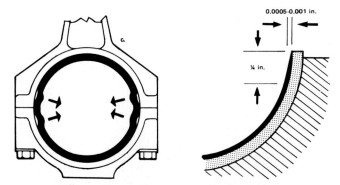

FIGURE 12-7 Bearing crush tends to deform bearing slightly near parting line when bolts are tightened (left). To offset this effect, most bearings are designed with some joint face relief (right). (Courtesy of Federal-Mogul Corporation.)

Oil Holes

Engine crankshaft bearings receive their oil supply through oil holes in the engine block. A matching oil hole in the upper bearing half admits oil between the bearing and bearing journal (Figure 12-8). The size and location of the oil holes in replacement bearings must match those of the original bearings installed by the engine manufacturer.

Oil Grooves

Oil grooves are provided in some bearing designs to increase oil distribution between the bearing and bearing journal (Figure 12-8). Oil grooves may also be used in some cases to increase oil throw-off to lubricate nearby engine parts, such as camshafts or cylinder walls. In some OHC engines oil grooves are used to direct oil to adjacent cam followers. Oil grooves are also used to drain oil away from the area of an oil seal in some designs (Figure 12-9).

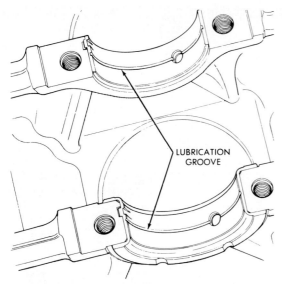

FIGURE 12-8 Oil holes and grooves in upper bearing halves. (Courtesy of Chrysler Canada Ltd.)

FIGURE 12-9 Diagonally grooved lower main bearing drains oil away from seal area. (Courtesy of Chrysler Canada Ltd.)

BEARING MATERIALS

Engine bearings consist basically of steel backing and one or more layers of lining material. There are three general types of bearing liner materials used in engine bearings: (1) lead-based babbit, (2) copper–lead alloy, and (3) aluminum alloy.

Lead-Based Babbit

This material consists of about 83% lead, 15% antimony, 1% tin, and 1% arsenic. This material is a light-duty bearing material and not in widespread use in today's engines. Lead-based babbit bearings have good corrosion resistance, embedability, and conformability, while fatigue resistance is relatively low.

Copper–Lead Alloy

This alloy consists of about 50% copper and 50% lead. The lead and copper in powder form are sintered (heated to cause fusion without melting) in strip form on the steel backing. Copper–lead bearings have better

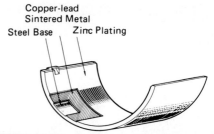

FIGURE 12-10 Typical bearing layers with tin-plated copper lead on steel back. (Courtesy of TRW, Inc., Automotive Aftermarket Division.)

fatigue resistance than babbit bearings but slightly less corrosion resistance, embedability, and conformability (Figure 12-10).

Aluminum Alloy

Aluminum alloy bearings are manufactured by using the roll-bonding process. Included in these alloys are such materials as copper, tin, lead, and silicon. Percentages of these materials vary depending on manufacturer's preference and the particular application of the bearings. Aluminum alloy bearings provide an excellent level of fatigue resistance, corrosion resistance, embedability, and conformability.

Bearing Overlays

Many alloy bearings are designed with an electroplated babbit overlay of from 0.0005 to 0.001 in. The babbit overlay increases corrosion resistance, surface action, embedability, and conformability.

BEARING CHARACTERISTICS

The operating characteristics of the various bearing materials are critical to good engine operation and long service life. The following bearing material characteristics are considered by all bearing manufacturers.

Embedability

Embedability is the ability of the bearing material to embed tiny particles of hard foreign material. If these hard particles do not become embedded in the bearing material they will scratch and score the bearing journal. The softness of the bearing material determines its embedability (Figure 12-11).

Conformability

The bearing material must have the ability to conform slightly to the surface irregularities of the bearing journal. To do this, the bearing material must flow slightly

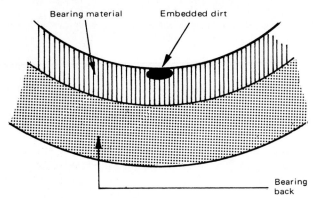

FIGURE 12-11 Soft bearing liner embeds small hard particles of foreign matter. (Courtesy of Federal-Mogul Corporation.)

under pressure to alter its shape to match the surface irregularity of the journal.

Corrosion Resistance

The bearing material must be able to resist the corrosion action of acids produced as by-products of combustion. The presence of these acids attacks the metal and causes etching, which leads to further damage.

Fatigue Resistance

The bearing material must have the ability or strength to resist fatigue failure. The greater the load-carrying strength of the bearing, the higher its fatigue resistance. All materials fail eventually under continued heavy loading. Engine bearings are designed to withstand heavy loads over a long period of time before fatigue failure occurs.

Surface Action

Good surface action is defined as the resistance to scoring when heavy loading breaks down the oil film between the bearing and the bearing journal. The resulting metal-to-metal contact can result in scoring damage. Certain metals provide better surface action than others in this respect.

CAMSHAFT BEARINGS

Camshaft bearings for in block camshafts are of full-round design. They are manufactured from round seamless metal alloy tube stock or from layered metal strips. The strips are cut, formed round, and secured with an interlock or butt-type seam (Figure 12-12). These bearings are press-fit into their bores and assume the correct inside diameter when installed. Some OHC engines use split-type precision bearing inserts

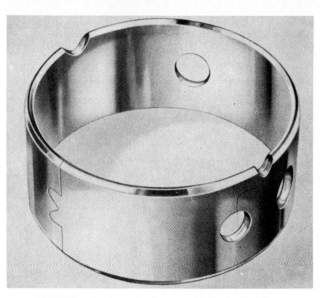

FIGURE 12-12 Typical camshaft bearing with an interlocking joint. (Courtesy of Federal-Mogul Corporation.)

similar to crankshaft bearings. Camshaft bearings are available in standard sizes for camshafts that have not been reground and in undersizes for reground camshafts.

BEARING REPLACEMENT GUIDELINES

Consideration must be given to bearing material, size, clearance, and end play when replacing engine bearings as follows.

Bearing Material

Many bearing manufacturers use a material code as a suffix to the part number to identify the bearing material. These codes vary between different bearing manufacturers. To determine the code information used consult with the bearing supplier. One example is shown in Figure 12-13. Generally, the type of bearing material used originally by the engine manufacturer should be used when replacing bearings.

Material Code	Meaning
AP	Steel back bearing, aluminum alloy, babbit overplate
AT	Solid aluminum alloy bearing
CA	Steel back bearing, copper alloy lined
CP	Steel back bearing, copper-alloy lined, lead–tin overplate
RA	Steel back bearing, aluminum alloy
SA or SB	Steel back bearing, babbit lined

FIGURE 12-13 Example of bearing material coding.

Bearing Sizes

Replacement main and connecting rod bearings are available in standard and undersize. Undersizes of 0.001, 0.002, 0.010, 0.020 and 0.030 in. are normally available. Semifinished main bearings of 0.060 in. undersize can be resized to fit the particular crankshaft being used. They can also be installed in the main bearing saddle bores and align bored to correct saddle bore misalignment.

To determine the bearing size needed requires measuring the crankshaft journals. If the original crankshaft is used, one or more bearing journals may be undersized. Be sure to mike all bearing journals accurately to determine the bearing sizes needed. If a reground crankshaft is being used, the correct set of bearings will be supplied with the crankshaft and bearing kit. If the bearings are ordered separately, the crankshaft must be measured to determine the bearing sizes required.

Sometimes bearings are packaged incorrectly or the package is improperly labeled as to bearing size, standard or oversize. The only way to check whether you have the correct bearing is to measure the bearing at the crown (Figure 12-14) and compare this figure with specifications in the bearing catalog.

Bearing Clearances

Typical main and connecting rod bearing clearances are shown in Figure 12-15. Normal main and connecting rod bearing clearance should not exceed 0.001 in. per inch of journal diameter and should not be less than 0.0005 in. per inch of journal diameter. Recommended crankshaft thrust bearing clearances are shown in Figure 12-16. Thrust washers (Figure 12-17)

Crankshaft Journal Diameter [in. (mm)]	Crankshaft End Clearance [in. (mm)]
2–2¾ (50–70)	0.004–0.006 (10–15)
2¹³⁄₁₆–3½ (71–89)	0.006–0.008 (15–20)
Over 3½ (over 89)	0.008–0.010 (20–25)

FIGURE 12-16 Recommended thrust bearing clearances.

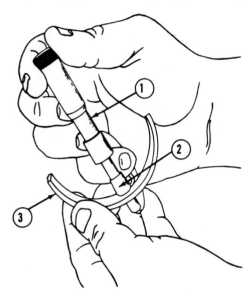

FIGURE 12-14 Measuring bearing shell thickness with a ball spindle micrometer: 1) micrometer; 2) ball; 3) bearing. (Courtesy of Federal-Mogul Corporation.)

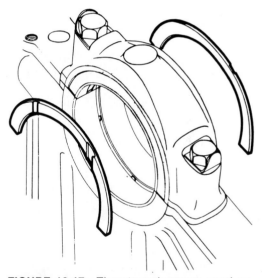

FIGURE 12-17 Thrust washers are used on some engines to control crankshaft end play. (Courtesy of Chrysler Canada Ltd.)

Shaft Size (in.)	Clearance (in.)			
	Babbit (SB)	Copper Alloy (CA)	20% Tin Aluminum (RA)	Overplated Lining (AP/CP)
2–2¾	0.001–0.002	0.002–0.003	0.001–0.002	0.001–0.002
2¹³⁄₁₆–3½	0.0015–0.0025	0.0025–0.0035	0.0015–0.0025	0.0015–0.0025
3⁹⁄₁₆–4½	0.002–0.003	0.003–0.004	0.002–0.003	0.002–0.003

FIGURE 12-15 Typical bearing clearances for different shaft sizes and bearing materials.

or a flanged main bearing control crankshaft end play. Flanged thrust bearings with extra material on the flanges are available for fitting to crankshafts that have had the thrust surfaces reground. Material is removed from the the flanges until the desired end play is achieved (Figure 12-18).

Camshaft bearing clearance is required for lubrication. Make sure that the cam bearings are correct for the camshaft being used—standard for a new cam-shaft and undersized to fit a reground camshaft with undersized journals. Endwise movement of the camshaft is controlled by thrust surfaces in many applications.

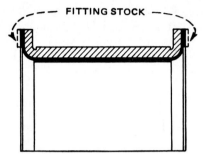

FIGURE 12-18 Flanged main bearing with extra material on thrust surfaces which can be reduced to achieve proper crankshaft end play. (Courtesy of Federal-Mogul Corporation.)

BEARING INSTALLATION PROCEDURE

For procedures for installing specific bearings and related engine components, refer to the appropriate chapter in this book.

1. Make certain that the bearing bores are in good condition, round, and not misaligned.
2. Do not touch the bearing surfaces since moisture from the skin contains acids that corrode the bearing metal.
3. Bearing and bearing bores must be clean and dry before installing the bearings. Wipe the back of the bearings and the bores with a clean, lint-free cloth.
4. Check the oil holes in the bearings and the bores. Oil holes or slots in the bearings must

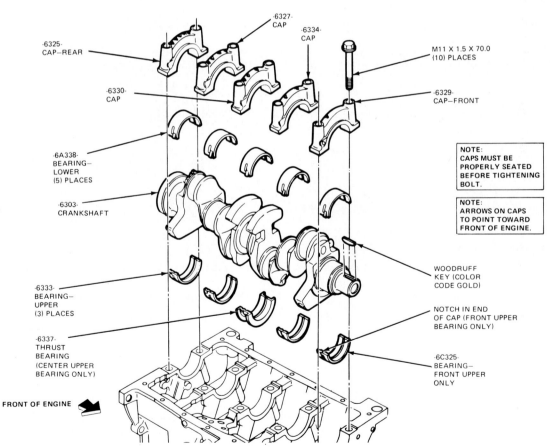

FIGURE 12-19 Typical main bearing installation procedure. (Courtesy of Ford of Canada.)

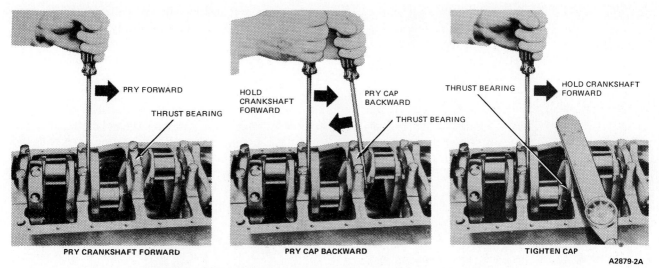

FIGURE 12-20 Aligning bearing thrust surfaces before tightening main bearing caps. (Courtesy of Ford of Canada.)

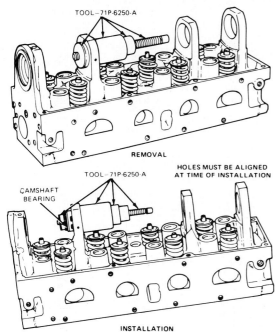

FIGURE 12-21 Replacing camshaft bearings in OHC engine cam towers. (Courtesy of Ford of Canada.)

be aligned with the oil holes in the block (Figures 12-19 and 12-20).

5. When installing cam bearings, be sure to use the correct-size driver or puller. Pull or drive the bearings in straight. A cocked cam bearing will distort as it is installed. Make sure that the correct-size cam bearing is installed in the appropriate bore (Figures 12-21 to 12-23).

6. Use plastic gauge to check rod and main bearing clearances.

7. Make sure that the bearings are properly positioned in their bores and fully seated.

8. Lubricate all bearing surfaces and seal lips liberally with engine oil or a good assembly lubricant.

9. When installing the crankshaft, lower it evenly and carefully into place.

10. Install the bearing caps (and bearings) and bolts, making certain they are not reversed, interchanged, or misaligned. Install the bolts finger-tight. Tap each cap lightly with

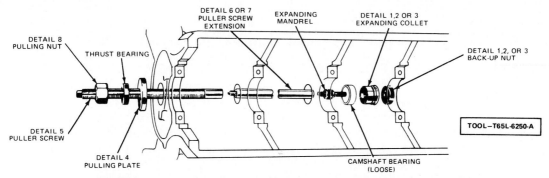

FIGURE 12-22 Installing cam bearings in-block with threaded puller. (Courtesy of Ford of Canada.)

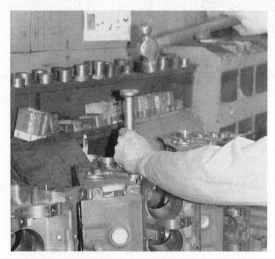

FIGURE 12-23 Installing cam bearings in-block with bearing driver and hammer. (Courtesy of Western Engine Ltd.)

a soft hammer to make sure that it is positioned properly and fully seated. Pry the crankshaft forward to align the bearing thrust surfaces. Then tighten the main bearing bolts in the recommended sequence and in steps to the specified torque.

11. When installing the piston assemblies, pull the connecting rod into place against the crankpin. Install the bearing caps with the cap and yoke numbers on the same side. Tighten the rod nuts to specifications using a socket that will not interfere with the rod cap while turning.

REVIEW QUESTIONS

1. Why are bearings needed in an engine?
2. Define "hydrodynamic lubrication."
3. What happens when the oil film breaks down and the bearing journal contacts the bearing?
4. Crankshaft bearings are:
   (a) half-round split-type precision inserts
   (b) of the friction, roller, and ball types
   (c) replaced with oversized bearings when worn
   (d) ovate to ensure proper lubrication
5. Bearing spread:
   (a) results from excessive loading
   (b) is a bearing design feature
   (c) is required when bearings shrink
   (d) occurs when bearings are improperly stored
6. Crankshaft end play is limited by:
   (a) a thrust control button
   (b) a flanged connecting rod
   (c) a flanged main bearing
   (d) none of the above
7. Bearing crush:
   (a) is an undesirable abnormality
   (b) is designed to overcome bearing spread
   (c) results from overloading
   (d) is a bearing design feature
8. What is the purpose of oil holes and grooves in crankshaft bearings?
9. Name three common bearing materials.
10. Bearing overlay:
    (a) occurs when the engine is stored too long

    (b) results from overloading
    (c) may be from 0.0005 to 0.001 in. thick
    (d) none of the above
11. List five good bearing characteristics and describe each one briefly.
12. True or false: Camshaft bearings are manufactured from a seamless metal tube or from layered metal strip.
13. True or false: All camshaft bearings are of the split precision insert design.
14. List the normal sizes of crankshaft bearing availability.
15. Why is bearing clearance required?
16. Crankshaft bearing clearance per inch of journal diameter should be:
    (a) 0.1 in.
    (b) 0.01 in.
    (c) 0.001 in.
    (d) 0.0001 in.
17. An undersized bearing compared to a standard bearing is:
    (a) thinner and has a larger ID
    (b) thicker and has a larger ID
    (c) thinner and has a smaller ID
    (d) thicker and has a smaller ID
18. True or false: Do not touch the bearing surface when replacing bearings.
19. When installing in block camshaft bearings, make sure that the _____ are aligned.
20. True or false: The main bearing inserts with the oil holes must be installed in the bearing cap.

CHAPTER 13

Gaskets, Sealants, and Seals

OBJECTIVES

To develop the ability to select and use the appropriate gaskets, seals, and sealants in a manner that meets manufacturers' specifications.

INTRODUCTION

Gaskets, sealants, and seals are used in engines to prevent leakage of fluids and gases. Wide variations in temperature and pressure create a hazardous environment in which these products must function. The properties of these products must be such that they will not fail even though they are subjected to these conditions over a long period of time. Here is a list of common engine gaskets and seals, most of which are shown in Figure 13-1.

- Oil pan drain plug gasket
- Oil pan side rail gaskets
- Oil pan end seals
- Rear main bearing cap side seals
- Rear main bearing oil seal
- Oil pump pickup tube gasket
- Oil pump cover plate seal
- Fuel pump mounting gasket
- Timing cover gasket
- Timing cover seal
- Distributor mounting gasket
- Water pump gasket
- Cylinder head gasket
- Valve stem seals

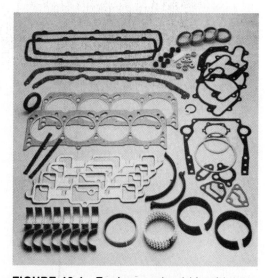

FIGURE 13-1 Engine overhaul kit with gaskets, seals, rings, and bearings. (Courtesy of Sealed Power Corporation.)

- Valve cover gasket
- Exhaust manifold gasket
- Exhaust pipe gasket
- Intake manifold side gaskets
- Intake manifold end seals
- Thermostat housing gasket
- EGR valve plate gasket
- Carburetor base gasket
- Air cleaner to carburetor gasket
- Air plenum gasket
- PCV valve grommet
- Oil filler cap gasket

GASKET PROPERTIES

Gasket manufacturers spend a great deal of time and money researching and testing new product combinations. Among the properties they design into their gaskets are the following.

1. *Compressibility and extrudability.* Gaskets must be able to flow and conform to minor surface irregularities. The machined surfaces of the parts being sealed by the gasket may have minor undulations and roughness that must be accommodated by the gasket.

2. *Resilience.* This is the ability of the gasket to maintain its sealing ability despite wide temperature extremes and engine vibration. If the gasket relaxes too much over time, bolting force and flange pressure are reduced, allowing leakage to occur.

3. *Impermeability.* The gasket must not allow the fluids or gases it is designed to contain to seep through the gasket material.

The type of gasket used for a particular application is determined by the following factors.

- Type of material to be confined: gases, liquids
- Pressure of confined material
- Finish or smoothness of mating parts
- High- and low-temperature extremes that will be encountered
- Clearance required between assembled parts

GASKET MATERIALS

Cork

Cork granules are made from the tough elastic outer tissue of the cork oak. These granules are bonded together with special glues and resins. During manufac-

ture the cork is produced in sheets of different thickness for different applications. Cork gaskets are used to contain liquids where pressures and bolt torque values are low and temperature variations are less extreme. Cork gaskets tend to dry out, shrink, and become brittle when stored too long. Cork has low impermeability and resilience and extrudes or cracks when bolting torque is too high. Cork gaskets can be made from cork granules of different size. Finer granules make for a firmer gasket (Figures 13-2 and 13-3).

FIGURE 13-2 Fine granule cork gasket. (Courtesy of Sealed Power Corporation.)

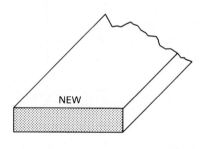

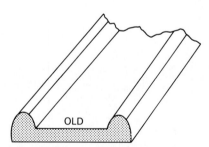

FIGURE 13-3 Used cork gaskets should not be used again since their sealing properties are lost. (Courtesy of FT Enterprises).

Rubber

Rubber gaskets are more impermeable and resilient than cork and can withstand higher temperatures. Rubber tends to be displaced under pressure, resulting in greatly reduced clamping pressure. Rubber gaskets and the surfaces to be sealed should be absolutely clean and dry to reduce displacement.

Cork and Rubber

Cork granules with a rubber bonding agent make an excellent gasket material with good impermeability, resilience, and compressibility.

Paper

Paper is used where high temperatures and pressures are not encountered. Paper gaskets are usually treated to resist the fluids encountered in engines.

Fiber

Asbestos and cellulose fibers are used, often in combination, for gaskets used where high temperatures and high pressure are present. Fiber gaskets may be reinforced at openings with a metal ring. Some fiber gaskets are bonded to both sides of a steel core.

Embossed Steel

Sheet steel embossed with ridges in the sealing areas is used where high pressures and temperature extremes are present. These gaskets are usually coated with tin or aluminum to provide additional sealing. The raised or embossed areas provide the compressibility and resilience required (Figure 13-4).

Other gasket designs are shown in Figures 13-5 to 13-9.

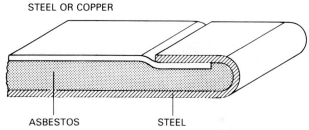

FIGURE 13-6 Sandwich gasket construction. (Courtesy of FT Enterprises).

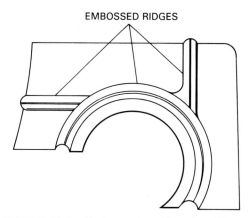

FIGURE 13-4 Embossed steel gasket. (Courtesy of FT Enterprises).

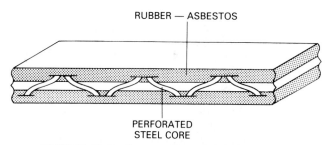

FIGURE 13-7 Composition gasket with perforated steel core. (Courtesy of FT Enterprises).

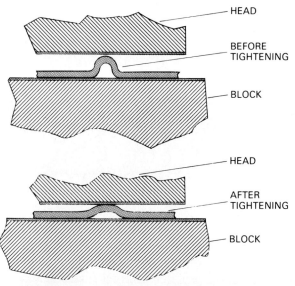

FIGURE 13-5 Embossed steel gasket before and after tightening of bolts. (Courtesy of FT Enterprises).

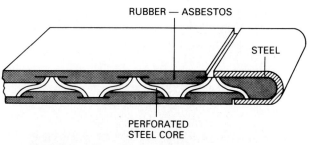

FIGURE 13-8 This composition gasket has a steel flange to seal against high pressures. (Courtesy of FT Enterprises).

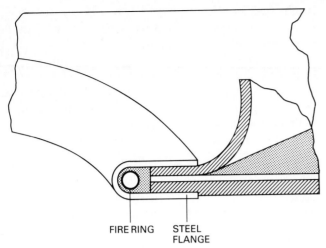

FIGURE 13-9 High sealing pressure is obtained with this fire ring gasket design. (Courtesy of FT Enterprises).

GASKET PACKAGES

Gaskets may be purchased individually or in sets. Gasket packages usually include a list of contents on the package label. Typical gasket packages for engines include the following with common abbreviations in brackets.

1. *Full set (FS)*: includes all required gaskets for a complete engine overhaul; may also include valve stem seals and front and rear oil seals

2. *Valve cover set (VS)*: gaskets for valve covers and seals for center bolt mounting where applicable

3. *Cylinder head set (HS)*: includes all gaskets and seals required to perform a valve grind

4. *Oil pan set (OS)*: side gaskets, end seals, and drain plug gasket for oil pan

5. *Timing cover set (TC)*: all gaskets and seal for front timing cover

6. *Water pump set (WP)*: all gaskets and seals required to replace water pump

7. *Manifold set (MS)*: all gaskets and seals required for mainfold replacement

HANDLING GASKETS

Gaskets, particularly sandwich and metal gaskets, should never be bent. The inner material is easily damaged by bending and straightening, which can result in leakage. Bent or straightened gaskets should not be used. Never reuse cylinder head gaskets since they

have been compressed and have conformed to minor surface irregularities. This causes them to lose their sealing ability. Gaskets that are not packaged should be stored flat. Storing them on edge causes them to bend out of shape and straightening a bent gasket can result in early gasket failure.

SEALANTS

There are two types of sealants used in automotive service. One type is used with gaskets and is applied directly to the gasket or to the mating surfaces to be sealed by the gasket. A great variety of material is available for this purpose. The manufacturer's service manual recommendations should be followed as to whether to use a sealer with the gasket, and if so, which type of sealer to use (i.e., hardening or nonhardening, compatibility, and the like). Another type of sealant used is the "form-in-place gasket material." There are two basic types of this material available. They are not interchangeable and should not be so used. These sealants come in tubes.

Aerobic Sealant (RTV Sealant)

Aerobic or room-temperature vulcanizing sealant cures when exposed to moisture in the air. It has a shelf storage life of one year and should not be used if the expiration date on the package has passed. Al-

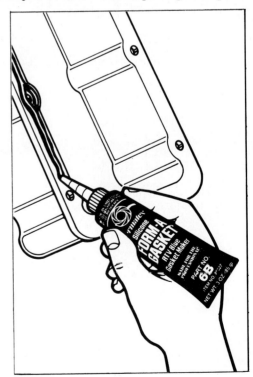

FIGURE 13-10 Using silicone RTV gasket maker on a valve cover. (Courtesy of Loctite Corporation.)

ways inspect the date on the package before use. This material is normally used on flexible metal flanges (Figure 13-10).

Anaerobic Sealant

This type of sealant cures in the absence of air (as when squeezed between two metal surfaces). It will not cure if left in an uncovered tube. This material is used between two smooth machined surfaces. It should not be used on flexible metal flanges.

Sealant and Gasket Maker Usage

Typical gasket sealants, adhesives, and gasket makers and their uses are shown on the accompanying chart.

Product/Part No.	Function	Features
Cut gasket sealants and adhesives		
Form-A-Gasket Sealant 1A, 1B, 1C, 1AR, 1BR	Hardening sealant permanently seals all cut gaskets and fills uneven surfaces. Dries fast, sets hard.	Works from −65°F (−55°C) to 400°F (204°C). Withstands pressures up to 5000 psi. Resists vehicle and shop fluids.
Form-A-Gasket sealant 2A, 2B, 2BR, 2C, 2D, 2E, 2F, 2AR	Nonhardening sealant seals and repairs all types of cut gaskets. Fills uneven surfaces. Dries slowly, sets to a pliable film.	Works from −65°F (−55°C) to 400°F (204°C). Withstands pressures up to 5000 psi.
Aviation Form-A-Gasket sealant 3D, 3F, 3H	Nonhardening brush-on sealant for easy application. Dries slowly, sets to a pliable film.	Works from −65°F (−55°C) to 400°F (204°C). Withstands pressures up to 5000 psi. Resists vehicle and shop fluids.
Indian Head gasket shellac 5J, 5H	Hardening, brush-on sealant for easy application. Dries fast. Sets hard.	Works from −65°F (−55°C) to 350°F (178°C). Withstands pressures up to 100 psi. Resists vehicle and shop fluids.
Tack & Seal sealant 9A, 9B, 9C, 9AR	Nonhardening, "clean hands" sealant that seals and holds all types of cut gaskets. Dries slowly; sets to a tacky, pliable film.	Works from −65°F (−55°C) to 400°F (204°C). Withstands pressures up to 5000 psi. Resists vehicle and shop fluids, including gasohol. Hands clean up with soap and water.
Super 300 sealant 83H, 83D	High-performance, brushable sealant for cork, paper, asbestos, and most metal gaskets. Dries fast, sets to a semiflexible seal.	Works from −65°F (−55°C) to 425°F (222°C). Withstands pressures to 5000 psi. Resists vehicle and shop fluids.
High tack adhesive/sealant 97BR, 97S, 98H, 98D, 99GA, 99MA	Gasketing adhesive sealant holds gasket in place during assembly to ensure correct positioning. Dries fast, sets to a semiflexible seal. Available in tube, brush, and aerosol formulas.	Works from −65°F (−55°C) to 450°F (232°C). Withstands pressures up to 5000 psi. Resists vehicle and shop fluids.
Copper Form-A-Gasket sealant 101H, 101MA	High-temperature gasket adhesive/sealant. Metallic copper particles help dissipate heat, prevent burnout, and improve heat transfer.	Works from −50°F (−45°C) to 500°F (260°C). Can be used in place of a cut gasket. Fills surface imperfections to .035."
High temperature Form-A-Gasket sealant 1372W	High-temperature gasket sealant seals and repairs all types of cut gaskets. Dries slowly, sets to a pliable film.	Works from −50°F (−45°C) to 600°F (316°C). Dries slowly, sets to a pliable film. Withstands pressures up to 5000 psi.
Gasket makers (RTV silicone and anaerobic)		
Blue silicone Form-A-Gasket sealant 6B, 6BR, 6C, 6M, 6SR	Blue universal RTV gasketing silicone replaces almost any cut gasket.	Effective from −80°F (−62°C) to 600°F (315°C), 600° intermittent exposure only. Resists regular and synthetic oil, antifreeze, grease, and transmission fluid.
Black RTV silicone 16B, 16C, 16BR, 16SR	Black, all-purpose RTV silicone for sealing and noncritical gasketing applications.	Effective from −80°F (−62°C) to 600°F (315°C), 600° intermittent exposure only. Resists regular and synthetic oil, antifreeze, grease, and transmission fluid.
Red high temp. silicone Form-A-Gasket sealant 26B, 26BR, 26SR, 26C	Red, high-temperature RTV silicone for all high-temperature gasketing applications.	Effective from −80°F (−62°C) to 650°F (315°C), 650° intermittent exposure only. Resists regular and synthetic oil, antifreeze, grease, and transmission fluid. Will replace almost any cut gasket.
Orange low volatile silicone Form-A-Gasket sealant 27BR	Orange RTV silicone for GM vehicles. Specially formulated, high-temperature, low-volatile RTV gasket maker designed for all oxygen sensor-equipped GM vehicles.	Effective from −80°F (−62°C) to 650°F (315°C), 650° intermittent exposure only. Resists regular and synthetic oil, antifreeze, grease, and transmission fluid. Specially designed for oxygen sensor-equipped GM engines. Low-volatile formula.

(continued)

Product/Part No.	Function	Features
Clear RTV silicone sealant 66B, 66BR, 66C	Clear, all-purpose RTV silicone sealant. Seals, insulates, and bonds glass, metal, plastics, fabrics, vinyl, and rubber.	Works up to 400°F (204°C). Resists aging, cracking, and shrinking. Dries tough and flexible.
Gasket Eliminator sealant 51517, 51531, 51580	Anaerobic gasket maker for precision rigid flange assemblies. Replaces cut gaskets. Use where OEM's specify "anaerobic" gasket.	Effective from −65°F (−54°C) to 300°F (150°C). Resists gasoline, gasohol, diesel fuel, regular and synthetic oil, grease, and transmission fluid. Meets OEM service requirements.

Gasketing aids

Gasket remover 4MA, 4K	Removes baked-on gaskets and gasketing adhesives and sealants. Prepares surface for new gasket and assembly.	Works in 10 to 12 minutes. Reduces scraping and sanding. Prevents damage to flanges.
O-ring splicing kit 0157, 90060, 90061, 90062, 90063	Makes custom O-rings on the spot.	Incudes cutting blade, splicing fixture and 12 ft of Buna-N cord stock. Requires Superbonder adhesive (49450)—not included.

Source: Loctite Corporation.

Suggested Applications	Description	OEM Interchange
Cut gasket sealants and adhesives		
Oil galley plugs. Rear axle cover gaskets. Freeze plugs. Assemblies that are semipermanent.	Hardening gasket sealant.	Ford D7AZ-19B-508A
Oil pan gaskets. Transmission pan gaskets. Timing cover gaskets. Assemblies that are frequently serviced.	Nonhardening sealant.	GM 1050026
Water pump gaskets. Heater valve gaskets. Coolant hose connections.	Nonhardening, brushable sealant.	
Thermostat gaskets. Water pump gaskets. Radiator hose connections.	Hardening sealant, brushable.	
Cut gaskets on vertical and hard-to-reach surfaces. Suggested applications: head gaskets, oil pan gaskets, engine seals, transmission seals, timing cover gaskets, transmission pan gaskets.	Nonhardening gasket sealant.	
Air compressor gaskets. Final drive assemblies. Valve covers. Diesel engine heads.	High-performance gasket sealant.	
Carburetor gaskets, intake manifold gaskets. Valve cover gaskets. Head gaskets. Oil pan gaskets. Gasketing adhesive. Differential cover gaskets. Rear axle gaskets.	Gasket sealant.	
Exhaust manifolds. Head gaskets. Exhaust flanges. Turbocharger flanges.	Copper gasket sealant.	
Exhaust manifold gaskets. Exhaust flanges and connections. Steam flanges and connections.	High-temperature gasket sealant.	
Gasket makers (RTV silicone and anaerobic)		
Oil pan, valve covers. Timing chain covers. Engine side covers. Transmission pans. Oil pumps. Differential housings.	RTV silicone.	Chrysler 4026070 Ford D6AZ-19562-A,B GM 1052289 1051435, 1052751
Water pumps. Thermostat housings. Windshields. Body seams.	RTV silicone.	Ford D6AZ-19562-B

(continued)

Suggested Applications	Description	OEM Interchange
Cross over manifolds. Exhaust manifolds. High-temperature cam covers. High-temperature timing covers.	High-temperature RTV silicone.	Chrysler 4206070 Ford D6AZ-19562-A,B

All GM RTV gasketing applications. Can be used to replace almost any cut gasket.	Specified by General Motors.	GM 1052434,
		   1052366
		   1052751,
		   1052734

Windshields, lights. Cab panels, seams, and roofs. Dash boards and firewalls. Rubber hose connections. Protects and repairs automotive wiring.	Silicone sealant.	Ford D6AZ-19562-A

Overhead cam housings. Cast metal timing chain and differential covers. Compressors and transmission assemblies.	Anaerobic sealant.	Chrysler
		   4205918, 4318029
		Ford E1FZ-19562-A
		GM 1052357
		   1051517, 1052756

Gasketing aids		
Cut gaskets. Silicone gaskets. Weatherstrip adhesive. Dried oil, grease, and paint.	Gasket remover.	
Works on $\frac{3}{32}$, $\frac{1}{8}$, $\frac{3}{16}$, and $\frac{1}{4}$ in. diameter applications.	O-rings.	Chrysler 00112 Ford 00112 GM 00112

(Chart Courtesy of Loctite Corporation.)

CYLINDER HEAD GASKETS

Cylinder head gaskets have the toughest sealing job in the entire engine. They must confine high combustion pressures, engine coolant, and engine oil under wide temperature ranges which cause the head and block to expand and contract. The expansion rate of aluminum cylinder heads is quite different from that of the cast-

FIGURE 13-11 Embossed steel head gasket set. (Courtesy of Chrysler Canada Ltd.)

FIGURE 13-12 Teflon-coated composition head gasket set. (Courtesy of Chrysler Canada Ltd.)

iron block, making the job even more difficult. The temperature it must withstand can range from below zero to 300 or 400°F in minutes from a cold start.

Cylinder head gasket designs currently used include the following: embossed steel, metal sandwich, and soft surface (Figures 13-11 and 13-12). The embossed steel gasket was described earlier under "Gasket Materials." The metal sandwich gasket consists of an asbestos core between two faces of steel or copper. The soft-surface gasket has a perforated steel core covered on both sides with treated asbestos. A stainless steel fire ring is used at each cylinder opening to seal combustion pressures. Another head gasket design has a no retorque feature. This eliminates the need for retorquing cylinder head bolts after initial break in. This gasket consists of a metal core with a Teflon-coated fiber face. A wire ring encased in metal surrounds each cylinder opening (Figures 13-5 to 13-9).

Effects of Varying Head Gasket Thickness

The thickness of an installed, compressed head gasket is a factor affecting the volume of the combustion chamber and therefore the compression ratio. An increase in installed head gasket thickness means an increase in combustion chamber volume and a lower compression ratio. Installed head gasket thickness is also a factor affecting the piston-to-head clearance as well as the piston-to-valve clearance. When replacing head gaskets, be sure to take these factors into con-

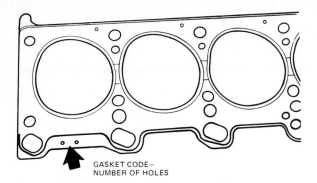

GASKET CODE—
NUMBER OF HOLES

HIGHEST PISTON PROTRUSION OF ALL 6 PISTONS mm	CYL. HEAD GASKET CODE NO. OF HOLES	THICKNESS OF CYL. HEAD GASKET mm
0.60 – 0.70	1	1.4
0.70 – 0.85	2	1.5
0.85 – 1.00	3	1.6

A8070-A

FIGURE 13-13 Typical head gasket thickness selection information. (Courtesy of Ford of Canada.)

sideration. Figure 13-13 shows one example of head gaskets available in varying thicknesses.

HEAD GASKET INSTALLATION GUIDELINES

1. The cylinder head and block gasket surfaces should not be too smooth or too rough. If they are too smooth, there is too little friction between them and the gasket, making it easier for combustion pressures to cause a blown gasket, especially between cylinders. A surface finish that is too rough results in poor sealing since the gasket cannot conform to the rough surface. A surface finish of 90 to 110 microinches (μm) is preferred. Sealing surfaces must be clean and free of any foreign material.

2. Check the gasket and locate the Top, Front, or This Side Up marking (Figure 13-14). If

Identification mark

FIGURE 13-14 Head gasket identification mark indicates "this side up." (Courtesy of Chrysler Canada Ltd.)

the gasket is not marked, it should be installed with the stamped ID number toward the head.

3. Check the gasket holes against the holes in the block and cylinder head. In some cases a water jacket hole may be covered by the gasket or the hole in the gasket may be much smaller than the water jacket hole. This aids in controlling and directing coolant flow in the block and head.

4. Be certain that all head bolt holes in the block are chamfered, clean, and do not contain any oil or coolant. Fluid in a blind bolt hole causes inaccurate torque readings and can create enough hydraulic force to crack the block.

5. Throughly clean all the head bolt threads.

6. Most gaskets do not require sealer to be used. Embossed steel gaskets may require a spray sealer to be applied to both sides. Follow the gasket manufacturer's recommendations.

7. Make certain that special headbolts or bolts of different length are installed in the proper location. Use guide bolts to keep the gasket in place and guide the head into place.

8. Tighten all head bolts in several steps and in the sequence specified by the engine manufacturer. See Figure 13-15 for typical installation instructions.

VALVE COVER GASKETS

There are two methods of sealing the valve covers to the head by using a gasket or by using RTV sealant. In some cases the valve cover gasket acts as a spacer to provide the necessary clearance between the rocker arms and the valve cover. Whether a gasket or a sealant is used, care must be taken that the installation will not cause interference between the valve covers and rocker arms. In some cases spacers may be necessary between the valve cover and cylinder head when sealant is used.

Make sure that the valve cover flanges are not damaged or distorted and that the sealing surfaces are clean and dry. Flange straightness can be checked by laying the valve cover flange side down on a flat surface. Check all around with a feeler gague. Minor surface irregularities are permissible. Distorted flanges must be straightened with a hammer and block of wood.

To install a gasket, apply a good gasket sealant to the valve cover flange. Allow the sealant to become tacky, then place the gasket in position on the valve cover. Allow to dry until the gasket is firmly stuck to the cover. Install the cover using load spreader wash-

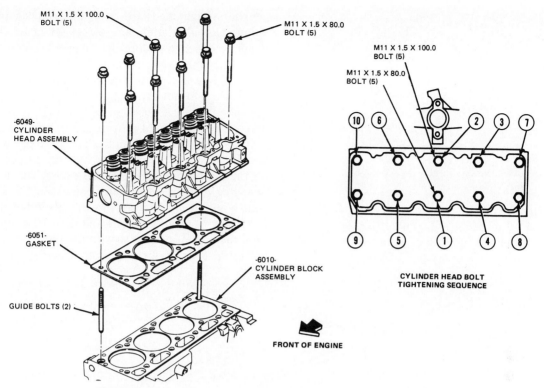

FIGURE 13-15 Head and gasket installation instructions for 2.3-liter HSC Ford engine. (Courtesy of Ford of Canada.)

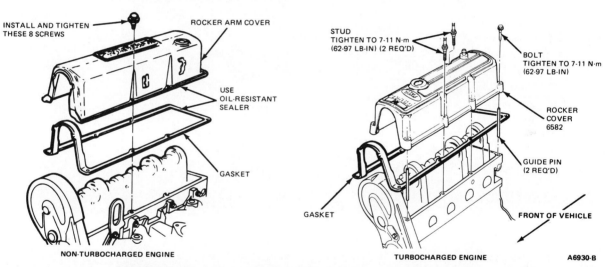

FIGURE 13-16 Valve cover installation instructions for Ford 2.3-liter engine. (Courtesy of Ford of Canada.)

ers if applicable. Tighten the bolts evenly in steps to specified torque. Do not overtighten. Check for leaks with the engine running (Figures 13-16 and 13-17).

FIGURE 13-17 Load-spreading washers are used on some engine valve covers. (Courtesy of FT Enterprises).

OIL PAN GASKETS

Oil pans are either stamped steel or die-cast aluminum. Both types should be checked for flange flatness. This can be done by placing the pan flange side down on a flat surface. Check between the flange and the flat surface with a feeler gauge. Minor deviations are acceptable. A badly distorted aluminum pan must be replaced. Bent flanges or pulled bolt holes on steel pans

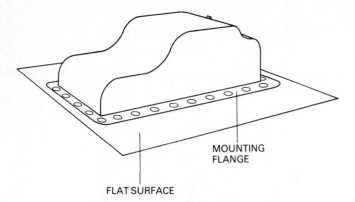

FIGURE 13-18 Mounting flange on oil pan must be flat and true. Straighten if necessary with a hammer and block of wood with pan on flat surface. (Courtesy of FT Enterprises).

can be straightened with a hammer and a block of wood. Do not flatten any ridges designed into the flange to increase sealing effectiveness (Figure 13-18).

Place the new gasket on the flange and check the fit, including the length, the bolt hole pattern, and the end seals. Do not attempt to stretch a gasket to fit. Shrunken cork gaskets can be restored by placing in lukewarm water until they expand to fit. Make sure that the gasket surfaces on the block and pan are clean and dry. Use a gasket sealer to hold the gasket in place on the pan. Make sure that the sealer has set before installing the pan. Wet sealer can cause the gasket to squish out around the bolts even before they are fully tightened. Start all the bolts and draw them up evenly until the pan is in contact with the block. Tighten the bolts evenly in uniform steps to specified torque. Do

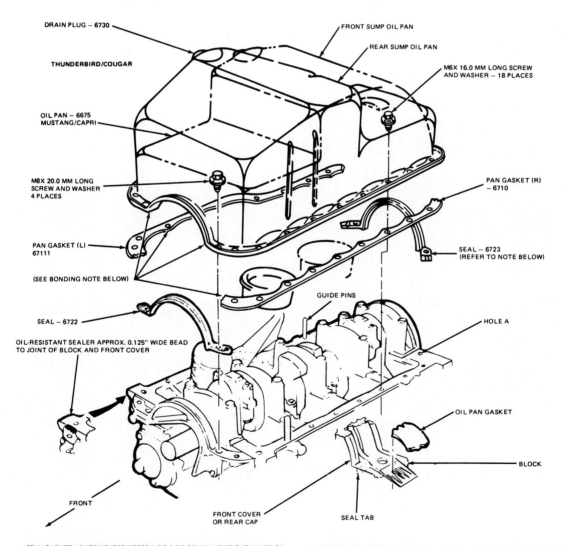

1. APPLY GASKET ADHESIVE (ESE-M2G52-A OR B OR EQUIVALENT) EVENLY TO OIL PAN SIDE GASKETS. ALLOW ADHESIVE TO DRY PAST WET STAGE, THEN INSTALL GASKETS TO OIL PAN.

2. APPLY SEALER (ESE-M4G195-A OR EQUIVALENT) TO JOINT OF BLOCK AND FRONT COVER. INSTALL SEALS TO FRONT COVER AND REAR BEARING CAP AND PRESS SEAL TABS FIRMLY INTO BLOCK. BE SURE TO INSTALL THE REAR SEAL BEFORE THE REAR MAIN BEARING CAP SEALER HAS CURED.

3. POSITION 2 GUIDE PINS AND INSTALL THE OIL PAN. SECURE THE PAN WITH THE FOUR M8 BOLTS SHOWN ABOVE.

4. REMOVE THE GUIDE PINS AND INSTALL AND TORQUE THE EIGHTEEN M6 BOLTS, BEGINNING AT HOLE A AND WORKING CLOCKWISE AROUND THE PAN.

FIGURE 13-19 Oil pan installation instructions for several engine models. (Courtesy of Ford of Canada.)

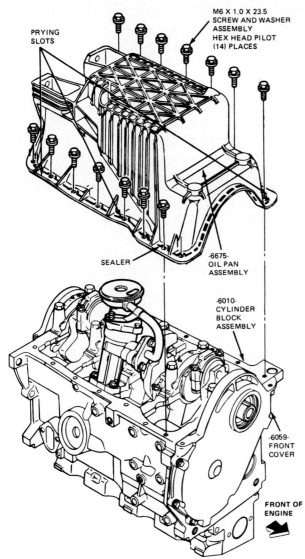

FIGURE 13-20 Typical die-cast aluminum oil pan installation using RTV sealant. (Courtesy of Ford of Canada.)

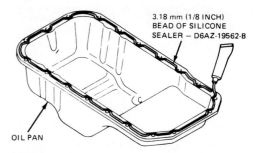

FIGURE 13-21 Typical application of RTV silicone sealer to oil pan. (Courtesy of Ford of Canada.)

the head combine the two manifolds. Intake manifold gaskets range from embossed steel to steel-cored fiber. Many V engines use separate rubber end seals on the intake manifold. A one-piece embossed steel pan-type intake manifold gasket is used on some V-8 engines. The pan serves to deflect oil away from the exhaust crossover. Many engines do not use exhaust manifold

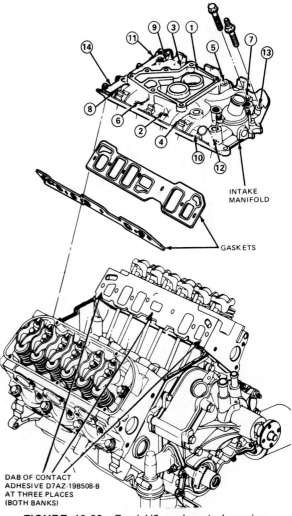

FIGURE 13-22 Ford V6 carbureted engine intake manifold installation. Rubber and seals are in place on the block. Bolt tightening sequence is indicated on manifold. (Courtesy of Ford of Canada.)

not overtighten since this causes bolt holes to pull and gaskets to crack. Check for leaks with the engine running (Figure 13-19).

Oil Pan Sealant

With all sealing surfaces clean and dry and flanges flat and true, apply a bead of gasket making RTV sealant in a pattern as shown in Figures 13-20 and 13-21. Install the oil pan as outlined above.

MANIFOLD GASKETS

V-type and cross-flow in-line engines have separate intake and exhaust manifolds. Some in-line engines with the intake and exhaust ports on the same side of

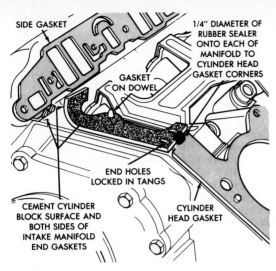

FIGURE 13-23 Special instructions for end gasket sealing on 318-CID Chrysler engine. (Courtesy of Chrysler Canada Ltd.)

gaskets. They rely on the close fit between parts and manifold openings that fit over smaller ports. The most common exhaust manifold gasket consists of a perforated steel core covered on both sides with asbestos.

When installing intake manifold gaskets on V engines, particular attention must be paid to sealing in the corners at the front and rear. Apply RTV sealer to these areas to ensure a leak-free seal. RTV sealer can also be used around the coolant passages. RTV sealer must not be used in a manner that will restrict intake passages, water passages, oil passages, or be drawn into the engine's cylinders. Tighten all bolts in several steps and in the recommended sequence to specified torque. Use antiseize compound on EGR mounting bolts to prevent seizure (Figures 13-22 to 13-24).

When installing exhaust manifold gaskets, make sure that the manifold is not warped. Resurface it if

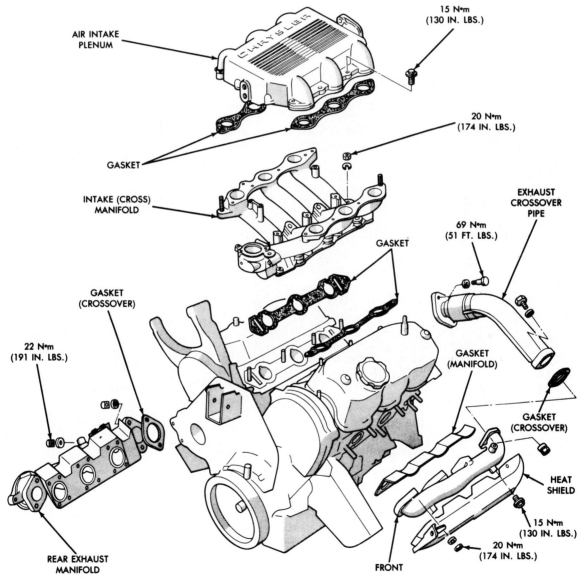

FIGURE 13-24 Intake manifold and plenum installation on Chrysler Mitsubishi 3.0-liter V-6 engine. Exhaust manifold mounting is also shown. (Courtesy of Chrysler Canada Ltd.)

necessary. When a gasket with asbestos on one side only is used, install the asbestos side toward the cylinder head. This allows the manifold to slide on the metal surface as it expands and contracts. Tighten all bolts down evenly and torque to specifications in several steps.

OTHER ENGINE GASKETS

Gaskets used with timing covers (Figures 13-25 and 13-26), water pumps, thermostat housings, oil pumps, and fuel pumps are usually of the high-density paper or fiber variety. Mounting bolt torque is considerably higher than it is for oil pan or valve cover bolts since these parts are subject to operating loads. The general procedure includes making sure that all surfaces are clean, flat, and dry. Check gasket fit before installation to make sure that all oil and coolant passages are accommodated. Threads must be clean and in good condition. It is good practice to use a good thread sealant to ensure sealing. Make sure that bolts of different length are placed in appropriate holes. Bolts that are too short may not reach the threaded hole or may catch just a few threads and strip. Bolts that are too long may bottom out in blind holes and cause serious damage as they are tightened. Tighten all bolts evenly in steps to specified torque.

OIL SEALS

Two types of oil seals are used in automotive engines: static and dynamic seals. Static seals are used to seal between stationary parts. Dynamic seals are used between a moving part and a stationary part. Dynamic

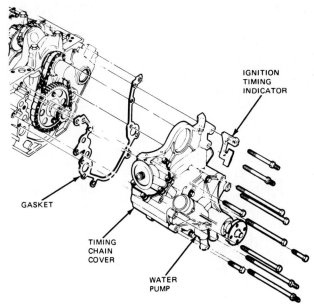

FIGURE 13-25 Timing cover assembly for 3.8-liter V-6 Ford engine. (Courtesy of Ford of Canada.)

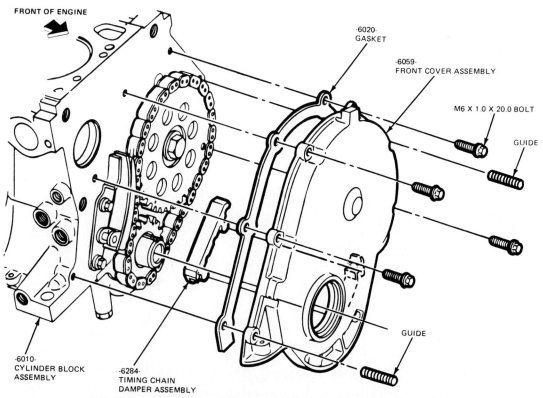

FIGURE 13-26 Timing cover installation on 2.3-liter HSC engine. (Courtesy of Ford of Canada.)

seals include the rear main bearing seal, the timing cover seal, and the valve stem seals. Static seals include O-ring seals used between the distributor housings and the block and between intake manifolds and cylinder heads in some engines. Seal designs include the rubber O-ring, the single- and double-lip metal-backed rubber (with or without a garter spring), the wick type, and the injection type.

Installing a Lip-Type Seal

1. Make sure that you have the correct seal. Check the ID, OD, and depth.

2. Check the seal bore to make sure that it is clean and in good condition.

3. Lubricate the sealing lip with the type of lubricant it will seal.

4. Protect the sealing lip during installation. Where applicable, use an installing sleeve to protect the lip.

5. Position the seal squarely in its bore with the sealing lip facing the fluid to be contained.

6. Apply uniform pressure to the seal case during installation and do not allow the seal to cock in the bore. Use the appropriate size and type of seal driver; do not apply hammer blows directly to the seal. Some seals require special installation tools. See Figure 13-27.

7. Make sure that the seal is fully seated in its bore. Check with a feeler gauge.

8. With split-type rear main lip seals, install them with the seal halves protruding so that the seal parting surface is offset from the main bearing cap parting surface (Figures 13-28 and 13-29).

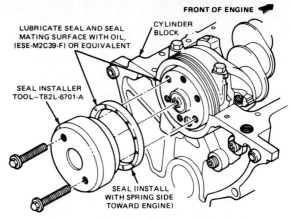

FRONT OF ENGINE

NOTE: REAR FACE OF SEAL MUST BE WITHIN 0.127mm (0.005-INCH) OF THE REAR FACE OF THE BLOCK

FIGURE 13-27 One-piece lip-type rear main bearing seal being installed with a special seal installer. (Courtesy of Ford of Canada.)

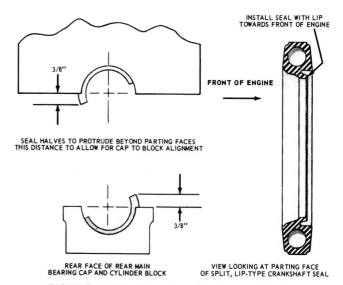

FIGURE 13-28 Two-piece lip-type seal is installed here with the seal parting line offset from the main bearing cap parting line. (Courtesy of Ford of Canada.)

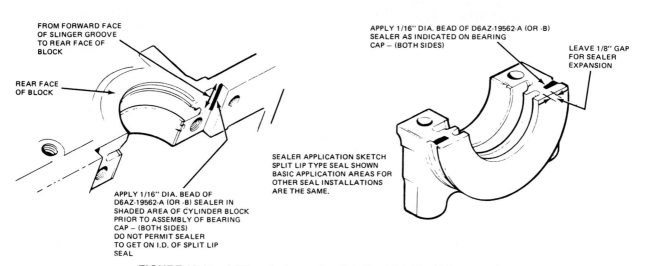

FIGURE 13-29 RTV sealer is used on this Ford 5.8-liter V-8 rear main bearing cap and block. (Courtesy of Ford of Canada.)

Installing a Wick-Type Seal

1. Soak the seal in engine oil.
2. Make sure that the seal grooves are clean.
3. Place the seal in the groove. Force the seal to seat in its groove by using a smooth round tool.
4. When the seal is fully seated, cut off the ends of the seal flush with the block surface and the main bearing cap mating surface (Figures 13-30 and 13-31).
5. Where side seals are used between the cap

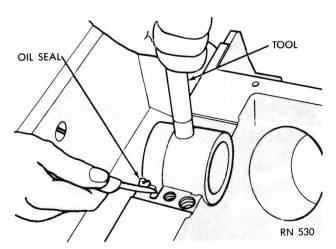

FIGURE 13-30 Seal installing tool is used to push rope seal fully into groove. Protruding ends of seal are cut off flush with parting line on block. (Courtesy of Chrysler Canada Ltd.)

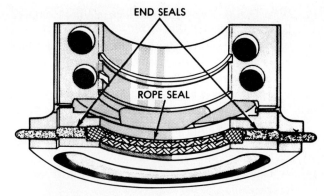

FIGURE 13-31 Rope- or wick-type rear main bearing seal lower half installed in main bearing cap. Rubber end seals are used to seal cap parting line to cylinder block. (Courtesy of Chrysler Canada Ltd.)

and block, make sure that they stay in position while tightening the main bearing cap bolts. Where no seals are used, apply a light coat of anaerobic sealer to the mating surface before installing.

Installing an Injection-Type Seal

Some engines use a sealing material that is injected into a groove at the rear main bearing. This includes Pinto, Capri, and Vega. The applicator resembles a hypodermic springe. The applicator plunger must be depressed slowly and evenly to avoid air entrapment. Sealing material is injected into a groove between the main bearing cap and engine block until completely filled.

REVIEW QUESTIONS

1. What are gaskets used for in an engine?
2. Why should gaskets be compressible and extrudable?
3. Why is resilience an important gasket property?
4. A gasket that is not impermeable:
   (a) allows fluids or gases to escape
   (b) prevents fluids or gases from escaping
   (c) cannot be compressed
   (d) is not resilient
5. The types of gasket materials currently used include:
   (a) paper, cork, metal, and asbestos
   (b) asbestos, cork, rubber, and metal
   (c) metal, paper, fiber, and asbestos
   (d) all of the above
6. The ability of a gasket to make a good seal depends on:
   (a) clamping pressure
   (b) surface irregularities
   (c) the temperature of the gasket
   (d) expansion and contraction rates
   (e) all of the above
7. How does an embossed steel gasket provide for compressibility?
8. Embossed steel gaskets are usually used:
   (a) where pressures and clamping forces are low
   (b) where pressures and clamping forces are high
   (c) where impermeability and resilience are not a factor
   (d) where low temperatures and low pressures are present
9. A room-temperature vulcanizing (RTV) sealant cures when:
   (a) clamped between two surfaces
   (b) exposed to moisture in the air
   (c) used with an epoxy
   (d) none of the above
10. Which gasket has the toughest job in an engine?
11. Describe the basic makeup of a "sandwich" gasket.
12. True or false: A used cylinder head gasket in good condition can be used again.

13. What effect do very smooth block and head surfaces have on a cylinder head gasket's ability to seal?

14. Why should there not be any fluid in cylinder head bolt holes in the block when installing the head?

15. Why is the hole in a head gasket often much smaller than the mating hole in the cylinder block water jacket?

16. What two methods are used to seal valve covers to the cylinder head?

17. Before installing a valve cover, the straightness of the valve cover _____ should be checked.

18. How should the condition of an oil pan flange be checked?

19. True or false: Many engines do not use exhaust manifold gaskets.

20. What type of gasket is most commonly used with exhaust manifolds?

21. Where should RTV sealer be used when installing intake manifolds on V engines?

22. What three types of rear main bearing oil seals are commonly used?

23. What three dimensions are critical to proper seal selection?

24. What is an injection seal?

CHAPTER 14

Oil Pumps and Oil Pans

OBJECTIVES

To develop the ability to:

1. Identify different types of lubrication systems, oil pumps, and oil pans.
2. Correctly diagnose lubrication system problems.
3. Inspect oil pumps to determine their serviceability.
4. Install oil pumps to industry standards.

LUBRICATION SYSTEM FUNCTION

The lubrication system of an engine is designed to perform the following tasks.

- Reduce friction to minimize wear and loss of power
- Help form a seal between the pistons, rings, and cylinders
- Help cool engine parts
- Clean engine parts to reduce the formation of sludge, carbon, and varnish
- Absorb shock and dampen the noise of moving parts

How successful the system is in performing these tasks depends on the ability to maintain a film of oil between the moving parts of an engine. Once this film breaks down there is metal-to-metal contact between parts, and friction, wear, noise, and temperatures all increase. This results in rapid parts failure. Lubricating oil is delivered to engine parts either by pressure from the oil pump or by splash from oil throw-off.

LUBRICATION SYSTEM COMPONENTS

The engine's lubrication system consists of the following components.

1. *Oil pump:* delivers oil under pressure to engine components
2. *Pressure relief valve:* prevents oil pressure from exceeding predetermined maximum pressure
3. *Oil pan:* provides a reservoir for a supply of lubricating oil (also closes bottom of engine crankcase)
4. *Oil filter:* traps larger particles suspended in oil as oil flows through the filter
5. *Oil passages:* in the cylinder block, crankshaft, connecting rods, pistons, cylinder heads, camshafts, lifters, pushrods, and rocker arms distribute the oil to the parts needing lubrication
6. *Positive crankcase ventilation (PCV) system:*

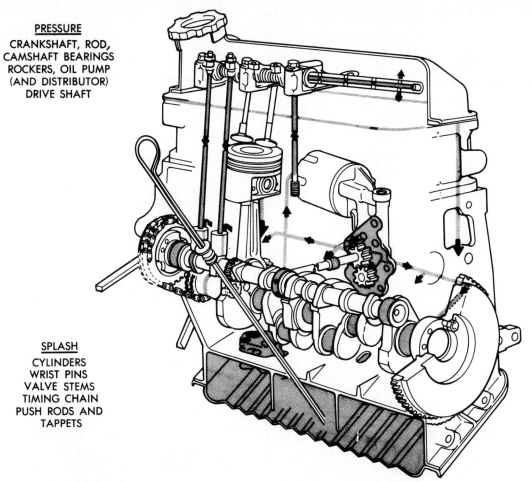

PRESSURE
CRANKSHAFT, ROD,
CAMSHAFT BEARINGS
ROCKERS, OIL PUMP
(AND DISTRIBUTOR)
DRIVE SHAFT

SPLASH
CYLINDERS
WRIST PINS
VALVE STEMS
TIMING CHAIN
PUSH RODS AND
TAPPETS

FIGURE 14-1 Typical lubrication system, showing parts that are pressure lubricated and splash lubricated. (Courtesy of Chrysler Canada Ltd.)

removes blowby gases from the crankcase and directs them to the engine's intake system, where they are fed to the engine cylinders

7. *Engine oil:* liquid lubricant used in the engine's lubrication system

See Figures 14-1 to 14-4 for examples of typical lubrication systems.

TYPES OF LUBRICATION SYSTEMS

The pressure-fed full-flow lubrication system is currently being used in automotive engines. In this system oil is fed under pressure to most of the essential moving parts of the engine. This includes the crankshaft and camshaft bearings, valve lifters, pushrods, and rocker arms. In some engines the piston pin is also pressure lubricated through a drilled passage in the connecting

rod. Other engine parts are lubricated through oil splash or by oil being squirted at them. The cam lobes and engine cylinders on some engines rely on oil splash for lubrication. Many engine designs use a squirt hole in the connecting rod to direct a spurt of oil to the cylinder wall at the moment when it is in the right position and the oil hole in the journal and rod are indexed. Oil drain back from the top of the cylinder head is usually designed to flow over the in-block camshaft. In some engine designs cylinder wall lubrication is provided by oil throw-off from the crankshaft.

Full-Flow System

In a full-flow system system (Figures 14-5 and 14-6) all the oil delivered by the oil pump must first pass through the filter before it reaches the bearings; only filtered oil reaches the bearings. When the filter gets plugged a bypass valve opens, allowing unfiltered oil to be supplied to the engine. This is the system used in current automotive engines.

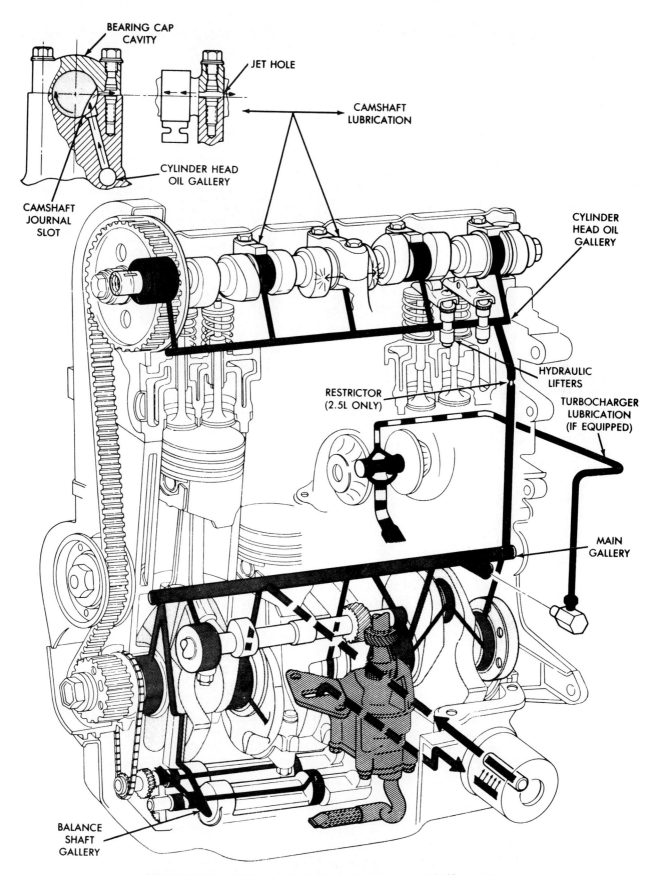

FIGURE 14-2 Oil flow path for typical pressure lubrication system for 2.2- and 2.5-liter engines with auxiliary shaft-driven oil pump. (Courtesy of Chrysler Canada Ltd.)

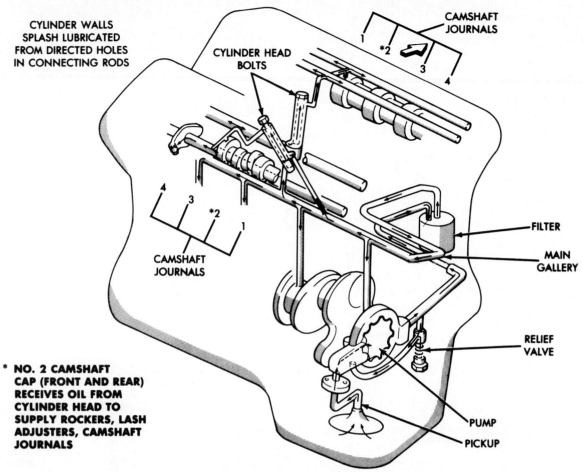

CYLINDER WALLS
SPLASH LUBRICATED
FROM DIRECTED HOLES
IN CONNECTING RODS

CYLINDER HEAD
BOLTS

CAMSHAFT
JOURNALS

CAMSHAFT
JOURNALS

FILTER

MAIN
GALLERY

RELIEF
VALVE

* NO. 2 CAMSHAFT
CAP (FRONT AND REAR)
RECEIVES OIL FROM
CYLINDER HEAD TO
SUPPLY ROCKERS, LASH
ADJUSTERS, CAMSHAFT
JOURNALS

PUMP

PICKUP

FIGURE 14-3 Pressure-fed lubrication system on V-6 engine with crankshaft-driven oil pump. (Courtesy of Chrysler Canada Ltd.)

Shunt System

The shunt system delivers both filtered and unfiltered oil to the engine bearings. A passageway parallel to the oil passage through the filter shunts oil past the filter at all times. This allows oil to be delivered to the bearings when the filter is plugged.

Bypass System

In the bypass system engine bearings are fed directly from the oil pump. A parallel passage leading to the oil filter directs some oil pump output to the filter. Filtered oil is returned to the sump.

OIL PUMP DESIGN AND OPERATION

There are two types of engine oil pumps: the gear type (with two externally toothed gears or with one external gear operating inside an internally toothed gear) and the rotor type. Oil pumps are driven by the crankshaft, camshaft, or intermediate shaft. The oil pump may be

driven directly by one of these shafts or through a set of gears and an oil pump drive shaft.

Engine oil pumps are classified as positive-displacement pumps since there is no direct path or opening between the pump inlet and outlet. As oil enters the pump it is trapped between gear teeth or rotor lobes. As the gears or rotors turn, the teeth or lobes come into mesh and squeeze the oil out from between them to the pump outlet. The inlet side of the pump is connected to a pickup tube that extends into the oil sump. A screen is attached to the bottom of the pickup tube to screen out larger particles of foreign matter (Figures 14-7 to 14-9).

Bearing clearances and metered oil holes restrict the oil flow out of the pump outlet. This restriction results in a pressure rise as the pump continues to force oil into the system. To limit maximum oil pressure, a pressure relief valve is used.

Pressure Relief Valve

The pressure relief valve consists of a ball or plunger type of valve held in position by a coil spring. When

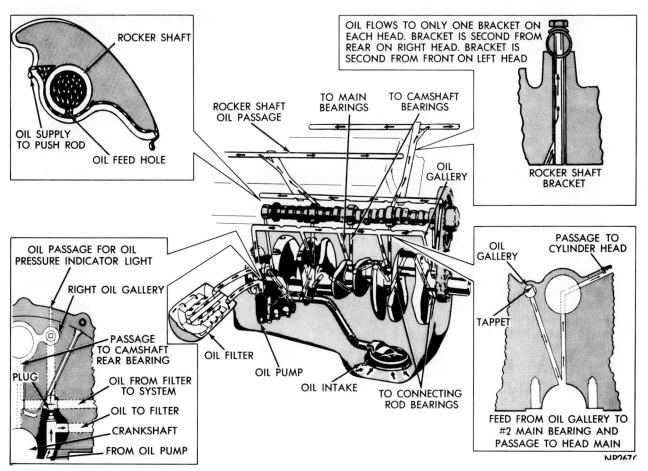

FIGURE 14-4 Chrysler 318-CID V-8 engine lubrication system. (Courtesy of Chrysler Canada Ltd.)

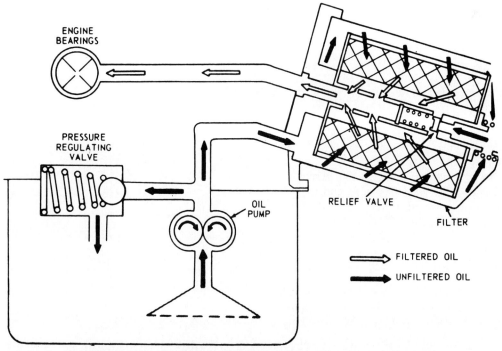

FIGURE 14-5 Full flow pressure lubrication system with full flow filter. (Courtesy of Ford of Canada.)

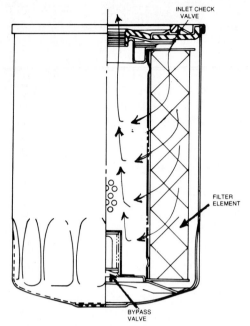

FIGURE 14-6 Cross section of full flow oil filter, showing check valve and bypass valve locations. (Courtesy of Chrysler Canada Ltd.)

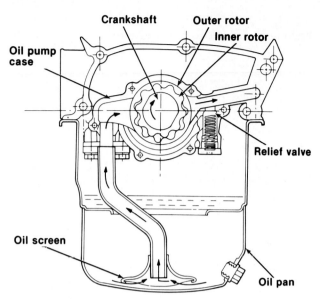

FIGURE 14-8 Crankshaft-driven rotor-type oil pump operation. (Courtesy of Chrysler Canada Ltd.)

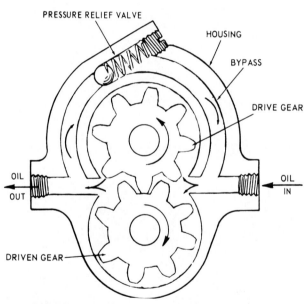

FIGURE 14-7 External gear-type oil pump operation with pressure relief valve and bypass passage shown. (Courtesy of Ford of Canada.)

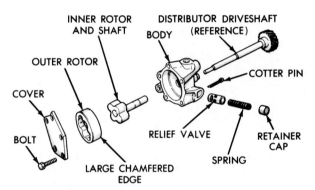

FIGURE 14-9 Exploded view of camshaft-driven rotor-type oil pump. (Courtesy of Chrysler Canada Ltd.)

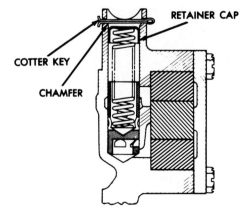

FIGURE 14-10 Typical plunger-type pressure relief valve in assembled position. (Courtesy of Chrysler Canada Ltd.)

seated the pressure relief valve keeps the pressure relief passage closed. As oil pressure rises in the system, the valve is unseated, allowing some oil to escape. This oil is redirected through the pressure relief passage to the inlet side of the pump. Maximum oil pressure is established by the tension of the relief valve spring. Increased spring tension increases maximum oil pressure. Reduced spring tension has no effect on the volume of oil a pump is able to deliver (Figure 14-10). Other factors that affect engine lubricating oil pressure are:

1. *Engine speed.* As engine speed increases, so does the speed at which the oil pump is driven. This increases the pump volume and pressure.
2. *Bearing clearances.* Increased bearing clearances allow more oil to escape, thereby reducing pressure buildup. This is why high-mileage engine oil pressures are usually low since wear increases bearing clearance.
3. *Internal clearance in the oil pump.* As the oil pump gears, rotors, and housing wear, more oil can escape to the inlet side of the pump due to the increased clearance between these parts. This reduces the ability of the pump to produce pressure. However, engine oil pumps are designed with some overcapacity, which allows some wear without affecting pressure requirements.
4. *Oil viscosity.* The viscosity of an engine oil is its fluidity—how easily it pours. Oil that is "thicker" (lower fluidity or higher viscosity) results in higher oil pressure.

Oil Pump Drives

Engine oil pumps are driven by the following methods:

1. By the bottom end of the distributor shaft, which is gear driven from the camshaft. The distributor shaft may connect directly to the top of the pump drive gear shaft or an intermediate shaft may be used between the oil pump and distributor shaft.
2. By a gear on the camshaft that meshes with a gear at the top of the oil pump drive shaft.
3. By the front of the engine crankshaft. In this case the oil pump drive gear or rotor is keyed to the crankshaft and the driven gear or rotor operates in a machined section of the timing chain cover.

OIL FILTERS

Oil filters are designed to trap larger particles of dirt, metal, or carbon delivered to it by the oil from the oil pump. Today's engines use the full-flow spin-on type of filter (Figure 14-6). The oil filter or filter adapter is equipped with a bypass valve. As oil flow through the filter becomes increasingly restricted, the bypass valve

opens to allow full oil delivery to the engine bearings. The bypass valve opens at a pressure differential of about 10 psi. Many filters are designed with a check valve that prevents the filter from draining when the engine is stopped. This provides more immediate lubrication after the engine is started. Oil filters must be changed periodically as recommended by the vehicle manufacturer to ensure good engine lubrication.

OIL COOLERS

Some turbocharged engines and diesel engines are equipped with an oil cooler (Figure 14-11). Since these engines produce more power and therefore more heat, an oil cooler is required to help maintain engine lubricating oil temperature. Increased oil temperatures increase the rate of carbon and varnish formation.

Oil coolers usually consist of a series of tubes or plates through which oil is circulated. Air or engine coolant is directed over the plates or tubes to remove the heat from them. In the event that engine failure results in metal particles circulating through the cooler, the cooler must be replaced when the engine is repaired. Trapped metal particles in the cooler would enter the lubrication system of the repaired engine and cause damage if the cooler is not replaced.

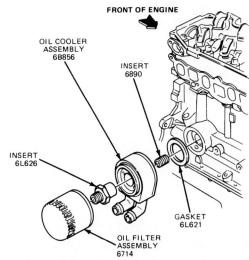

FIGURE 14-11 Oil cooler assembly on Ford turbocharged 1.6-liter engine. (Courtesy of Ford of Canada.)

OIL PUMP PROBLEMS

Most engine rebuilders supply either a new or rebuilt oil pump with a rebuilt or remanufactured engine. Although oil pumps generally last as long as the engine, oil pump problems do sometimes occur. Some of these are: (1) stuck pressure relief valve, (2) broken relief valve spring, (3) cracked or leaking pickup tube, (4) scored and worn gears or rotors, (5) pump drive shaft failure, (6) cracked pump casting, and (7) sludge formation.

Stuck Pressure Relief Valve. There is very limited clearance between the relief valve plunger and the plunger bore. Foreign material entering the pump can become jammed between the plunger and bore preventing normal movement of the relief valve. A relief valve stuck in the open position results in little or no oil pressure (Figure 14-12). A relief valve stuck in the closed position causes excessively high oil pressure and may cause the oil filter to rupture from the pressure.

Broken Relief Valve Spring. A broken relief valve spring results in little or no engine oil pressure.

Cracked or Leaking Pickup Tube. This can cause a fluctuating oil pressure gauge or flickering oil pressure indicator light. Air leaking into the oil pump pickup tube causes aeration and bubbles, which usually results in noisy hydraulic valve lifters. Aeration can also cause the pressure relief valve to vibrate and fail. Cracks or leakage in the pickup tube can be caused by improper installation methods, a dented oil pan, or a cracked pump housing (Figure 14-13). Aeration can also be caused by too high or too low an oil level.

Scored and Worn Gears or Rotors. Abrasives such as metal particles, dirt, carbon, cleaning, or machining residues can score the gears or rotors and pump housing. Scored parts wear rapidly and result in excessive clearances between gears or rotors and the pump housing. Excessive pump internal clearances result in reduced oil pressure (Figure 14-14).

Pump Drive Shaft Failure. When a large enough piece of foreign material enters the pump, it becomes wedged between the gears or rotors, causing the pump to stop turning. This results in a twisted or broken pump drive shaft and of course no oil pressure (Figure 14-15).

Cracked Pump Casting. Incorrect tightening of pump mounting bolts or tightening the pump in a misaligned position can cause the pump housing to crack (Figure 14-16). The pump mounting flange must fit snugly against the engine block with no foreign particles in between before the mounting bolts are tightened. Bolts must be tightened alternately a little at a time to specified torque.

FIGURE 14-12 Relief valve plunger stuck open due to foreign material entering the pump. (Courtesy of TRW, Inc., Automotive Aftermarket Division.)

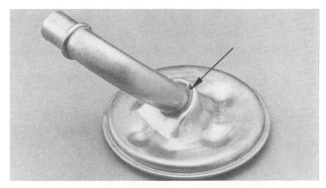

FIGURE 14-13 Cracked oil pickup tube connection. (Courtesy of TRW, Inc., Automotive Aftermarket Division.)

FIGURE 14-14 Badly worn oil pump gears. (Courtesy of TRW, Inc., Automotive Aftermarket Division.)

Sludge Formation Sludge forms in the engine due to poor servicing of the lubrication system, water, dirt and antifreeze contamination, and poor crankcase ventilation (Figure 14-17). Sludge can restrict the oil pickup screen and restrict the oil pump intake. Severe restriction causes starving of the oil pump and severe damage to the oil pump gears or rotors.

FIGURE 14-15 Oil pump drive shaft failure caused by pump lockup due to piece of metal lodged between pump lobes. (Courtesy of TRW, Inc., Automotive Aftermarket Division.)

FIGURE 14-16 Oil pump casting broken due to pump misalignment as mounting bolts were tightened. (Courtesy of TRW, Inc., Automotive Aftermarket Division.)

FIGURE 14-17 Severe sludge formation in crankcase. (Courtesy of TRW, Inc., Automotive Aftermarket Division.)

CHECKING OIL PUMP CLEARANCES

Both gear and rotor-type pumps are measured similarly to check clearances. The relationship of the pump parts to each other must be maintained during this pro-

cedure if the pump is to be used again. Do not mark pump parts with a center punch. Use a felt pen or chalk to mark parts for correct reassembly. Some pumps have the gears or rotors marked during manufacture. Procedures for measuring pump clearances are illustrated, and results should be compared to limits specified in the service shop manual (Figures 14-18 to 14-22). The oil pump should be assembled properly lu-

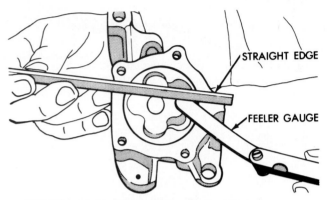

FIGURE 14-18 Measuring rotor clearance with straightedge across pump housing. (Courtesy of Chrysler Canada Ltd.)

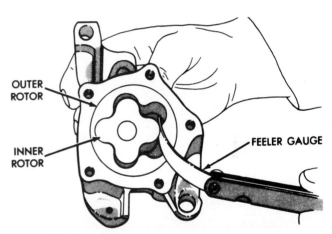

FIGURE 14-19 Measuring rotor tip clearance. (Courtesy of Chrysler Canada Ltd.)

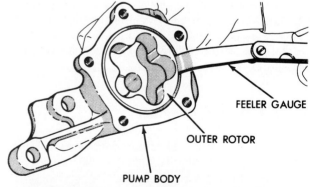

FIGURE 14-20 Measuring outer rotor to housing clearance. (Courtesy of Chrysler Canada Ltd.)

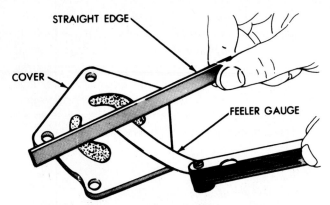

FIGURE 14-21 Checking pump cover flatness and wear. (Courtesy of Chrysler Canada Ltd.)

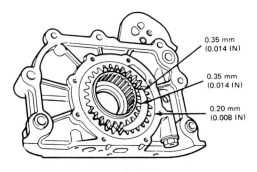

0.35 mm
(0.014 IN)

0.35 mm
(0.014 IN)

0.20 mm
(0.008 IN)

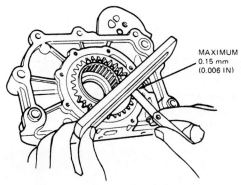

MAXIMUM
0.15 mm
(0.006 IN)

FIGURE 14-22 Typical clearance tolerances for crankshaft-driven gear-type oil pump. (Courtesy of Ford of Canada.)

FIGURE 14-23 Priming the pump with oil. (Courtesy of TRW, Inc., Automotive Aftermarket Division.)

FIGURE 14-24 Mounting the oil pickup tube and screen. (Courtesy of TRW, Inc., Automotive Aftermarket Division.)

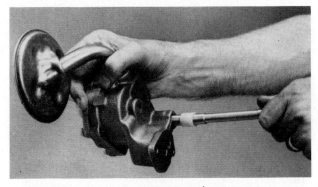

FIGURE 14-25 Installing the pump drive connector. (Courtesy of TRW, Inc., Automotive Aftermarket Division.)

bricated and primed to ensure immediate lubrication on engine startup. The correct thickness of the cover plate gasket must be used (where applicable) to ensure correct gear or rotor tolerances and adequate oil pump pressure.

INSTALLING THE OIL PUMP

Certain precautions must be observed when installing the oil pump, including the following. See Figures 14-23 to 14-27 for typical procedures.

1. Follow the instruction sheet supplied with the oil pump.

2. Be sure that mounting surfaces on the pump and on the engine block are clean and flat.

3. A new pickup tube and screen should be used. The old tube will not fit as well in the new pump if it is of the press-in type. The wire mesh of the screen may be frayed or broken, allowing foreign material to bypass the screen or the bypass valve under the screen strap may be faulty.

4. Be sure that the mounting surfaces of the

FIGURE 14-26 Tightening the pump mounting bolts. (Courtesy of TRW, Inc., Automotive Aftermarket Division.)

6. Be sure to use the intermediate shaft retainer. If left off or installed incorrectly, it can allow the intermediate shaft to fall into the oil pan.

7. If more than one relief valve spring is supplied with the pump, make sure that you use the correct spring for the engine being serviced. Use of an incorrect spring results in improper oil pressure levels.

ENGINE OIL PANS

Two types of engine oil pans are used on automotive engines: the stamped steel oil pan (Figure 14-28) and the die-cast aluminum oil pan (Figure 14-29). The

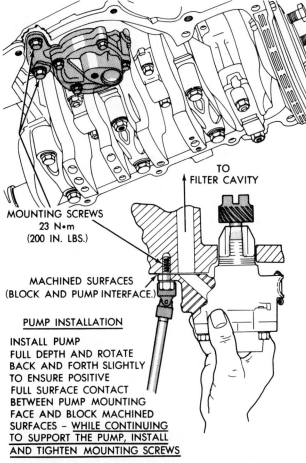

TO
FILTER CAVITY

MOUNTING SCREWS
23 N•m
(200 IN. LBS.)

MACHINED SURFACES
(BLOCK AND PUMP INTERFACE.)

PUMP INSTALLATION

INSTALL PUMP
FULL DEPTH AND ROTATE
BACK AND FORTH SLIGHTLY
TO ENSURE POSITIVE
FULL SURFACE CONTACT
BETWEEN PUMP MOUNTING
FACE AND BLOCK MACHINED
SURFACES – WHILE CONTINUING
TO SUPPORT THE PUMP, INSTALL
AND TIGHTEN MOUNTING SCREWS

FIGURE 14-27 Typical oil pump mounting instructions. (Courtesy of Chrysler Canada Ltd.)

bolt-on-type pickup tube are clean and flat to avoid any entry of air. Make sure that the gasket center hole is punched out.

5. Make sure that there is sufficient intermediate shaft end play. Intermediate shaft pressure against the oil pump gears cause them to be forced against the pump cover, thereby ruining the pump.

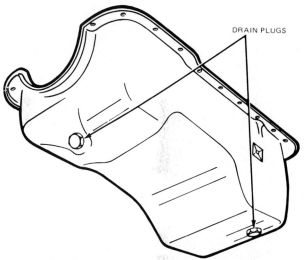

DRAIN PLUGS

FIGURE 14-28 Typical stamped steel oil pan. This design has two drain plugs since there are two sump areas. (Courtesy of Ford of Canada.)

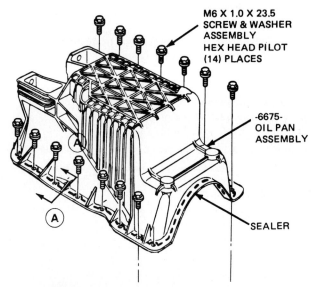

M6 X 1.0 X 23.5
SCREW & WASHER
ASSEMBLY
HEX HEAD PILOT
(14) PLACES

-6675-
OIL PAN
ASSEMBLY

A

SEALER

A

FIGURE 14-29 Die-cast aluminum oil pan. Note reinforcing ribbing. (Courtesy of Ford of Canada.)

stamped steel pan is most commonly used. The oil pan is designed to enclose the bottom of the engine crankcase and is provided with a sump that contains a supply of lubricating oil. This is called a wet sump.

Some racing engines use a dry sump with a separate supply of oil which is not affected by the angle of the engine or by high-speed turns, which can starve a conventional oil pump. Severe braking causes oil to move forward in a conventional wet oil pan, while high-speed turns cause oil to move to the right or left side. Baffles in conventional wet sump oil pans reduce the possibility of starving the oil pump on production engines.

Oil pan configurations include designs with the sump at the rear, front, or center of the pan. Engines with the Y block design use oil pans that are much shallower than those for engines where the lower edge of the block is parallel to the crankshaft centerline. Most oil pans have only one drain plug; however, some engines have two.

Many oil pans are designed to fit around the curved bottom of the rear main bearing cap at the rear and the timing cover at the front. This requires two side gaskets and two end seals for sealing against the engine. See Chapter 13 for details on oil pan and gasket installation procedures.

REVIEW QUESTIONS

1. List five tasks that a lubrication system performs.
2. When the oil film between moving parts breaks down:
   (a) friction and wear increase
   (b) the temperature of parts increases
   (c) engine noise increases
   (d) all of the above
3. The pressure fed full _____ lubrication system is currently used in automobiles.
4. How is cylinder wall lubrication achieved in an engine?
5. How are the cam lobes of an in-block camshaft lubricated?
6. True or false: In the full-flow lubrication system all the oil flows through the bearings before it is filtered.
7. What is meant by the term "positive displacement"?
8. Automobile engine oil pump designs include the following:
   (a) plunger and rotor
   (b) vane and plunger
   (c) rotor and vane
   (d) gear and rotor
9. True or false: Restriction in the engine lubrication system causes oil pressure to rise.
10. Oil bypassing the pressure relief valve:
    (a) is fed back to the oil pan
    (b) is directed to the pump intake
    (c) is directed to the pump outlet
    (d) is fed to the oil filter
11. What effect does engine speed have on oil pump volume and pressure?
12. How do increased bearing clearances affect engine oil pressure?
13. List three ways in which engine oil pumps can be driven.
14. Engine oil pressure drops as:
    (a) bearing clearances are reduced
    (b) internal clearances in the pump increase
    (c) the oil viscosity is increased
    (d) none of the above
15. What is the function of an oil filter?
16. What type of oil filter is commonly used in today's engines?
17. Why is a bypass valve needed in the filtering system?
18. What type of automotive engine may require an oil cooler?
19. Describe the effects of a stuck relief valve in an oil pump.
20. Engine oil pressure drops when the:
    (a) relief valve is stuck closed
    (b) relief valve spring breaks
    (c) oil viscosity increases
    (d) none of the above
21. What causes oil pump gears or rotors to wear?
22. What causes an oil pump drive shaft to break?
23. Name the internal clearance checks required when checking oil pump wear.
24. What effect does a crack in the pickup tube of an oil pump have on pump operation?
25. What two types of oil pans are used on automotive engines?

CHAPTER 15

Cylinder Head Reconditioning

OBJECTIVES

1. To develop an understanding of the reasons for cylinder head reconditioning.
2. To develop the ability to perform the following cylinder head reconditioning procedures to meet industry standards.

   - Resurface cylinder heads and manifolds.
   - Repair threaded spark plug holes.
   - Ream valve guides to specific oversize.
   - Replace replaceable valve guides.
   - Replace bronze valve guides.
   - Knurl valve guides.
   - Bronze wall valve guides.
   - Hone valve guides.
   - Replace press-fit and threaded rocker arm studs.
   - Replace press-fit rocker arm studs with threaded studs.
   - Machine valve guides for positive valve stem seals.

RESURFACING CYLINDER HEADS AND MANIFOLDS

The cylinder head gasket surface and the intake and exhaust manifold mounting surfaces should be checked for erosion or other damage and checked for straightness. Surfaces that are not flat and straight should be resurfaced. To check for straightness use a straightedge and feeler gauge as outlined in Chapter 3. Figures 15-1 and 15-2 illustrate cylinder heads awaiting

FIGURE 15-1 Cylinder heads stacked in storage awaiting cleaning. (Courtesy of Western Engine Ltd.)

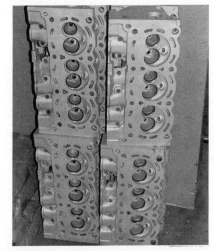

FIGURE 15-2 After cleaning, cylinder heads are ready for reconditioning. (Courtesy of Western Engine Ltd.)

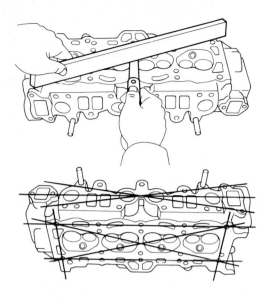

FIGURE 15-3 Cylinder head gasket surface and manifold mounting surfaces must be checked for flatness with a straightedge and feeler gauge. (Courtesy of Ford of Canada.)

FIGURE 15-4 Longitudinal leveling of a cylinder head. Leveling is achieved by adjusting the screw jacks under the head. (Courtesy of Jasper Engine and Transmission Exchange.)

FIGURE 15-5 Leveling across the head surface. (Courtesy of Western Engine Ltd.)

reconditioning. Figure 15-3 shows how to check cylinder head surface flatness.

Caution Regarding Compression Ratios

To maintain equal compression ratios between cylinders, the same amount of metal must be removed across the full length of the head. If excess material has been removed from one end of the head, the compression ratio at that end will be higher. To maintain the same compression ratio and detonation characteristics for all cylinders, both heads on V engines must be machined the same amount.

Caution Regarding Leveling the Head

Cylinder head and block surfacing equipment is designed to keep the cutting head and cylinder head support frame parallel. For this reason the cylinder head should always be leveled to the cutting head. Leveling the head actually means positioning the head surface to be parallel to the cutting head or grinder surface. A bubble-type level is used for the purpose (Figures 15-4 and 15-5). To ensure accuracy the level must be calibrated on a surface plate. First place the level on the reference surface and note the bubble position. Next place the level on the exact same spot in the reversed position. If reversing the level does not yield the same reading, the reference surface is not level. To correct, adjust the reference surface one-half the error indicated and adjust the level the other half. Repeat the checks in both level positions and readjust until the same reading is obtained in either level position.

SURFACING METHODS

There are two basic methods used for stock removal from cylinder heads and manifolds: (1) milling and (2) grinding.

Milling Machine

Milling machines vary somewhat in design, but all use a rotating cutting head with a series of sharp carbide cutting tools mounted near its outer edge. These cut-

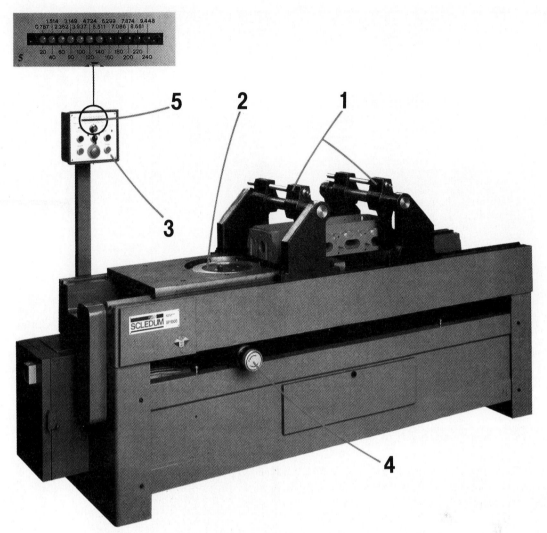

FIGURE 15-6 Bottom cutting cylinder head milling machine. (Courtesy of Storm Vulcan Co.)

ting tools shave small amounts of metal from the head surface (Figures 15-6 to 15-10). Most machines provide several cutting head speeds as well as varying feed rates to obtain the desired rms finish on both cast-iron and aluminum heads. Milling machines are designed to remove material from the top of a piece of work or from the bottom. Machines that remove material from the bottom require a mirror to inspect the work from underneath while the work is always visible with machines that cut from the top of the workpiece. Setting up the milling machine requires the following steps.

1. Level the cylinder head until the cutters touch at all four corners.
2. Move the cutting head across the cylinder head to find the low spot. This will determine how much stock must be removed to clean up the head.
3. Adjust the cutting depth and milling speed to produce the desired rms finish.

Milling or Grinding Aluminum. Milling or grinding aluminum heads requires slightly different procedures to achieve good results. The rotating speed, the depth of cut, and the rate of feed are usually reduced. A cutting depth of no more than 0.001 in. may be required. Special cutting tools for aluminum may also be required. The head surface should be sprayed with silicone spray or coated with wax to prevent aluminum from sticking to the cutting bits.

Rotary Grinding Machine

Grinding machines use a rotary grinding head with an abrasive wheel. The depth of cut, the rotating speed of the abrasive wheel, and the rate of feed are adjustable to achieve the desired rms finish and to accommodate both cast-iron and aluminum heads (Figure 15-11). There are two basic methods of mounting the head on rotary grinders. One design uses the rollover fixture method and the other uses the jackstand method of

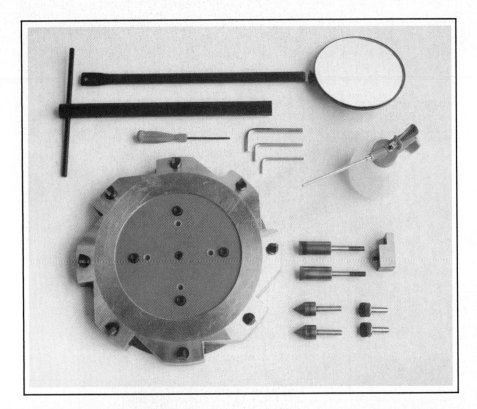

FIGURE 15-7 Cutting head and related parts for machine shown in Figure 15-6. Mirror is used to inspect bottom of head. (Courtesy of Storm Vulcan Co.)

FIGURE 15-8 Cylinder head in position against cutters. (Courtesy of Storm Vulcan Co.)

head mounting. With the rollover fixture the head can be bolted to the cradle via the manifold bolt holes to surface the head gasket surface. To surface the manifold gasket surface, the head is bolted to the cradle via the cylinder head bolt holes. With the jackstand fixture the head is mounted on jackstands at each corner. While the mounting methods differ, the procedural steps are the same for both machines:

1. Mount the cylinder head.
2. Level the cylinder head.
3. Set the cutting depth.
4. Select the desired rotating speed.
5. Select the desired rate of feed (traverse speed).
6. Resurface the head.

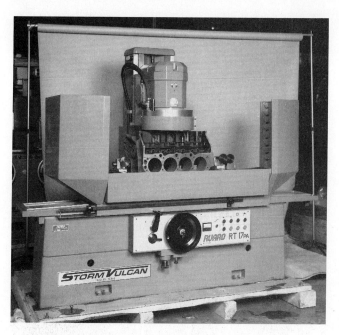

FIGURE 15-9 Surface milling machine that works on the top of the head or block being milled. (Courtesy of Storm Vulcan Co.)

FIGURE 15-10 Cylinder head on rollover fixture positioned to check manifold mounting surface with dial indicator to determine amount of stock to be removed in relation to combustion side of head. (Courtesy of Storm Vulcan Co.)

Dressing the Grinding Wheel. To keep the grinding wheel cutting properly, it must be kept clean and its surface trued. To clean the grinding wheel, hold a wire brush against the grinding surface while it turns. Although procedures vary with machine design, the typical procedure for dressing the stone is as follows.

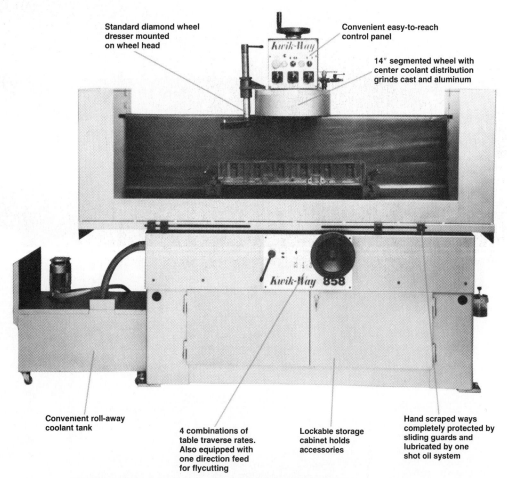

Standard diamond wheel dresser mounted on wheel head

Convenient easy-to-reach control panel

14″ segmented wheel with center coolant distribution grinds cast and aluminum

Convenient roll-away coolant tank

4 combinations of table traverse rates. Also equipped with one direction feed for flycutting

Lockable storage cabinet holds accessories

Hand scraped ways completely protected by sliding guards and lubricated by one shot oil system

FIGURE 15-11 Rotary surface grinder grinds cast iron or aluminum. (Courtesy of Kwik-Way Manufacturing Company.)

FIGURE 15-12 The diamond dresser is mounted on the grinding head and swivels inward for stone dressing. (Courtesy of Kwik-Way Manufacturing Company.)

Bolt the dressing attachment securely to the traverse table. Position the table (or grinder) so that the dresser clears the grinding surface. Turn on the motor that drives the grinding wheel. Carefully raise the table (or grinder) so that the dresser just meets the abrasive surface. Traverse right and left so that the dresser has crossed the grinder surface completely. Turn on the coolant. Increase the dresser depth by 0.005 in. Carefully traverse right and left to dress the stone surface completely. Readjust for a second or third cut if needed. Figure 15-12 illustrates a stone dresser that remains attached to the grinding head. To dress the stone with this unit, simply pivot the stone dresser to swing across the face of the grinder. A depth adjustment is provided.

Belt Grinder

Belt grinders use a motor-driven endless abrasive belt supported between two rollers (Figure 15-13). Belt tension is adjustable. Both cast-iron and aluminum heads and manifolds can be resurfaced wet or dry. Wet grinding is dust free, is cleaner, and produces a better finish. The surface to be ground is placed on the belt grinder

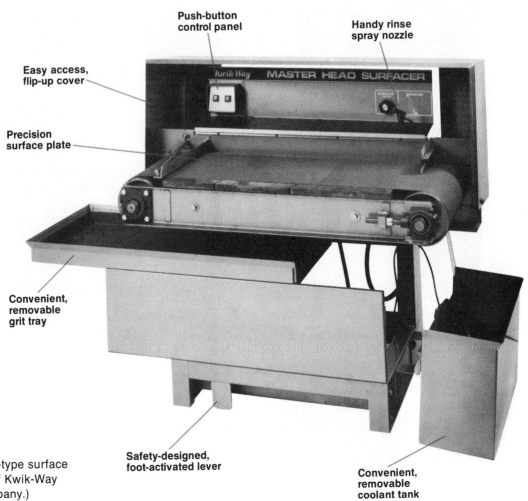

FIGURE 15-13 Belt-type surface grinder. (Courtesy of Kwik-Way Manufacturing Company.)

FIGURE 15-14 Grinding the combustion side of a head on a belt grinder. (Courtesy of Kwik-Way Manufacturing Company.)

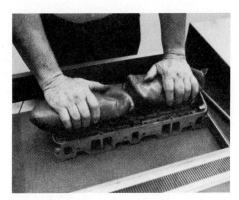

FIGURE 15-15 Grinding the intake manifold side of a V-8 head on a belt grinder. (Courtesy of Kwik-Way Manufacturing Company.)

FIGURE 15-16 Grinding an exhaust manifold on a belt grinder. (Courtesy of Kwik-Way Manufacturing Company.)

and held in place by hand (Figures 15-14 to 15-16). The machine is stopped to make a visual inspection of the surface and to check flatness with a straightedge and feeler gauge. During grinding, additional pressure can be applied by the use of 25-lb bags (Figure 15-17). Grinding belts with 40, 50, 60, and 80 grit are available to achieve the desired rms finish. Belt tension is adjustable (Figure 15-18).

FIGURE 15-17 Using two 25-lb bags to increase grinding pressure on belt grinder. (Courtesy of Kwik-Way Manufacturing Company.)

FIGURE 15-18 Adjusting the belt tension on a belt grinder. (Courtesy of Kwik-Way Manufacturing Company.)

PORT ALIGNMENT ON V ENGINES

As stock is removed from either the cylinder head or block on V engines, the heads will move closer together and downward. This reduces the dimensions between the intake manifold mounting surfaces. Consequently, the intake manifold will not go down far enough for ports and bolt holes to be aligned. Usually, metal removal of 0.010 in. or more from either the heads or block at surface A (Figure 15-19) will require removing metal from the heads at surface B. On manifolds that form the lifter valley cover, metal may also have to be removed from the block at surface C (Figure 15-20).

The following method can be used to calculate the amount of stock that must be removed at surface B after stock removal from surface A (Figure 15-20) on 90° V engines.

1. Check the angle of surface A against surface B with a protractor (Figure 15-21). The amount by which this angle is less than 90° is the head angle in degrees.

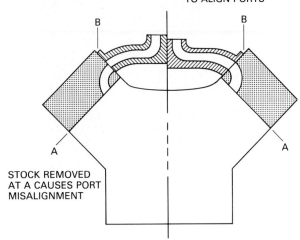

FIGURE 15-19 Stock removed from head or block at surface A requires stock removal from both heads at surface B for port alignment on 90° V engines. (Courtesy of FT Enterprises).

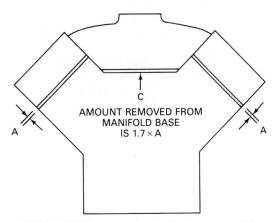

FIGURE 15-20 On manifolds that seal at the front and rear of the block, stock removal from the block at C is required as indicated here. (Courtesy of FT Enterprises).

2. Multiply the amount removed at surface A by the appropriate head angle factor in the following table to determine the amount to be removed at surface B.

Head Angle (degrees)	Factor
5	1.1
10	1.2
15	1.4
20	1.7
25	2.0
30	3.0

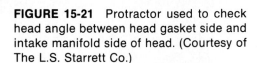

FIGURE 15-21 Protractor used to check head angle between head gasket side and intake manifold side of head. (Courtesy of The L.S. Starrett Co.)

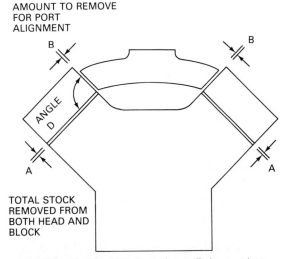

FIGURE 15-22 Head angle at D is used to calculate amount of stock removal required at B. (Courtesy of FT Enterprises).

Example: If the angle formed between surface A and surface B is 80°, the head angle factor is 10° (90° − 80° = 10°). If 0.015 in. was removed from surface A, calculate as follows: 0.015 × 1.2 = 0.018 in. This amount must be removed from the manifold surface of both cylinder heads for ports and bolt holes to align (Figure 15-22).

A quick method is to use the gauge shown in Figure 15.23. This gauge indicates the factor by which the amount removed from surface A is to be multiplied to determine how much stock to remove from surface B.

On manifolds that seal at the front and rear of the block, material must also be removed at surface C of the cylinder block to prevent the manifold from bottoming against the block. The amount to be removed may be calculated as follows. Assuming 0.015 in. of stock was removed from surface A, simply multiply by a factor of 1.7 (0.015 × 1.7 = 0.025 in.). This is the amount to be removed at surface C in this example. This will provide for end gasket compression and prevent the manifold from bottoming on the block (see Figure 15-20).

FIGURE 15-23 Quick check head angle gauge provides multiplication factor to calculate the amount of stock removal required on the intake side of the head. (Courtesy of Storm Vulcan Co.)

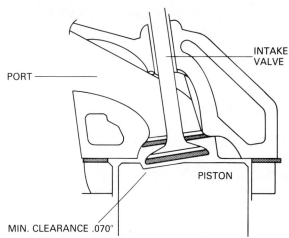

FIGURE 15-24 Minimum piston-to-valve clearance must be maintained. (Courtesy of FT Enterprises.)

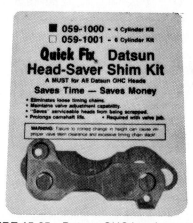

FIGURE 15-25 Datsun OHC head saver shim kit for shimming cam towers. (Courtesy of Silver Seal Products Company, Inc.)

MAXIMUM STOCK REMOVAL

There is a limit to how much stock can safely be removed from the cylinder heads. Removing too much stock can cause the following conditions.

1. Stock removal from the cylinder heads increases the compression ratio. Removing from 0.060 to 0.080 in. of stock increases the compression ratio by approximately 1. For example, a compression ratio of 8.0:1 would be increased to 9.0:1. Increasing the compression ratio increases the potential for detonation.

2. Excessive stock removal can cause piston-to-valve contact. A minimum of 0.070-in. piston-to-valve clearance must be maintained (Figure 15-24).

3. Stock removal from the cylinder heads changes the position of the rocker arms on pushrod engines. The plungers in the hydraulic lifters will also be forced farther down, and if stock removal is excessive cause them to bottom out. Shimming under the rocker arm pedestals will correct this condition. Shim thickness should be one-half the amount of stock removed from the heads. The other way to correct this problem is to use shorter pushrods.

4. On OHC engines the cam bearing bore centerline will be lowered due to stock removal from the combustion chamber side. On Datsun and Mercedes-Benz OHC engines with removable cam towers, shimming under the towers will restore the camshaft-to-crank-

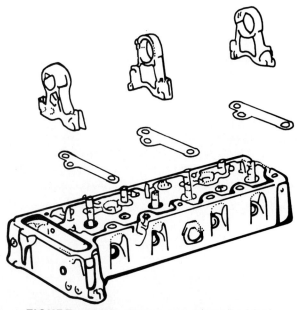

FIGURE 15-26 Shim arrangement for Mercedes-Benz cam towers. (Courtesy of Silver Seal Products Company, Inc.)

shaft centerline dimension. Simply use shims equal in thickness to the amount of stock removed from the head (Figures 15-25 and 15-26).

STRAIGHTENING ALUMINUM HEADS

Since aluminum cylinder heads expand at twice the rate of the cast-iron block to which they are bolted, warpage of the head is not uncommon. This generally occurs by the head becoming bowed upward in the middle. Warpage can be severe, approaching as much as 0.050 in. With overhead cam engines machining the head surface still leaves the cam bearing bores misaligned. On engines with removable cam bearing pedestals, shimming under the pedestals can correct alignment. Another method is to use semifinished bearings and align honing them to proper size.

Straightening the cylinder head corrects the warpage as well as the cam bearing bore alignment. However, using a press to straighten the head cold most often cracks the head. The head can be straightened by bolting the stripped head to a 2-in.-thick steel plate (Figure 15-27) through two of the center bolt holes with shims at either end of the head. Shims are placed between the head and plate at each end. Shim thickness is half the total amount of head warpage. The center bolts are tightened to no more than 25 ft-lb, leaving the ends of the head free to slide with expansion. Overtightening the bolts can cause the head to crack. The head is then placed in an oven heated to 500°F and left there for 4 hours. The oven is then shut off and left to cool slowly with the head still in place. After removal, a minor resurfacing is all that is required.

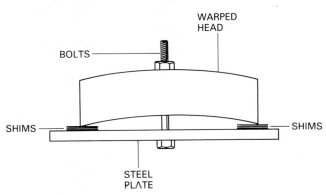

FIGURE 15-27 Warped head in place on steel plate with shims at each end ready for straightening. (Courtesy of FT Enterprises.)

VALVE GUIDE SERVICE

Valve guide service must always be performed before any seat grinding is done. The guide is the base from which seat reconditioning is done. If the guide is worn,

it is not possible to recondition the seat properly. Valve guide service includes measuring the valve stem to guide clearance to determine the amount of valve guide wear. If wear and stem to guide clearance are excessive, one of the following methods may be used to correct the condition.

1. Remove and replace replaceable guides.
2. Ream the guides to specified oversize and use oversized stem valves.
3. Knurl and ream the guides to standard dimensions.
4. Boring and sleeving integral guides.
5. Bronze walling guides.

Valve Stem-to-Guide Clearance

Valve stem-to-guide clearance is an important factor in proper valve operation. Too little clearance can result in scuffing, scoring, and valves sticking in the open position. Pistons can strike valves that are stuck open. Too much clearance allows the valve to cock in the guide. A cocked valve seats only on one side, causing leakage and burning of valves and seats. Since exhaust valves run hotter and expand more than intake valves, they require more stem-to-guide clearance. Typical intake valve stem clearance ranges from 0.001 to 0.003 in., while exhaust valve stem clearance is from 0.002 to 0.004 in. When guides are worn enough to increase valve clearance by 50% or more over specifications, the guides must be replaced or reconditioned. See Chapter 3 for details on measuring valve guide wear.

REPLACING REMOVABLE VALVE GUIDES

On engines equipped with replaceable guides (Figure 15-28) the worn guides are removed and new guides installed. The worn guides can be removed with a press or by using a driving tool and hammer. The new guides are pressed or driven into place. On engines equipped with aluminum cylinder heads and replaceable guides, worn guides are often reconditioned rather than replaced. This eliminates any possible damage to the head that guide replacement can easily cause on the softer aluminum head material. The following guidelines can be used for guide replacement.

1. Measure the installed height of the worn guide before removal. This is the amount the guide protrudes on the spring side of the head. Use this figure for proper depth of new guide installation (Figure 15-29).
2. Drive or press the guide out from the spring side or the port side as specified in the manual. Driving the guide out from the port side

FIGURE 15-28 Valve guide drivers. (Courtesy of Easco/KD Tools, Inc.)

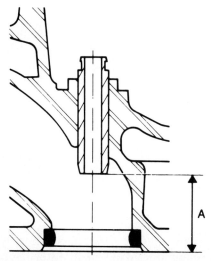

FIGURE 15-29 Installation height of guide is shown at A. (Courtesy of Chrysler Canada Ltd.)

may be more difficult because of the buildup of hard carbon deposits. On aluminum heads carbon residues on the guide OD can damage the guide bore in the head. On some engines the guides cannot be removed from the spring side because of guide taper. In this case they must be driven out from the port side (Figure 15-30). Consult the appropriate service manual for the correct procedure.

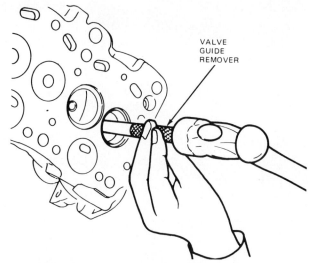

FIGURE 15-30 Driving out a valve guide. (Courtesy of Ford of Canada.)

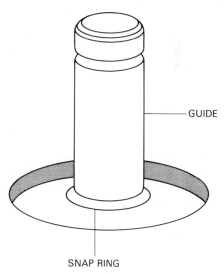

FIGURE 15-31 On some Toyota engines the guide is held in place by a snap ring. The guide must first be broken off at the snap ring groove before removal. (Courtesy of FT Enterprises.)

3. On some Toyota aluminum head engines the protruding part of the guide on the spring side must be broken off at the snap ring with a brass punch and hammer before guide removal. Next heat the head gradually to 200°F (93°C), then drive the guide out toward the combustion chamber side using a guide driver and hammer. Reheat the cylinder head to the same temperature, then drive the new guide in from the spring side of the head until the snap ring contacts the cylinder head (Figure 15-31).

4. Coat the new guide with the recommended installation lubricant.

FIGURE 15-32 Typical replacement valve guides. (Courtesy of Sealed Power Corporation.)

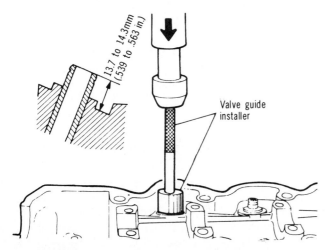

FIGURE 15-33 Installing a valve guide with a press. See the appropriate service manual for installation depth specifications. (Courtesy of Chrysler Canada Ltd.)

5. Press or drive the new guide into place to the depth as measured in step 1 (Figures 15-32 and 15-33). If the guide is pressed in too far, airflow in the port will be affected and more rapid stem wear will result. If the guide extends too far above the spring seat, the spring retainer may bottom out against the guide.

6. After installing the guide it should be reamed to remove any high spots that may be present (Figure 15-34). Be sure that you use the correct reamer type and size and do not enlarge the guide diameter.

Guide Beveling

The top of the guide can be beveled to shed oil if positive valve stem seals are not being used. To bevel the guide, install the appropriate size pilot tightly in the guide. Adjust the cutter blade in the cutting tool to just clear the pilot. Turn the beveling tool while applying downward pressure. Stop cutting when the taper nears the pilot at the top of the guide.

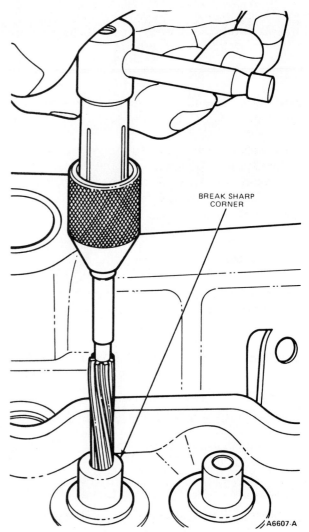

FIGURE 15-34 Reaming the newly installed valve guide. (Courtesy of Sealed Power Corporation.)

REPLACING BRONZE GUIDES

Due to the softness of bronze guides, they present a special problem during replacement. Special precautions are required to avoid guide distortion and swelling when trying to remove them. For bronze guide removal and installation a press is better than a hammer and driver. Hammering can easily swell the exposed

FIGURE 15-35 Bronze valve guide insert. (Courtesy of Sealed Power Corporation.)

FIGURE 15-36 Valve guide reamer has pilot to keep reamer in alignment. (Courtesy of Easco/KD Tools, Inc.)

guide and damage the guide bore in the head. To remove the guide, cut threads on the inside and install a bolt from the spring side. Press the guide out by pressing against the end of the bolt inside the guide. To install a new bronze guide, preheat the cylinder head evenly in an oven, then install the guide to the correct depth. A new bronze guide insert is shown in Figure 15-35.

REAMING WORN GUIDES

If oversized stem valves are available for the engine being serviced, the guides can be reconditioned by reaming to the appropriate oversize and installing new, oversized stem valves. Be sure to use only the kind of reamers specially designed for guide reaming (Figure

Diameter Fractional	Diameter Decimal	Diameter Metric
¼"	.250	6.35
	.2756	7.00
9/32"	.2812	7.14
5/16"	.3125	7.94
5/16" + .003	.3155	8.00
5/16" + .005	.3175	8.06
·5/16" + .015 (21/64")	.3275	8.32
5/16" + .030	.3425	8.70
11/32"	.3437	8.73
11/32" + .003	.3467	8.81
11/32" + .010	.3537	8.98
	.3543	9.00
11/32" + .015 (23/64")	.3587	9.11
11/32" + .020	.3637	9.24
.373 Std.	.373	9.48
11/32" + .030	.3737	9.50
3/8"	.375	9.58
.373 + .003	.376	9.55
3/8" + .003	.378	9.60
3/8" + .005	.380	9.65
.373 + .015	.388	9.86
3/8" + .015 (25/64")	.390	9.91
.373 + .020	.3937	10.00
3/8" + .030	.405	10.29
7/16"	.4375	11.11
7/16" + .015	.500	12.70
½"	.500	12.70
9/16"	.5625	14.29
5/8"	.625	15.88

FIGURE 15-37 Typical range of valve guide reamer sizes available. (Courtesy of Easco/KD Tools, Inc.)

15-36). Reaming is a method of removing metal from the inside of the guide. Reaming produces the desired guide bore diameter and keeps the sides of the bore parallel throughout the length of the guide. The size (diameter) of the reamer determines the amount of metal that will be removed. Replacement oversize stem valves are generally available in standard oversizes of 0.003, 0.005, 0.015, and 0.030 in. Original-equipment manufacturers sometimes use oversizes other than standard such as 0.001 or 0.013 in. oversize.

To ream a guide, select a reamer size that will produce the correct hole size for the clearance required with the selected oversize valve stem diameter (Figure 15-37). It is best not to try to remove more than 0.005 in. of metal per reamer pass. If more metal must be removed, start with a smaller reamer and enlarge the guide in progressive steps. Always turn the reamer in a clockwise direction while applying slight downward pressure. Never reverse the reamer direction since this will destroy its cutting edge. Make sure that the reamer does not wobble during the operation. A wobbling reamer produces a bell-mouthed guide. After reaming, measure to make sure that the desired stem-to-guide clearance with the new oversized stem valve has been achieved. Reaming worn guides produces a new wear surface with minimum time and effort; however, the necessity to use new oversized stem valves adds to the cost.

KNURLING VALVE GUIDES

Proper valve stem-to-guide clearance can be restored by knurling the guides. Knurling is a process of cold metal displacement. As the knurling tool is pressed firmly into the inside surface of the guide, metal is raised on each side of the tool as it forms a groove. A series of closely spaced spiral ridges and grooves is formed throughout the guide bore. After knurling, the guide is reamed to the desired finish diameter. Stem-to-guide clearance for knurled guides is generally one-half the normal clearance specifications. The reduced clearance is possible since the grooves in the knurled guide carry lubricating oil.

There are two general methods used for guide knurling. One method uses a tool that has a tapered spirally grooved knurling area at one end and a square drive at the other. This method will recondition guides with up to 0.007 in. wear. The other method uses a knurling arbor with a rotating knurling wheel. Both methods use a portable electric drill motor and speed reducer to drive the knurling tool (Figures 15-38 to 15-40).

With the first method the proper-size knurling arbor is dipped in special knurling lube. The arbor is then installed into the guide from the combustion chamber side of the head. Start driving the arbor by hand to ensure straight entry into the guide. Complete the knurling by driving the arbor right through the guide with the drill motor and speed reducer. After knurling, ream the guide to the desired finish diameter to provide the specified clearance. After reaming,

FIGURE 15-40 Good-quality knurling oil must be used when knurling guides. (Courtesy of Silver Seal Products Company, Inc.)

FIGURE 15-38 Valve guide knurling arbor. (Courtesy of Kwik-Way Manufacturing Company.)

FIGURE 15-39 Using a speed reducer and electric drill motor to drive the knurling arbor. (Courtesy of Red River Community College.)

clean the guide with a guide cleaning brush to remove any metal chips.

Using the second knurling method proceed as follows (Figures 15-41 to 15-44).

1. Place the drill guide in the spring side of the guide to be knurled. Using the $\frac{3}{16}$ in. drill and drill stop, drill a starting hole at the edge of the guide. The hole should be drilled $\frac{3}{16}$ in. deep. This hole provides a starting point for the knurling wheel.

2. Select the proper-size knurling arbor for the guide. Install the smallest-diameter knurling wheel into the arbor by placing one end under the staked marks in the arbor and the other end under the retainer spring. (Measure the knurling wheel with a micrometer to ensure starting with the smallest one first.)

3. Lubricate the knurling wheel with the special high-pressure knurling lubricant. Insert the arbor into the guide with the drive end up and the knurling wheel positioned in the drilled starting hole.

4. Install the drive shaft into the arbor. Place the speed reducer in the drill motor and tighten the chuck. Place the drill motor and speed reducer on the drive shaft. While holding the speed reducer, turn on the drill motor keeping slight downward pressure on the arbor as it passes through the guide.

5. Select the same-size reamer as the valve stem diameter and ream the guide using the drill motor and speed reducer to drive the reamer. Using the same-size reamer as the valve stem diameter provides the required clearance since this is built into the reamer. Clean the guide with a guide brush after reaming.

FIGURE 15-41 Guide knurling kit with speed reducer, knurling arbors, knurling wheels, reamers, and brushes. (Courtesy of Red River Community College.)

BEFORE

AFTER

FIGURE 15-43 Cross section of valve guide before knurling (left) and after knurling and reaming (right). (Courtesy of Hastings Manufacturing Company.)

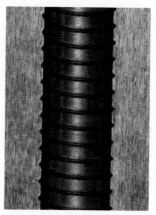

FIGURE 15-44 Cross section of roller knurled and single-stroke diamond-reamed valve guide using the Sunnen method. (Courtesy of Sunnen Products Company.)

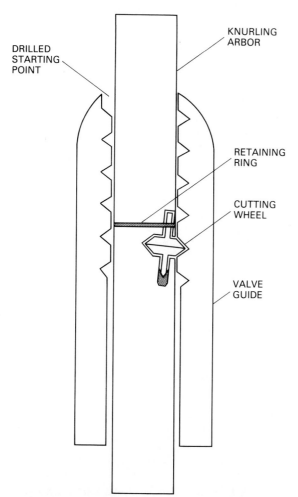

DRILLED STARTING POINT

KNURLING ARBOR

RETAINING RING

CUTTING WHEEL

VALVE GUIDE

FIGURE 15-42 Cross section of knurling arbor in valve guide. (Courtesy of FT Enterprises.)

6. If knurling with the smallest knurling wheel leaves the guide with too much clearance, use the next-larger-size knurling wheel and repeat steps 3 to 5.

Note: Positive-type valve stem seals should not be used with knurled guides which have 0.001 in. or less stem-to-guide clearance. Insufficient stem lubrication could result causing scuffing, scoring, and sticking of the valves. If positive valve stem seals are to be used, ream the guides to provide normal specified stem to guide clearance.

REPAIRING INTEGRAL GUIDES WITH THIN-WALLED INSERTS

Worn integral valve guides can be restored to original bore diameters by reboring and installing thin-walled guide inserts. The guide bore is enlarged by boring to a size slightly smaller than the OD of the guide insert to provide a press-fit. The new thin-walled guide inserts are then pressed into place. After installation the

FIGURE 15-45 Head shop shown here is capable of performing complete head reconditioning. (Courtesy of Sunnen Products Company.)

FIGURE 15-46 Thin-wall guide installing equipment. (Courtesy of Kwik-Way Manufacturing Company.)

FIGURE 15-47 Drilling fixture in place to prepare head for thin-wall guide installation. (Courtesy of Kwik-Way Manufacturing Company.)

guide insert must be reamed to the desired finish diameter to provide the specified stem-to-guide clearance. The entire process is done on specialized head reconditioning equipment as shown in Figures 15-45 to 15-47. This equipment is designed to ensure that the original guide bore centerline is maintained.

BRONZE GUIDE INSERTS

Bronze guide inserts can be used in badly worn guides to restore stem-to-guide clearance. Bronze inserts have the advantage of increased wear resistance, good lubrication, and heat dissipation. Reduced stem-to-guide clearance is also possible and can be as low as 0.0005 in.

The spiral bronze guide insert resembles a thread repair insert (Figure 15-48). The guide is tapped all the way through with a special tap. Next the bronze insert is screwed into place by hand with a special installing tool (Figure 15-49). The excess portion of the insert is unwound down to the guide and cut off, leaving a $\frac{1}{32}$ in. tailpiece. A bushing retainer is then installed over the guide to hold the insert in place (prevent it from turning) while reaming. The tail of the insert must engage one of the serrations of the holding tool. A special reaming oil is used to lubricate the guide bore. Using the broaching tool provided in the kit, dip the bulb of the broach in the reaming oil. Drive the broach completely through the guide with a soft hammer. Using the special drill motor, drive the lubricated reamer through the guide with slight downward pressure. Remove the bushing holder. Install the cutoff tool in the guide and cut the insert down flush with the guide. Clean the finished guide with a guide cleaning brush.

Honing Valve Guides

Valve guides may be honed to finish size instead of using a reamer. Honing produces an excellent surface finish to the guide bore, and it can be done to a closer tolerance than reaming. A special honing device with expanding stones driven by a slow-speed drill motor is used (about 350 rpm). A dial-type bore gauge is used to check bore diameter during the honing process (Figures 15-50 and 15-51).

FIGURE 15-48 Valve guide with cutaway section showing bronze insert in place at left. Bronze insert is shown on right. (Courtesy of Winona Van Norman Machine Company.)

← 1

← 2

FIGURE 15-49 Installing tool with bronze guide insert in place. Hold knurled handle (1) and turn thimble (2) counterclockwise until the upper tang breaks off. (Courtesy of Winona Van Norman Machine Company.)

FIGURE 15-50 Guide sizing using the Sunnen single-stroke sizing tool. (Courtesy of Sunnen Products Company.)

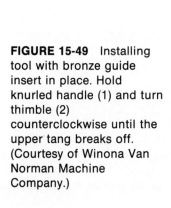

FIGURE 15-51 Checking finished bore size of valve guide. (Courtesy of Sunnen Products Company.)

HEAD SHOPS

Cylinder head reconditioning equipment called a head shop is designed to provide complete cylinder head reconditioning. Figure 15-45 is an example of this type of equipment. Many head shops are available with an air float table or working head that makes centering the work fast and easy. Once centered, the table or head locks in place. Jobs this type of equipment is normally capable of performing are shown in Figures 15-52 to 15-54 and listed here.

• Boring integral valve guides
• Reaming valve guides
• Cutting off damaged integral guides
• Installing valve guides
• Counterboring valve guides
• Machining integral guides for seals
• Knurling and reaming valve guides
• Machining for valve seat inserts
• Machining spring seats for larger OD springs
• Drilling, tapping, and facing for replacing press fit rocker studs with threaded studs

ROCKER ARM STUD SERVICE

On many pushrod engines, individually mounted rocker arms are mounted on studs that are screwed into the head or (on some older engines) are a press-fit in the head. There are three versions of these studs, as shown in Figure 15-55. The threaded type screws into a threaded hole in the head and is locked into place by a locknut or jamming taper. The conventional press-

FIGURE 15-52 Leveling (left), tapping for rocker studs (center), and drilling out broken studs (right). (Courtesy of Kwik-Way Manufacturing Company.)

FIGURE 15-53 Cutting off stud bosses for screw in studs (left), machining for positive valve stem seals (center), and machining for oversized OD valve spring seats (right). (Courtesy of Kwik-Way Manufacturing Company.)

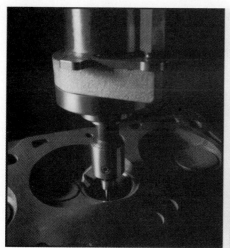

FIGURE 15-54 Machining all three valve seat angles at once (left), machining for valve seat insert (center), and drilling and reaming a guide (right). (Courtesy of Kwik-Way Manufacturing Company.)

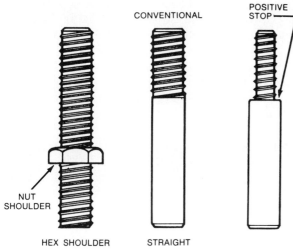

CONVENTIONAL

POSITIVE STOP

NUT SHOULDER

HEX SHOULDER

STRAIGHT

FIGURE 15-55 Three types of rocker arm studs. (Courtesy of Ford of Canada.)

fit stud is used on engines with the valve train adjustment provided by an interference thread-type adjusting nut. The positive-stop press-fit stud is used on engines with no valve train adjustment at the rocker arm.

Replacing Press-Fit Rocker Arm Studs

Rocker arm studs that show wear due to rocker arm contact, have damaged threads, are broken, or have started to pull out of the head must be replaced. Damaged studs can be replaced with new oversized press-fit studs or with screw-in studs. To replace press-fit studs, proceed as follows.

1. Measure stud height on a stud that has not started to pull out of the head. Record this

measurement and use it later to ensure that replacement stud height is the same.

2. Remove the faulty stud by one of the following methods.
   a. Use a stud removal tool (Figure 15-56).
   b. Slide several nuts onto the stud and place a flat washer on top of the nuts. This arrangement must be high enough to allow a nut screwed onto the stud to exert pressure against the washer and stacked nuts when the stud nut is turned down on the stud. Add washers if necessary and repeat until the stud is pulled out.

3. Ream the stud hole to the size specified for the oversized replacement stud. Oversized studs are usually 0.003 or 0.006 in. larger than standard. A 0.002- to 0.003-in. interfer-

FIGURE 15-56 Rocker arm stud puller. (Courtesy of Silver Seal Products Company, Inc.)

Rocker Arm Stud Service **257**

ence fit is required. Do not try to install an oversized replacement stud in a standard-sized hole since cracking may occur.

4. Coat the replacement stud with a locking compound and drive or press the new stud into the head to the specified depth.

Pinning Press-Fit Studs

Pinning press-fit studs prevents stud pullout. After stud installation the stud boss and stud are drilled to the size specified in the stud pinning kit. After drilling, a roll pin is installed flush with the stud boss (Figure 15-57).

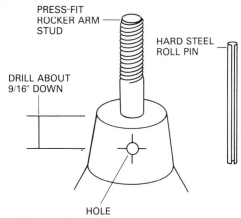

FIGURE 15-57 Pinning a press-fit rocker arm stud. (Courtesy of FT Enterprises.)

Screw-In Stud Replacement

Worn or damaged screw-in studs can be removed as follows.

1. Loosen the locknut (if so equipped).
2. Use a stud remover to remove the stud or use two nuts locked together on the stud to turn out the stud.
3. Chase the threads with a tap and remove all dirt and metal chips from the hole.
4. Apply a good hardening type of sealer to the replacement stud threads and screw the stud into the head to the specified depth. Install the locknut and tighten to specified torque. On tapered shoulder screw-in studs, simply tighten the stud to the specified torque to lock it into place.

Replacing Press-Fit Studs with Threaded Studs

Threaded rocker arm studs are available to replace press-fit studs. Screw-in studs prevent stud pullout. Drilling and tapping the stud holes is required (Figures 15-52 and 15-53). This operation must be done with careful attention to ensure maintaining the same stud centerline. An angled stud affects valve train action. For studs with a tapered locking shoulder, simply coat the stud threads with a hardening sealer, install, and tighten to specified torque. For studs with a locknut the stud boss must be milled flat an amount equal to the thickness of the locknut before installing the stud and locknut.

VALVE STEM SEAL FUNCTION AND DESIGN

Engine oil is pumped to the rocker arms and valve stem tips for lubrication. Rapid rocker arm and spring movement cause drain-off oil to be thrown around under the valve covers. Valve stem seals are used to prevent this oil from being drawn past the intake valve and into the combustion chamber on the intake stroke. They are also used on the exhaust valves to prevent oil from being drawn past the exhaust valve stem during the exhaust stroke. High-velocity exhaust gases rushing past the port end of the guide create a low-pressure area at the guide. Without exhaust valve stem seals, oil will be drawn into the exhaust stream. Several different methods are used to control oil flow past the valve stems as follows.

1. Tapered or chamfered valve guides at the spring end of the guide aid in draining oil away from the valve stem area (Figure 15-58).

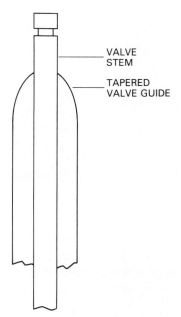

FIGURE 15-58 Tapered guide sheds oil away from valve stem. (Courtesy of FT Enterprises.)

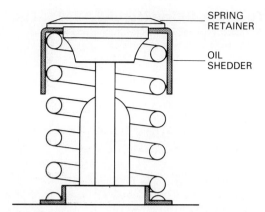

FIGURE 15-59 Cup-type oil shedder fits under spring retainer. (Courtesy of FT Enterprises).

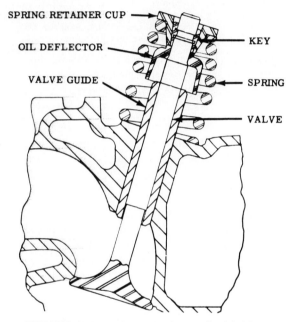

FIGURE 15-60 Umbrella-type oil shield on exhaust valve and positive seal on intake valve. (Courtesy of Chrysler Canada Ltd.)

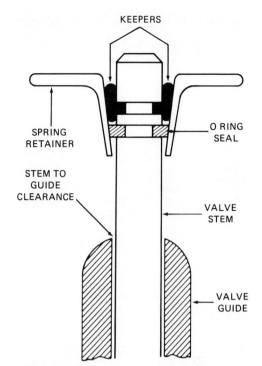

FIGURE 15-61 O-ring valve stem seal. (Courtesy of FT Enterprises.)

FIGURE 15-62 Positive valve stem seals: top, rubber; center, Teflon and rubber; bottom, Teflon. (Courtesy of Silver Seal Products Company, Inc.)

2. A metal cup oil shedder that fits under the valve spring retainer deflects oil away from the valve stem (Figure 15.59).

3. Umbrella-type rubber oil shedders that fit tightly on the valve stem and shed oil away from the stem. The umbrella moves up and down with the valve (Figure 15-60).

4. A rubber o-ring seal placed in a groove on the valve stem that creates a seal between the valve stem and spring retainer. This prevents oil from running down the valve stem into the guide (Figure 15-61).

5. A positive valve seal that fits tightly on top of the valve guide and seals against the valve stem. This type of seal provides the most effective oil control (Figures 15-62 and 15-63).

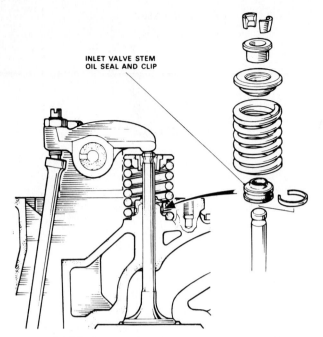

INLET VALVE STEM
OIL SEAL AND CLIP

FIGURE 15-63 Positive valve stem seal in place on guide. (Courtesy of Chrysler Canada Ltd.)

MACHINING GUIDES FOR POSITIVE TYPE SEALS

Several manufacturers produce positive valve stem seals, including Hastings Manufacturing Company, Dana Corporation (Perfect Circle), Sealed Power Corporation, and Raymond (Associated Spring Corporation). A variety of designs have been produced over the years. All of them require precise machining of the top of the valve guide. These seals are often used to control oil consumption and are effective even on slightly worn guides. However, proper stem-to-guide clearance should be restored, regardless of the type of seal used.

FIGURE 15-64 Tool for machining valve guides for positive seals. (Courtesy of Silver Seal Products Company, Inc.)

Guide machining is done with a special cutting tool (Figure 15-64), installed over a pilot placed in the valve guide. The cutting tool can be driven by a slow-speed electric drill motor, or it can be used in cylinder head reconditioning equipment as follows with the cylinder head in position (Figure 15-65).

1. To Check Guide Height

A. Install retainer on valve.
B. Measure from top of wear pattern to bottom of retainer (or variation of stem dia.) If this is less than 3/16" the guide height must be reduced the necessary amount to provide 3/16" clearance.

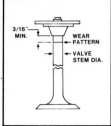

2. To Check Outside Diameter of Valve Guide

A. Measure valve guide O.D. if greater than specified above, guide must be machined.
B. Specified O.D. must extend at least 3/8" below top of guide. If less, machine until tool contacts top of guide.
C. Flat on top of guide must be at least 1/16". If less, machine to produce flat.
D. Guide must be machined if eccentricity exceeds .010" except when using the all rubber 2-ring valve seal.

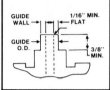

FIGURE 15-65 Instructions for machining valve guides for positive seals. (Courtesy of Silver Seal Products Company, Inc.)

1. Insert the machining tool in the chuck (Figure 15-53). Rotating speed should be about 350 to 450 rpm.
2. Lubricate the pilot on the cutting tool and insert it into the guide.
3. Switch on the motor drive and apply medium downward pressure to ensure that cutting takes place. Machine the guide to the specifications for the particular type of seal to be used. Do not allow the tool to bounce on the guide.
4. Repeat the process for the remaining guides.

Typical Positive-Seal Installation

Refer to Figures 15-66 and 15-67.

1. Place a plastic installation cap on the end of a valve stem. Lightly lubricate the cap. If the cap extends more than $\frac{1}{16}$ in. below the lower groove on the valve stem, remove the cap and cut off the excess length. (The cap prevents the sharp edges of the valve stem grooves from cutting the valve seal.)
2. Start the valve seal carefully on the cap over the valve stem. Hold your thumbs against the white seal insert to avoid dislodging it and push the seal down until the seal jacket touches the top of the valve guide. Remove the installation cap.
3. Use the installation tool to complete installation of the seal on the guide. If the tool is

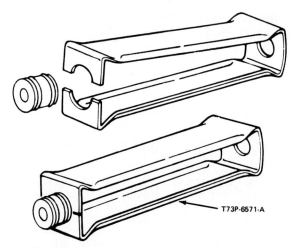

FIGURE 15-66 Installation tool for positive valve stem seals. (Courtesy of Ford of Canada.)

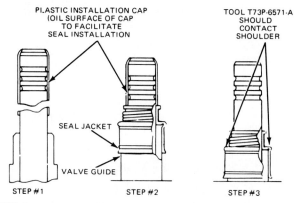

STEP #1 — WITH VALVES IN HEAD. PLACE PLASTIC INSTALLATION CAP OVER END OF VALVE STEM.
STEP #2 — START VALVE STEM SEAL CAREFULLY OVER CAP. PUSH SEAL DOWN UNTIL JACKET TOUCHES TOP OF GUIDE.
STEP #3 — REMOVE PLASTIC INSTALLATION CAP. USE INSTALLATION TOOL-T73P-6571-A OR SCREWDRIVERS TO BOTTOM SEAL ON VALVE GUIDE.

FIGURE 15-67 Installing positive valve stem seal. (Courtesy of Ford of Canada.)

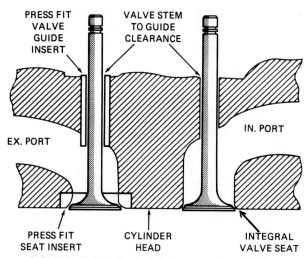

FIGURE 15-68 Typical integral and replaceable valve seats. (Courtesy of Prentice-Hall Inc., from Automotive Principles and Service by Thiessen and Dales.)

FIGURE 15-69 Valve seat inserts. (Courtesy of Chrysler Canada Ltd.)

not available, place two small screwdrivers about 90° from the gap in the lower retaining ring and push the ring down over the valve guide until the seal is flush with the top of the guide. (*CAUTION:* Do not push the seal down any farther than flush with the guide or the seal will open up.)

4. Install seals on the remaining valves.

VALVE SEAT INSERT SERVICE

Valve seats may be replaceable inserts or an integral part of the head (Figure 15-68). Damaged or worn seat inserts are replaced with new inserts. Integral seats

may be repaired by machining and installing seat repair inserts. Seat inserts may be hardened cast-iron alloy, chrome steel, or stellite material. Stellite seats are the hardest and most durable but also the most expensive. Stellite provides high corrosion resistance and a low wear rate (Figure 15-69).

Oversized valve seat inserts can be used in heads where original seat inserts have loosened. The head is machined to provide an interference (negative) fit and the oversized seat installed. Seat inserts are generally available in the following oversizes: 0.002, 0.005, 0.010, 0.015, 0.020, 0.030, 0.040, and 0.060 in. If a seat insert is to be installed to repair a damaged integral seat, the old seat area must be counterbored by a machining process. If a damaged seat insert is to be replaced, the old insert must be removed, the counterbore cleaned up and machined to oversize, and the new insert installed. The set up procedure for machining and the installation procedure is the same for both.

Removing a Valve Seat Insert

One way to remove a seat insert is to use a special curved pry bar. The tool tip is inserted to engage the lower edge of the seat insert while force is applied to the other end to pry the seat out. The seat may crack during this procedure, so eye protection should be used. Another method is to use a puller specially designed for seat insert removal as shown in Figure 15-70.

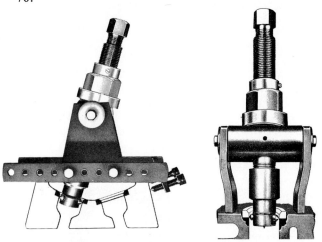

Angle Pull   Straight Pull

FIGURE 15-70 Removing a seat insert with a special puller. Tool is adjustable for angled (left) or straight pull (right). (Courtesy of Winona Van Norman Machine Company.)

Counterboring for a Valve Seat Insert

A variety of equipment types and designs is available for counterboring. Equipment ranges from hand operated devices to state of the art head reconditioning equipment with counterboring capability (Figures 15-71 and 15-72). The hand-operated equipment is relatively inexpensive and can be used by small shops; however, the procedure is much more time consuming than when it is done with more sophisticated equipment.

The diameter of the counterbore is determined from the insert diameter minus the interference fit required. Typical interference fit requirements for different insert sizes are given here.

Insert OD	Interference Fit
1–2	−0.002–0.004
2–3	−0.003–0.005
3–4	−0.004–0.006

Aluminum heads usually require an additional 0.001 to 0.002 in. interference fit. Although these are typical, the recommendations of the insert manufacturer should be followed, taking into consideration the

Valve Seat Insert Tool Versatility-Plus

Accurate Reliable — Rigid

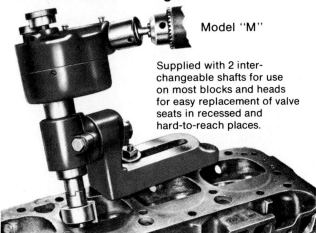

Model "M"

Supplied with 2 interchangeable shafts for use on most blocks and heads for easy replacement of valve seats in recessed and hard-to-reach places.

FIGURE 15-71 Valve seat insert counterboring tool. (Courtesy of Kwik-Way Manufacturing Company.)

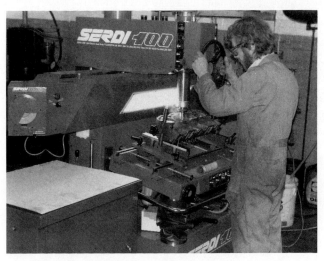

FIGURE 15-72 Counterboring for a valve seat insert on a cylinder head reconditioning machine. (Courtesy of Western Engine Ltd.)

insert material and design and whether they are to be installed into an aluminum or cast-iron head. The counterbore depth must also be considered, depending on the seat insert height dimension (Figure 15-73).

The following are the usual steps in counterboring for a valve seat insert.

1. Insert the pilot into the valve guide.
2. Level the head and lock it into position.

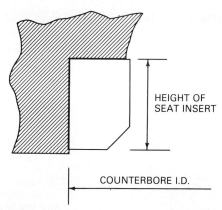

FIGURE 15-73 Dimensions required for counterboring head for seat insert. (Courtesy of FT Enterprises).

FIGURE 15-74 Close-up of cutter ready for counterboring. (Courtesy of Western Engine Ltd.)

3. Install the counterboring device into the drive head.
4. Adjust the cutters to the desired size to provide a negative fit for the seat insert (Figure 15-74).
5. Adjust the depth of cut. Use the seat insert as a gauge.
6. Cut the counterbore to the depth setting.
7. Remove all metal cuttings from the head and recheck the depth of cut.

Installing a Valve Seat Insert

1. Insert the proper-size pilot into the valve guide.
2. Shrink the seat insert in dry ice or by applying Freon from a spray can. (*Caution:* Avoid skin contact since instant freezing of the skin can occur. Use eye protection.) Shrinking the insert makes it easier to install.
3. If the insert is to be installed in an aluminum head, the head should be preheated to about 250°F to expand the counterbore.
4. Place the insert on the counterbore with the chamfered outer edge down. The chamfer aids installation and helps prevent shaving metal from the counterbore sides.
5. Place the insert driver onto the pilot.
6. With the insert centered over the counterbore drive or press it into place. Make sure that it bottoms in the counterbore. Remove the insert driver from the pilot (Figures 15-75 and 15-76).

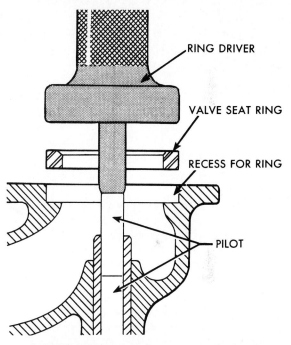

FIGURE 15-75 Installing a valve seat insert. (Courtesy of Ford of Canada.)

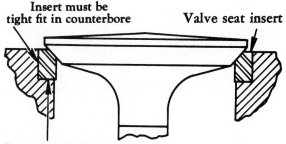

FIGURE 15-76 Conditions that must be met by properly installed seat insert. (Courtesy of Ford of Canada.)

7. If required, the seat insert may be staked, peened, or rolled in place. This prevents the insert from dropping out in case it loosens. A seat insert that drops into the cylinder causes severe piston, head, and valve damage. Place the rolling tool into the drive head of the head reconditioning equipment. Place enough downward pressure on the tool to roll the cylinder head metal over the edge of the seat insert. An alternative method is to peen metal over the edge of the insert (Figure 15-77).

8. After the insert is installed, the seat must be ground to the proper angle and seat width.

REPAIRING SPARK PLUG HOLE THREADS

Damaged threads in a spark plug hole can be repaired by using a special thread repair insert. This procedure is described in Chapter 6.

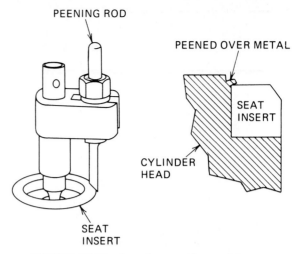

FIGURE 15-77 Peening tool is used to peen cylinder head metal over seat insert. (Courtesy of Sioux Tools Inc.)

REVIEW QUESTIONS

1. What two methods are used to resurface cylinder heads?
2. As material is removed from the cylinder heads or block on a V engine:
   (a) the heads move farther down when assembled
   (b) the heads move closer together when assembled
   (c) the intake ports are no longer aligned
   (d) all of the above
   (e) none of the above
3. Removing too much stock from the cylinder heads:
   (a) decreases the compression ratio
   (b) increases piston to valve clearance
   (c) raises the cam bearing bore centerline on OHC engines
   (d) all of the above
   (e) none of the above
4. Worn integral valve guides can be restored by:
   (a) reaming the guides and installing oversize stem valves
   (b) knurling
   (c) boring and sleeving
   (d) any of the above
5. What are the possible effects of too little valve stem-to-guide clearance?
6. Excessive stem-to-guide clearance causes:
   (a) poor valve seating
   (b) pistons to strike valves
   (c) valves to stick
   (d) none of the above
7. Why do exhaust valves require more stem-to-guide clearance?

8. After installing a new valve guide, it should be_____ to remove any high spots.
9. What is the purpose of beveling valve guides on the spring side?
10. Knurling is a process of cold metal_____.
11. The knurling tool is driven by an electric drill motor and _____.
12. After knurling, the guide must be_____.
13. True or false: A bronze guide insert resembles a thread repair insert.
14. Name three types of rocker arm studs.
15. How can a press-fit stud that has become loose in the head be repaired?
16. What method can be used to lock press-fit rocker arm studs to the head?
17. What is the purpose of valve stem seals?
18. Name three types of valve stem seals.
19. How can a badly worn integral valve seat be repaired?
20. Why should a plastic cap be placed on the valve stem before installing a positive stem seal?
21. How can a valve seat insert that has become loose in the head be repaired?
22. How should the size of the counterbore be determined when machining a head for a valve seat insert?
23. How can the valve seat insert diameter be reduced to aid in installation?
24. After installing a valve seat insert, it should be_____ and_____.

CHAPTER 16

Valve Train Components

OBJECTIVES

1. To develop an understanding of the function, design, and operation of valve train components.
2. To develop the ability to inspect, service, or replace valve train components, including rocker arms, pushrods, valve springs, retainers, and locks.

INTRODUCTION

Valve train components include the lifters, pushrods, rocker arms, valve springs, spring retainers, and valve locks. Because of the close relationship between the lifters and camshaft these are discussed separately in Chapter 11.

VALVE TRAIN REQUIREMENTS

The demands placed on the valve train are rigorous because of the speed at which these components operate, the rapid change of direction, the heat and friction involved, and the high operating loads imposed. The materials used in these components and their weight are critical design factors. Each valve in a four-stroke cycle engine is opened and closed 1500 times at an engine speed of 3000 rpm. This requires reversing the direction of travel of the pushrod, rocker arm spring, and retainer (Figure 16-1). If the valve train is

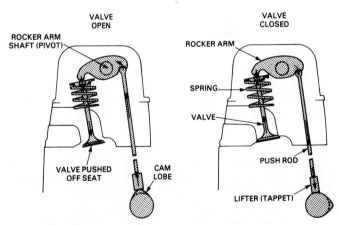

FIGURE 16-1 Valve train action in a pushrod engine. (Courtesy of Ford of Canada.)

too heavy or the valve springs are too weak, the valve train will not follow the cam lobe contour. This results in valve float and high impact seating of the valves and valve train with consequent rapid wear and possible breakage of parts. Valve train components are designed to meet these rigorous conditions.

PUSHROD FUNCTION AND DESIGN

Pushrod Function

Pushrods are designed to transmit camshaft action from the valve lifters to the rocker arms on overhead valve engines with in-block camshafts. Pushrods differ in design and material according to particular engine applications (Figure 16-2).

FIGURE 16-2 Common types of pushrods. (Courtesy of Sealed Power Corporation.)

Pushrod Design

Most engines use tubular pushrods because they are lighter and provide a convenient passage for oil flow to lubricate the rocker arms (Figure 16-3). Some engines using solid pushrods have other provision for rocker arm lubrication usually through a passage through the head and block. Tubular pushrods have steel inserts at each end. These inserts vary in size and shape as shown in Figures 16-2 and 16-4. Ends are usually spherical and convex or concave. The radius of the sphere may be the same at both ends on some pushrods but is different on others.

Pushrod Materials

Tubular pushrods are usually made from seamless steel tubing with hardened steel inserts welded to each end for wear resistance.

Pushrod Length

Pushrod length is determined by engine design and is usually fixed. Adjustable-length pushrods are available for some applications (Figure 16-5). Selective-length

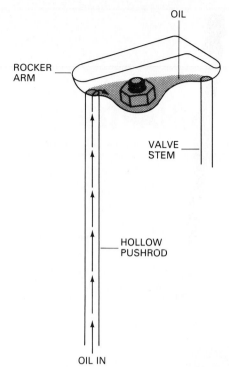

FIGURE 16-3 Tubular pushrod provides path for oil flow to lubricate rocker arm. (Courtesy of FT Enterprises).

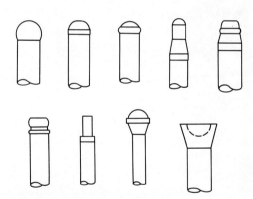

FIGURE 16-4 Pushrod end shapes vary considerably. (Courtesy of FT Enterprises).

pushrods are available for some big block FORD engines with nonadjustable valve trains. These are available in standard length, $\frac{1}{16}$ in. oversize and $\frac{1}{16}$ in. undersize.

Pushrod Guide Plate

Guide plates are used on some engines to keep pushrods in place (Figure 16-6). This prevents sideways movement of both the pushrod and rocker arm. On other engines pushrod alignment is maintained by the pushrod holes in the cylinder head.

FIGURE 16-5 Adjustable length pushrod. (Courtesy of Chrysler Canada Ltd.)

FIGURE 16-6 Pushrod guide plate. (Courtesy of Hastings Manufacturing Company.)

PUSHROD INSPECTION

Inspect both ends of the pushrods for wear, chipping, or ridging. Check the shank of the pushrod for contact wear from the guide plate or head. Check for straightness by rolling the pushrod by hand over a true flat surface. Maximum bend should not exceed 0.010 in. Replace damaged, worn, or bent pushrods. Do not straighten and use bent pushrods since they will bend again. Replace worn guide plates.

ROCKER ARM FUNCTION AND DESIGN

Rocker Arm Function

Rocker arms are designed to transmit and reverse pushrod movement. Upward movement of the pushrod causes downward movement of the valve. Rocker arms may also be designed to provide an increase in valve movement compared to pushrod movement. The difference is expressed as rocker arm ratio.

Rocker Arm Ratio

The rocker arm pivot is usually off-center toward the pushrod end. A rocker arm with the valve contact area $1\frac{1}{2}$ times farther away than the pushrod contact has a rocker arm ratio of 1.5:1 (Figure 16-7). This would provide a valve lift of 0.4125 in. with a camshaft lobe lift of 0.275 in. ($1.5 \times 0.275 = 0.4125$). Rocker arm ratios vary considerably with various makes and models of engines. Common ratios range from 1.5:1 to 1.75:1.

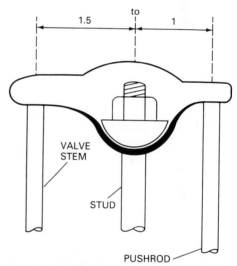

FIGURE 16-7 Rocker arm ratio. (Courtesy of FT Enterprises).

ROCKER ARM MATERIALS

Cast Rocker Arms

Cast rocker arms are made by pouring molten metal into a mold. After removal from the mold they are machined in the pivot area, the pushrod contact area, and the valve stem contact area. A valve lash adjustment screw is provided on some designs. In some cases bushings are used in the rocker arm pivot. Cast rocker arm materials include cast iron, die cast aluminum, and magnesium alloys. Wear inserts of harder material are often used at the valve stem end (Figures 16-8 and 16-9).

FIGURE 16-8 Cast iron rocker arm with oil groove and adjustment screw. (Courtesy of Sealed Power Corporation.)

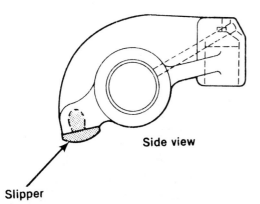

Side view

Slipper

FIGURE 16-9 Die-cast aluminum rocker arm with wear pad (slipper) at valve stem end and bore for hydraulic lash adjuster at pushrod end. (Courtesy of Chrysler Canada Ltd.)

FIGURE 16-10 Stamped steel rocker arms marked R for right and L for left. Right and left rocker arms are not interchangeable. (Courtesy of TRW, Inc., Automotive Aftermarket Division.)

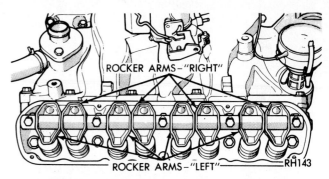

FIGURE 16-11 Right and left rocker arms are used on many engines. (Courtesy of Chrysler Canada Ltd.)

Stamped Steel Rocker Arms

Stamped steel rocker arms cost less to produce and are lighter than cast-iron rocker arms. Steel plate cut to shape is stamped into form with metal dies under very high pressure and controlled temperature. After stamping, wear surfaces are usually surface hardened (Figures 16-10 and 16-11).

ROCKER ARM MOUNTING METHODS

Rocker arms are mounted in a manner that allows them to pivot or "rock." Several methods are used, as follows.

Shaft-Mounted. All the earlier overhead valve automobile engines used shaft mounting of the rocker arms. Many of today's engines still use this method. Individual rocker arms are positioned on the shaft. Spacers and springs are often used to keep them in place on the shaft (Figure 16-12). Rocker arms are all the same design on some engines, while on others they are different for right- and left-hand positions for each cylinder. There may be only one shaft for each bank of cylinders, or there may be two—one for the intake valves and one for the exhaust valves. Rocker arm shafts are tubular and are drilled on one side to provide for lubrication of the rocker arms. The lubrication holes usually face down to lubricate the loaded side (Figure 16-13). Pedestal supports are used to mount the shafts on the head.

T-Bar-Mounted. Some engines use a support with a short stub shaft on each side to support two rocker arms (Figure 16-14). The support is bolted to the cylinder head. A plate with retainer clips keeps the rocker arms in place.

Stud-Mounted. Stud-mounted rocker arms have a ball pivot which slides over the stud and fits in

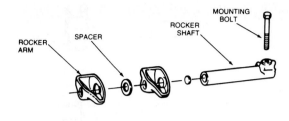

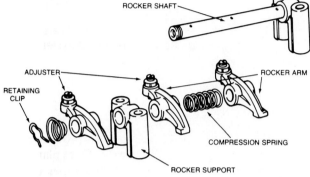

FIGURE 16-12 Examples of shaft-mounted rocker arms. (Courtesy of Chrysler Canada Ltd.)

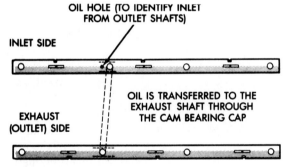

FIGURE 16-13 Tubular rocker shafts have a plug in each end and holes to provide rocker arm lubrication. (Courtesy of Chrysler Canada Ltd.)

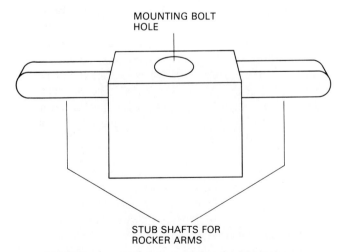

FIGURE 16-14 T bar is used on some engines to mount rocker shafts. (Courtesy of FT Enterprises.)

FIGURE 16-15 Stud-mounted rocker arm with ball pivot. Stud may be press-fit or threaded into head. (Courtesy of TRW, Inc., Automotive Aftermarket Division.)

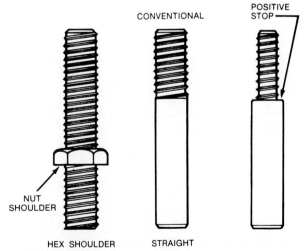

FIGURE 16-16 Three types of rocker arm studs. (Courtesy of Ford of Canada.)

a seat in the rocker arm. On others the ball pivot is integral with the rocker arm. A nut keeps the pivot and rocker arm in place (Figure 16-15). The rocker arm stud is either a press-fit in the head or it is screwed in. Some studs are designed with a shoulder or positive stop. The adjusting nut is tightened to specified torque against this stop. On other designs the threads on the nut are an interference fit and provide for valve lash adjustment (Figures 16-16 and 16-17).

Pedestal-Mounted. Pedestal-mounted rocker arms may be mounted singly or in pairs. The pedestal is cast into the head and the fulcrum is attached to it by a stud and nut or by a bolt. The fulcrum seats inside the rocker arm, holding it in place (Figures 16-18 and 16-19).

Overhead Camshaft Rocker Arms. Engines with overhead camshafts may have rocker arms, or the camshaft may act directly on the valves through

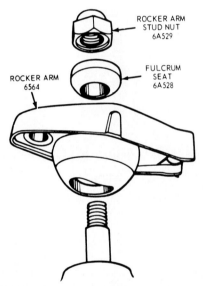

FIGURE 16-17 Positive-stop stud-mounted rocker arm with self-locking nut. (Courtesy of Ford of Canada.)

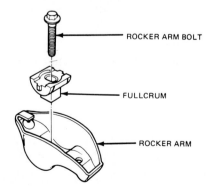

FIGURE 16-18 Pedestal-mounted rocker arm. Pedestal also acts as fulcrum. (Courtesy of Ford of Canada.)

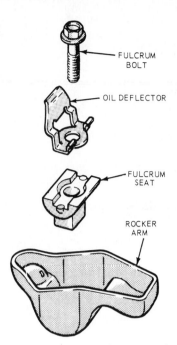

FIGURE 16-19 Pedestal-mounted rocker arm with oil deflector. (Courtesy of Ford of Canada.)

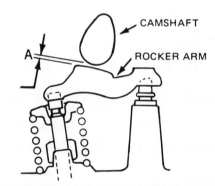

FIGURE 16-20 Mechanical OHC valve train and rocker arm. (Courtesy of Ford of Canada.)

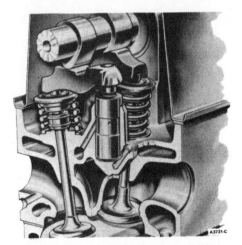

FIGURE 16-21 Hydraulic lifter OHC valve train and rocker arm. (Courtesy of Ford of Canada.)

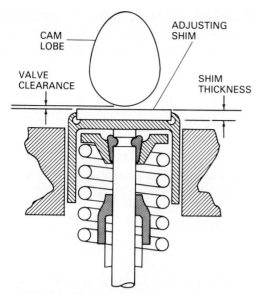

FIGURE 16-22 Direct-acting camshaft has no rocker arm. (Courtesy of FT Enterprises).

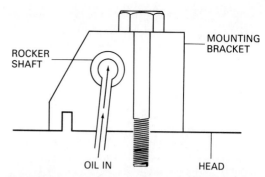

FIGURE 16-23 Drilled passage in head and rocker shaft mounting bracket feed oil to rocker shaft and rocker arms. (Courtesy of FT Enterprises.)

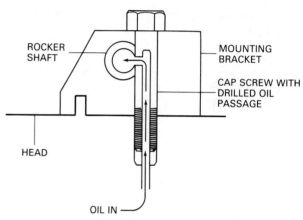

FIGURE 16-24 Drilled passage in head and cap screw feed oil to rocker shaft. (Courtesy of FT Enterprises.)

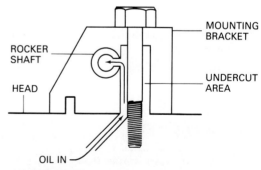

FIGURE 16-25 Undercut area in mounting bracket feeds oil from head to rocker shaft. (Courtesy of FT Enterprises.)

bucket-type cam followers. Rocker arms on OHC engines pivot at the lifter end, while the cam lobes act on a wear pad near the middle of the rocker arm. Retaining springs or clips are used to keep the rocker arms in place (Figures 16-20 and 16-22).

ROCKER ARM LUBRICATION

Rocker arms require lubrication to reduce friction and wear. Lubrication must occur at the valve stem end, the pivot, and the pushrod end. Rocker arm lubrication is provided in one of the following ways.

1. Through a hollow pushrod. Oil is pumped up to the rocker arm through a hollow pushrod and a hole in the rocker arm. Oil flows from the hole to the pivot and the valve stem tip (Figure 16-3).
2. Through a drilled passage in the block and cylinder head to a rocker shaft mounting bracket. The hole in the rocker shaft and mounting bracket must be aligned to provide oil flow (Figure 16-23).
3. Through a drilled passage in the block and a drilled cap screw that attaches a rocker shaft mounting bracket to the head. The specially drilled bolt must be installed in the position on the head, which has the oil passage for oil flow to occur (Figure 16-24).
4. Through an undercut area around the cap screw in the rocker shaft mounting bracket. The specially undercut mounting bracket

must be positioned on the head where the oil supply hole is located (Figure 16-25).
5. Through an oil supply line (tubing) that leads from an oil gallery in the block up through the head and is connected to the hollow rocker shaft (Figure 16-26).

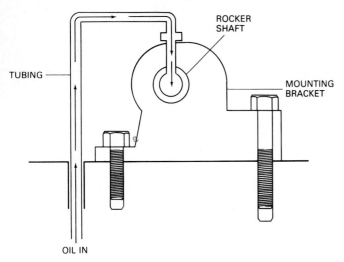

FIGURE 16-26 Oil supply line feeds oil from engine oil gallery to rocker shaft. (Courtesy of FT Enterprises.)

ROCKER ARM VALVE TRAIN ADJUSTMENT METHODS

Valve train adjustment is provided at the rocker arms on some engines. Several methods are used as follows.

1. An adjusting screw at the pushrod end of the rocker arm. The adjustment is maintained either by an interference type thread or by a locknut.
2. An adjusting nut on the rocker arm stud. An interference thread fit keeps the nut from losing its adjustment.
3. A positive stop rocker stud. The stud nut tightens against this stop on the stud, providing a fixed valve train adjustment. Selective-length pushrods are used to vary the adjustment.

See Chapter 19 for valve train adjustment.

ROCKER ARM SERVICE

Rocker Arm and Pivot Inspection

See Chapter 5 for detailed failure analysis of valve train parts. Inspect the rocker arms for wear or damage at the pushrod seat, the valve stem tip contact area, and the pivot area. Look for wear, cracks, chipping, and ridging. Check for galling in the pivot area. On stud-mounted rocker arms, inspect the pivot ball or pivot seat for similar wear or damage. On shaft-mounted systems inspect the lower side of the rocker shaft for galling, wear, scoring, or other obvious damage. Badly worn or damaged rocker arms, pivots, or shafts should

be replaced. Some cast types of rocker arms can be refaced at the valve stem end if otherwise in good condition. The tip radius must be maintained when refacing rocker arms.

It is important to remember that valve train components wear into each other during break-in. If these wear patterns are disturbed by interchanging of parts or by replacing some parts and not the others, rapid wear may occur between mismated wear surfaces. The life of these parts is greatest when the rocker arms, pushrods, and lifters are all replaced or reconditioned to provide well-matched wear surfaces.

Refacing Rocker Arms

The valve stem end of worn cast rocker arms can be ground to restore a good wear surface. An attachment on the valve grinding machine is used to hold the rocker arm in place while maintaining the tip radius as it is moved back and forth across the side of the grinding wheel (Figure 16-27). To reface a rocker arm, proceed as follows.

1. Dress the side of the stem grinding wheel on the valve refacer.
2. Install the rocker arm grinding fixture.
3. Mount the rocker arm on the fixture in a manner which will ensure that the wear surface will remain parallel.
4. Direct coolant flow to the grinding surface.
5. Slowly swing the fixture back and forth to grind the stem end of the rocker arm while applying light pressure to keep the rocker arm in contact with the stone. Maximum metal removal should not normally exceed 0.010 in., to avoid removing the hardened surface. If softer metal is exposed, rapid wear will take place.

Worn stamped steel rocker arms should be replaced.

FIGURE 16-27 Refacing a cast rocker arm. (Courtesy of Winona Van Norman Machine Company.)

Inspecting Rocker Arm Stud Nuts

The interference fit between the internal threads of the adjusting nut and the external threads on the stud prevent the nut from loosening or tightening during operation. This thread fit results in high turning effort. If this turning effort is reduced due to wear, the nut may self-adjust and must be replaced. The stud may also have to be replaced. Positive-stop stud nuts do not have the interference thread fit. These nuts should be inspected for thread condition and for damage or wear to the beveled positive stop surface. If the bevelled surface is more than $\frac{1}{16}$ in. wider than the original, it should be replaced. If it shows evidence of fracturing, it should be replaced.

VALVE SPRING FUNCTION AND MATERIALS

Valve Spring Function

The function of the valve spring assembly is to ensure that the valve train follows the cam lobe profile during both valve-opening and valve-closing action. Spring pressure tries to keep the valve closed. The rate of valve opening and closing is controlled by the shape of the cam lobe as it rotates. Once the valve train has accelerated to its maximum valve opening speed, its kinetic inertia is established. At very high engine speeds this kinetic inertia may be great enough to open the valve past its normal maximum-open position. This is called valve float. When this happens the pistons may strike the valves that are too far open and cause severe engine damage.

Valve spring pressures must be great enough to keep the valve train in contact with the cam lobe profile during opening and closing of the valves. If valve spring pressures are too low, valve float, valve bounce, and lifter pump-up may occur. If valve spring pressures are too high, rapid wear will take place on valve train components, valve seats, and valves.

Valve Spring Materials

Valve springs are wound from high-quality carbon steel, chrome vanadium, or chrome silicon steel, depending on load stresses and temperatures to be encountered. Valve springs are usually shot peened and stress relieved during manufacture to increase their fatigue life. Chrome vanadium and chrome silicon are normally used in high-performance and turbo-charged engines. The cross-sectional shape of the valve spring material may be round or ovate. A typical valve spring is shown in Figure 16-28.

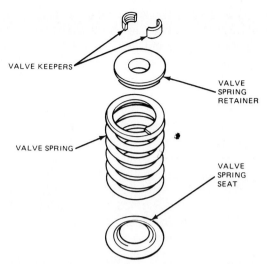

FIGURE 16-28 Valve spring and related parts. (Courtesy of Ford of Canada.)

VALVE SPRING VIBRATION CONTROL

The rapid compression and extension of valve springs during valve opening and closing sets up spring vibrations that must be controlled. The frequency and intensity of these vibrations, called spring surge, can result in spring breakage, rapid seat recession, rocker arm, and pushrod failure. Spring surge prevents proper spring action and allows valve train inertia to overcome spring tension causing valve float. Valve spring designs with uniform pitch and coil diameters are most subject to surge and require external surge control. Spring surge is controlled or dampened by one of the following methods.

Vibration Dampers. Vibration dampers control spring surge by friction. A reverse wound coil, wound from flat material fitted inside the spring with an interference fit, provides the friction to stop spring coil vibrations (Figure 16-29). A tubular steel insert positioned inside the valve spring is used in some applications as a damper.

FIGURE 16-29 Flat reverse-wound friction-type spring dampener inside valve spring. (Courtesy of Chrysler Canada Ltd.)

Dual Springs. A second smaller spring is placed inside the larger valve spring. The two springs, having coils of different size, have vibrations of different frequency and intensity when acting independently. Acting together, they ensure that at no time will the valve be unsprung. Dual springs of this type are designed to provide normal spring tension since excessive spring pressures result in rapid valve system wear and failure. The inner spring in the dual spring setup may be reverse wound or wound in the same direction as the outer spring (Figure 16-30).

Variable Pitch Springs. Variable pitch springs have unevenly spaced coils. The space between the coils is usually reduced at the base end of the spring. Unevenly spaced coils use different lengths of spring stock for each coil. This results in each coil having a different frequency and intensity of vibration. Spring surge at different engine speeds therefore occurs only in a small section of the spring, and never over the

entire length of the spring. Consequently, there is always some spring control over the valve (Figure 16-31).

Varying OD. Springs with progressively smaller-diameter coils toward the top maintain some spring control over the valve at different engine speeds since different coil ODs provide different frequencies and intensity of spring vibrations (Figure 16-31).

Color Coding of Valve Springs

Valve springs are often painted different colors to distinguish between different spring types. Colored springs with one or more colored stripes are also used. There is no uniform color coding of springs among manufacturers. Each manufacturer uses its own particular method of color coding. The most accurate method of spring identification is by the wire diameter, number of coils, coil diameter, free height, and spring rate (spring pressure at specified height).

Valve Spring Seats

The spring seat may be machined into the cylinder head or it may be a separate steel cup type of insert. The head may be machined flat or recessed to fit the spring. Recessing the seat prevents the spring from sliding across the seat surface. Aluminum cylinder heads require steel seat inserts to prevent wearing the head at the spring base.

VALVE SPRING INSPECTION AND TESTING

Valve springs should be inspected visually for etching, corrosion, pitting, base wear, and distortion. Damaged, worn, or distorted springs should be replaced. Valve springs should be measured for free length and squareness. Springs should all be within $\frac{1}{16}$ in. of each other and not more than $\frac{1}{16}$ in. out of height specifications (Figure 16-32). A spring that is shorter than specified has collapsed and should be replaced. Each spring should also be checked for squareness. An out-of-square spring will pull sideways on a valve, causing poor valve seating and excessive guide and stem wear. Springs between 2 and $2\frac{1}{2}$ in. in height should not be out of square more than $\frac{1}{16}$ in. (Figure 16-33).

Valve springs must have the proper spring tension to ensure proper functioning of the valves. Excessive spring tension increases cam lobe wear, valve train wear, and valve face and seat wear. Insufficient valve spring tension causes poor valve action and hydraulic lifter pump-up. Valve spring pressures are tested at two spring heights: the valve-closed (seated) spring height and the valve-open spring height. Ob-

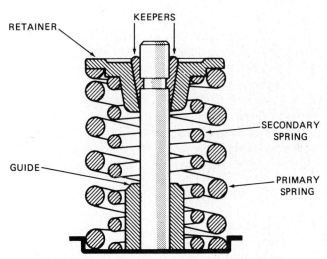

FIGURE 16-30 Reverse-wound secondary spring damper. (Courtesy of FT Enterprises.)

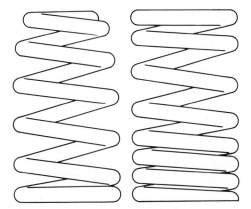

FIGURE 16-31 Variable-rate springs. (Courtesy of FT Enterprises).

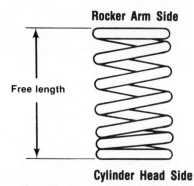

FIGURE 16-32 Free height of valve spring is spring height when it is relaxed. (Courtesy of Chrysler Canada Ltd.)

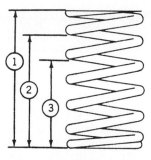

FIGURE 16-34 1, Valve spring free height; 2, valve-closed spring height; 3, valve-open spring height. (Courtesy of Ford of Canada.)

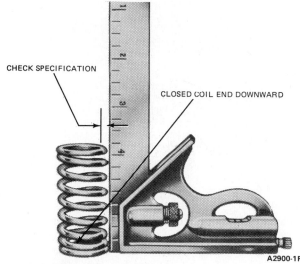

FIGURE 16-33 Checking spring squareness. (Courtesy of Ford of Canada.)

FIGURE 16-35 Testing valve spring pressure. (Courtesy of Ford of Canada.)

viously, the compressed height of a valve spring is less with the valve open as compared to its height when the valve is closed (Figure 16-34).

Before testing the springs, adjust the spring tester to read zero. Obtain spring pressure specifications from the appropriate service manual. To check the valve-closed spring pressure, place the spring on the tester base plate. Loosen the height-setting clamp. Turn the handle until the height pointer registers with the specified test height on the height scale. Tighten the height-setting clamp at this point. Test all remaining identical springs at this height. Repeat the tests for these springs at the valve-open spring height. Springs that fail the pressure tests must be replaced. In many cases intake and exhaust valve springs on a given engine are identical; however, this is not always the case. Be sure to abide by the specifications for both intake and exhaust valve springs (Figure 16-35).

Valve Spring Retainers

Valve spring retainers provide a flat square seating surface for the top of the valve spring. They also form part of the device that locks the valve stem to the spring retainer (Figure 16-28). Retainers for production engines are made from good-quality steel. Stepped retainers are used for dual spring installations (Figure 16-30). A conical section in the center of the retainer provides the wedging action under spring pressure that keeps the split locks in place. Lightweight aluminum alloy or titanium retainers are available for high-performance engines.

SPLIT LOCKS OR KEEPERS

Valve keepers or split locks have a tapered outer surface that wedges into the coned retainer. The inner surface usually has one or more ridges that fit into grooves in the valve stem. Spring tension keeps the valve, keepers, and spring retainer tightly locked together. The split lock keepers provide an easy method of assembly and disassembly of the valve and spring. Other keeper designs are also used, as shown in Figure 16-36.

A careful and thorough visual inspection of the keepers, spring retainer, and grooves in the valve stem must be made. Check for wear, nicks, and metal shear. Always replace split lock keepers in pairs. Combining a worn lock with a new one can cause the spring retainer to cock and break the valve stem at the keeper groove.

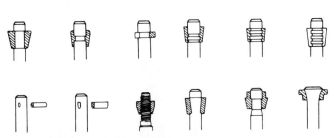

FIGURE 16-36 Various valve keeper designs. (Courtesy of Chrysler Canada Ltd.)

VALVE ROTATORS

Valve rotators are used to improve valve action and increase valve and seat life. During the transition from leaded fuel to unleaded fuel, some engines with softer valve and seat materials actually experienced increased valve and seat wear. The use of better valve and seat materials since then has reduced this problem. Other benefits derived from valve rotation include the following.

- More uniform valve head temperatures
- More uniform valve stem and tip wear
- More uniform valve stem lubrication
- Reduced valve stem deposits
- Better sealing between the valve face and seat
- Reduced valve burning and distortion

There are three types of valve rotators: (1) the rotovalve (also known as the free type or release type), (2) the rotocoil (coil spring type), and (3) the rotocap (ball and ramp type).

Rotovalve or Free-Type Rotator

The free-type rotator consists of a cup-type cap that fits over the valve stem tip. In operation the rocker arm pushes down on the cup. The cup pushes the keepers and the spring retainer down. This momentarily frees the valve of any spring pressure and allows it to turn in response to normal engine vibration. As the valve continues to open, the cup bottoms against the valve. At the maximum valve-open position, as the movement is reversed the valve is again momentarily free to turn (Figure 16-37).

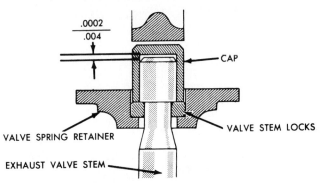

FIGURE 16-37 Free type valve rotator. (Courtesy of Ford of Canada.)

This type of rotator depends on proper clearance dimensions between the valve stem tip and the bottom of the cup. If clearance is excessive, high impact loads will be imposed on the assembly causing breakage. If clearance is insufficient, the valve will not release and rotation cannot occur. Clearance can be increased by removing material from the rim of the cup. When servicing this type of assembly, the best procedure is to replace the cups, keepers, and spring retainers.

Rotocoil

The rotocoil rotator imparts positive rotation to the valve. It consists of a coil spring inside the rotator body. When the valve opens, pressure applied to the sides of the spring causes its coils to tilt or flatten (Figure 16-38). Friction between the spring and rotator causes the retainer body, keepers, and valve to rotate each time the valve is opened. When the valve closes, the spring resumes its relaxed shape.

Rotocap

The rotocap rotator provides positive valve rotation every time the valve is opened. A series of spring-loaded balls and ramps and a belleville spring (dished washer-like spring) combine to cause the balls to move down and up the ramps as the valve is opened and

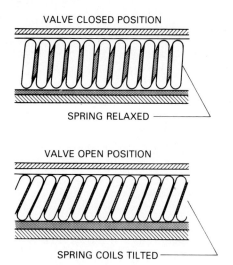

FIGURE 16-38 Rotocoil type valve rotator operation. (Courtesy of FT Enterprises).

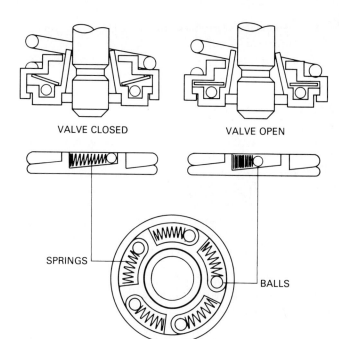

FIGURE 16-39 Rotocap valve rotator operation. (Courtesy of FT Enterprises).

closed. When the valve is opened the belleville spring flattens and pushes the balls down the ramps. This imparts rotation to the rotator body, keepers, and valve. When the valve closes, pressure from the belleville spring is removed from the balls, allowing the ball return springs to push the balls up the ramp again (Figure 16-39).

Rotator Replacement

Valve rotators should be replaced after approximately 100,000 miles (160,000 km). To check rotator operation

before disassembly, place a paint mark across the top and down on the collar of the rotator. Run the engine for a few minutes at 1500 rpm. If the rotator is working, the paint marks will no longer align. Movement may take place in either direction.

REVIEW QUESTIONS

1. At 3000 rpm each valve in a four-stroke cycle engine opens and closes:
   (a) 3000 times per minute
   (b) 300 times per minute
   (c) 1500 times per minute
   (d) 150 times per minute

2. Valve float can be caused by valves that are too light. Valve float can be caused by weak valve springs. Which statement is correct?
   (a) the first
   (b) the second
   (c) both are right
   (d) both are wrong

3. Valve float can cause:
   (a) high-impact valve seating
   (b) rapid valve train wear
   (c) both (a) and (b)
   (d) neither (a) nor (b)

4. What is the function of a pushrod?

5. Tubular pushrods are made from _____ steel tubing.

6. Pushrod alignment in the engine is maintained by _____ plates or holes in the _____.

7. The rocker arm pivot is usually off-center toward the (pushrod/valve) end.

8. A valve lift of 0.420 in. and a cam lobe lift of 0.280 in. means that the rocker arm ratio is:
   (a) 2:1
   (b) 2.5:1
   (c) 1.2:1
   (d) 1.5:1

9. Rocker arms are either _____ steel or cast.

10. Wear surfaces on steel rocker arms are usually _____.

11. Why are "rocker" arms so named?

12. How many rocker shafts are there on an in-line engine?

13. True or false: Some engines have left and right rocker arms while others have rocker arms that are all the same design.

14. On engines with T-bar-mounted rocker arms, how many T bars are required on an eight-cylinder engine?

15. Some rocker arm studs have an adjusting nut with interference fit _____.

16. True or false: Engines with overhead camshafts do not require pushrods but must have rocker arms.

17. List four methods of rocker arm lubrication.

18. Name three ways in which valve train adjustment is provided on different engines.

19. True or false: Worn stamped steel rocker arms should be refaced or replaced.

20. What is the function of valve springs?

21. Weak valve springs can cause:
    (a) valve float
    (b) valves opening farther than normal at high engine speed
    (c) hydraulic lifter pump-up
    (d) all of the above
    (e) none of the above

22. Valve spring surge is controlled by:
    (a) nonuniform spring coil diameters
    (b) dual springs
    (c) vibration dampers
    (d) any of the above
    (e) none of the above

23. What is valve spring surge?

24. True or false: Valve spring seats may be integral with the head or replaceable inserts.

25. Valve springs are normally tested on a spring tester for:
    (a) surge
    (b) vibration
    (c) etching
    (d) all of the above
    (e) none of the above

26. Valve springs should be checked for free _____ and for _____ with a square.

27. What are the effects of too much valve spring pressure?

28. What effect do weak valve springs have on engine operation?

29. Valve springs that are weak should be:
    (a) shimmed to increase the pressure
    (b) adjusted
    (c) replaced
    (d) aided by a secondary spring

30. Valve springs are held under compression by the spring _____ and valve _____.

31. List four benefits provided by the use of valve rotators.

32. Name three types of valve rotators.

33. How can valve rotator operation be checked?

CHAPTER 17

Valve and Seat Service

OBJECTIVES

1. To develop an understanding of the function, design, and operation of intake and exhaust valves and seats.
2. To develop the ability to perform the following service operations to meet industry standards:
   - Inspect the valve seats.
   - Grind the valve seats.
   - Narrow the valve seats.
   - Inspect and measure the valves.
   - Dress and chamfer the valve tip.
   - Grind the valves.
   - Measure and correct valve stem height.

INTAKE AND EXHAUST VALVES

Function and Design

The function of the intake and exhaust valves is to open and close the intake and exhaust ports in response to cam action. Typical valves and valve nomenclature are shown in Figure 17-1. A variety of valve keeper designs are shown in Figure 17-2. Intake and exhaust valves are designed to operate over a long period of time with relatively little attention and few problems. Valve heads may be domed, flat, dished, or recessed (Figure 17-3). Domed valves have the most heat capacity, while the dished and recessed valves have the most flexibility. Valve heads are sufficiently heavy for good heat capacity, yet light enough not to cause valve float very readily. Valve face and seat angles are usually 45°, while some are 30°. Seating angles are required to provide a positive seal. The wiping action of angled seating surfaces helps clear minor carbon particles that could prevent proper sealing. The wedging action of the angled surfaces also contributes to better sealing. Valves are made of high-grade steel alloy for long service life.

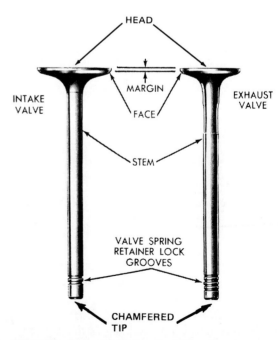

FIGURE 17-1 Intake and exhaust valves. (Courtesy of Chrysler Canada Ltd.)

FIGURE 17-2 Different types of keepers or valve locks. (Courtesy of Chrysler Canada Ltd.)

Production engines have been designed with two valves per cylinder for many years. Many are still being produced with only one intake and one exhaust valve per cylinder. In this design the exhaust valve is normally smaller in head diameter than the intake valve. Given the limited space in the combustion chamber, this is the best that can be done with the two-valve arrangement since the intake valve handles only slow-moving gases, while the exhaust gases are forced out by the fast-rising piston in the cylinder.

Since the ability of an engine is largely dependent on its ability to "breathe," any design change that enhances this ability will result in increased power. Any such design change, however, is always an increased cost of production. One such change is the addition of a turbocharger, which pushes more air into the cylinder. Another method is to use three or four valves per cylinder. As can be seen in Figure 17-4, the least-space-efficient arrangement is the two valve per cylinder design and the most space efficient is the four-valve design. One manufacturer has even produced a five valves per cylinder design.

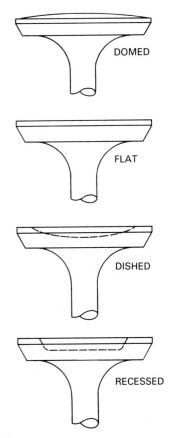

FIGURE 17-3 Valve head designs. (Courtesy of FT Enterprises).

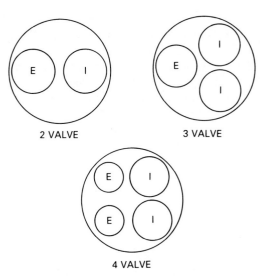

FIGURE 17-4 Two, three, and four valves per cylinder arrangements. (Courtesy of FT Enterprises.)

Materials

Intake and exhaust valve materials include alloy steel valves with an aluminized face and chrome stem, silchrome valve with an aluminized face, austenitic steel with aluminized face and chrome stem, and SAE 21-2 steel with a nickel-plated face as well as other alloys. Stellite is also a popular material for valves. On some engines, exhaust valves with sodium-filled stems are used for better valve cooling. At operating temperature, the sodium is liquefied. Valve movement causes the sodium to transfer heat from the head of the valve to the valve stem and then to the valve guide.

The four valves per cylinder arrangement has been used in competition engines for many years, for obvious reasons. This more expensive technology has also been used for many years in the engines of many higher-priced cars. Improvements in production technology and the desire to squeeze every bit of performance out of even the smaller four-cylinder engines have made this design more popular in standard production-line engines. The improved horsepower/weight ratio is a definite advantage. The three- and four-valve designs

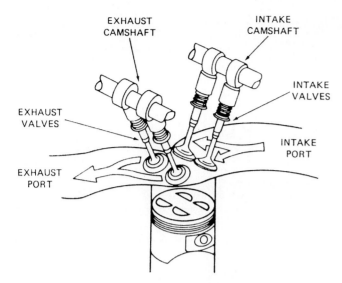

FIGURE 17-5 Four valves per cylinder with double overhead camshafts (DOHC). (Courtesy of Prentice-Hall, Inc., from *Automotive Principles and Service* by Thiessen and Dales.)

are used in a cross-flow arrangement using two cam-shafts (Figure 17-5). In the four-valve design one cam-shaft operates the intake valves and the other operates the exhaust valves. This results in more complex cam-shaft mounting and camshaft drive requirements. The four-valve design provides greater total valve and port capacity for any given cylinder bore size. Advantages include greater volumetric efficiency resulting from the increased flow rate, central spark plug location for better combustion, lower emissions, better performance, and increased fuel economy.

VALVE AND VALVE STEM DESIGN

Tapered Valve Stems

Some engines are equipped with valves that have tapered stems. The tapered stem provides increased stem-to-guide clearance near the valve head. This increased clearance reduces scuffing of the stem in this area. Usually, the stem is 0.001 in. (0.025 mm) less in diameter at the head end of the stem than it is at the tip end.

Canted Valves

Canted valves are positioned in the head at different angles. This allows changes in combustion chamber design and allows the use of valves with larger heads. The objectives in valve positioning include lower emissions, better engine breathing, and increased fuel economy (Figures 17-6 and 17-7).

Swirl-Finished Valves

A swirl-finished valve has a precisely dimensioned radius in the fillet area. A special grinding procedure is used to remove all surface irregularities. After grinding the valve is swirl polished, leaving a pattern of lines that circle the underside of the valve head and curve inward and down into the stem (Figure 17-8). The procedure aligns the swirl lines with the grain of the valve material and with the stress loads on the valve. This makes the valve much stronger.

Chrome Valve Stems

Hard chroming valve stems increases their ability to withstand higher temperatures and reduces valve stem wear. However, if stem lubrication is borderline, increased guide wear is possible with chromed valve stems.

Aluminized Valves

Aluminized valves have a dull gray unpolished appearance (Figure 17-9). This results from fusing an iron aluminum alloy material to the base metal of the valve, over which small particles of pure aluminum are fused. This material must not be removed. Removing the aluminized layer reduces the service life of the valve. The soft aluminum material improves valve sealing.

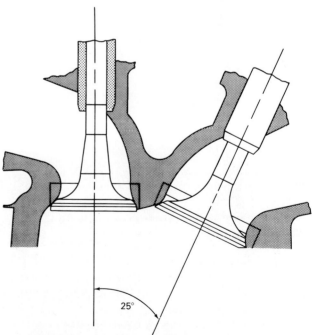

FIGURE 17-6 Intake valve is canted 25° in this valve arrangement. (Courtesy of FT Enterprises).

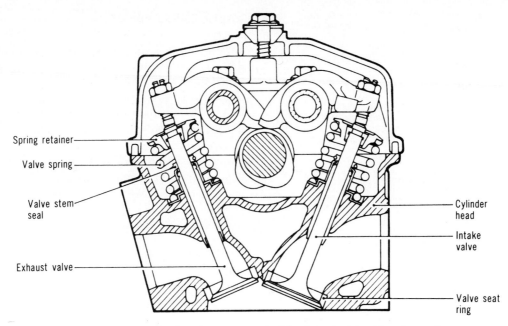

FIGURE 17-7 Both valves are canted in this design. (Courtesy of Chrysler Canada Ltd.)

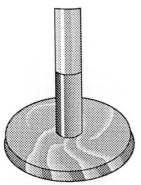

FIGURE 17-8 Swirl finished valve. (Courtesy of Chrysler Canada Ltd.)

FIGURE 17-9 Normal valve (left) and aluminized valve (right). (Courtesy of TRW, Inc., Automotive Aftermarket Division.)

Hardfacing Valves

To improve wear resistance and the effects of high temperatures, some valves are coated with a very hard metal (Figure 17-10). These metals include stellite, nichrome, and endurite. They are deposited on the valve by a welding process. In the trade, most of these valves are simply called stellite valves. Some valves are designed with a hardened steel tip welded to the valve stem to reduce wear (Figure 17-11).

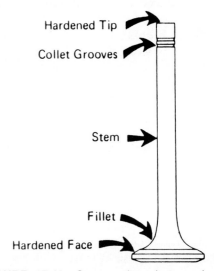

FIGURE 17-10 Some valves have a hardened face and tip. (Courtesy of Chrysler Canada Ltd.)

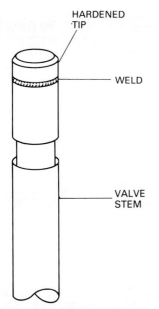

FIGURE 17-11 A hardened tip is sometimes welded to the valve stem end. (Courtesy of FT Enterprises.)

Sodium Valves

Sodium valves are designed with an oversized hollow stem that extends into the valve head (Figure 17-12). This space is half-filled with a metal called sodium. Sodium changes from a solid to a liquid at 208°F. After liquefying the sodium splashes back and forth in the valve (due to valve action), transferring heat from the valve head to the valve stem and guide. This reduces the operating temperature of the valve.

FIGURE 17-12 Sodium-filled valve. Note the extra-thick stem. (Courtesy of Sealed Power Corporation.)

VALVE TEMPERATURES

Exhaust valve temperature may reach approximately 1300 to 1500°F (704 to 815°C). This means that they are in fact running red hot while being blasted with hot exhaust gases with a blowtorch effect. Good heat transfer is therefore essential (Figure 17-13). It is important that valves be fully seated when they are closed. The exhaust valve is closed approximately two-thirds of the time while the engine is running. It is during this time that a large part of the heat transfer from the valve head to the seat takes place. The heat from the seat is transferred to the engine coolant. The remaining heat transfer takes place from the valve stem to the valve guide to the engine coolant (Figure 17-14). Intake valve temperatures are considerably lower than exhaust valve temperatures. Incoming air–fuel gases cool the intake valve while the valve is open.

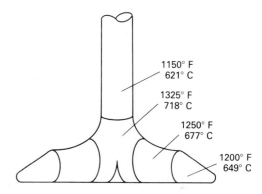

FIGURE 17-13 Typical exhaust valve temperatures. (Courtesy of FT Enterprises).

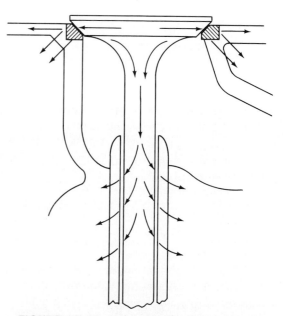

FIGURE 17-14 Heat transfer from valve to seat and guide. (Courtesy of FT Enterprises).

VALVE SEATS

The valve seat has a great deal to do with good heat transfer. A valve seat that is too narrow will not absorb sufficient heat from the valve head and will wear both seat and valve more rapidly. A seat that is too wide is not able to clear carbon particles as readily; it will, therefore, not seal as well and can cause valve seat burning. Many engines have integral valve seats that are induction hardened. Other engines have replaceable alloy-steel valve seat inserts pressed into machined recesses in the cylinder head. Integral valve seats are machined into the head casting. Aluminum cylinder heads have steel seat inserts since aluminum is unable to withstand the extreme conditions to which seats are subjected. Typical valve seat dimensions are shown in Figure 17-15.

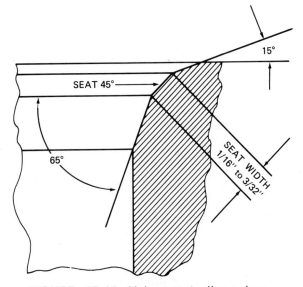

FIGURE 17-15 Valve seat dimensions. (Courtesy of FT Enterprises.)

Valve and Valve Seat Angles

Valve and seat angles are normally either 30 or 45°. The 30° angle is used on some engine intake valves and seats since it provides a slight increase in airflow at low valve lift. The 30° seat however has less wedging action and therefore seals less effectively. The 45° angle provides the greater overall benefits of wedging, such as better self-cleaning and sealing (Figure 17-16).

Seat Width

Seat width is an important factor in heat transfer and self-cleaning action. A seat that is too wide has greater heat transfer ability but less self-cleaning action and less ability to seal. A seat that is too narrow will seal well but has poor heat transfer capacity and will wear

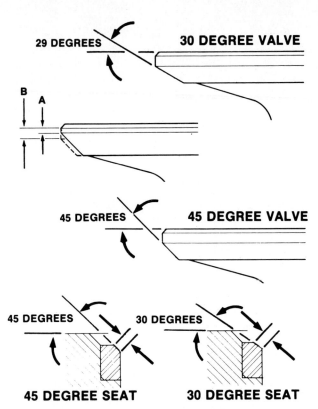

FIGURE 17-16 Common valve and seat angles. (Courtesy of Ford of Canada.)

more rapidly. In general, intake valve seats can be slightly narrower than exhaust seats since the exhaust valves run hotter than intake valves. Valve seat width specifications vary between different makes and models of engines; however, a general rule is that intake seat width should be from 0.060 to 0.080 in. and exhaust seat width from 0.080 to 0.100 in.

Interference Angle

A difference in the angle of the valve face and seat of from $\frac{1}{2}$ to $1\frac{1}{2}°$ is usually provided to improve the valve sealing. The difference in angles is provided on the valve face, which is $\frac{1}{2}$ to $1\frac{1}{2}°$ less than the seat angle

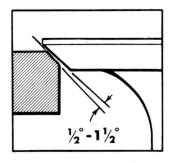

FIGURE 17-17 Interference angle. (Courtesy of Sioux Tools, Inc.)

(Figure 17-17). This ensures positive sealing at the combustion chamber side of the seat, where combustion pressures try to get between the valve and seat when the valve is closed. With no interference angle provided, the possibility of a negative interference angle (open on the combustion chamber side of the seat) exists, allowing combustion pressures to enter and unseat the valve (Figure 17-18).

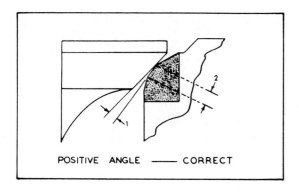

POSITIVE ANGLE —— CORRECT

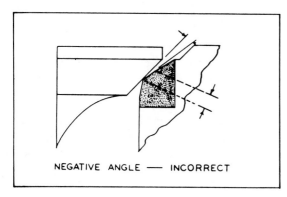

NEGATIVE ANGLE —— INCORRECT

FIGURE 17-18 Positive and negative interference angles. (Courtesy of Chrysler Canada Ltd.)

VALVE, SEAT, AND GUIDE WEAR

Normal Wear

Valves, seats, and guides will wear from normal friction, heat, and pressures. Valve stems and valve guides will wear, resulting in increased valve stem-to-guide clearance. Valve seats and valve faces will also eventually wear from continued opening and closing of the valves. Rocker arm action causes both rocker arms and valve stem tips to wear. The rocking action of the rocker arm also creates a slight side pressure on the valve stem, which contributes to valve stem and guide wear. By-products from normal combustion, such as carbon, acids, and moisture, also have a bearing on the wear of these parts.

Even in normal use, the intake and the exhaust valves absorb a fantastic amount of punishment. During a quick 15-minute trip to the shopping center, a single valve may open and close 10,000 times. Extreme temperatures scorch it many times each second. Violent explosions and powerful spring tension pound the red hot valve head. Hot gases under tremendous pressure swirl past it. Carbon deposits form on the face, preventing the valve from seating properly or cooling efficiently. As a result, the valves (particularly the exhaust valves) become pitted, burned, warped, and grooved. No longer concentric with the valve seat, they leak compression and fail to dissipate heat. Engine efficiency and economy nosedive.

The valve seat also wears. Hot gases burn it. Carbon particles that retain heat pit it. The valve guide wears in a corresponding position to the valve stem. Between stem and guide, carbon residues form which cause the valves to stick. To ensure top performance, the valve face, valve seat, and valve guide must be reconditioned. But not only the valve faces wear. The valve stem travels a mile or more in its guides during that short shopping trip. It wears at the top of the guides and at the bottom. Valve tips also wear and must be trued.

Abnormal Wear

The rate of wear on valves, valve seats, and valve guides will increase as a result of any abnormal conditions that may exist in the engine that affect valve operation. An overly rich air–fuel mixture can cause some dilution of lubricating oils, and increased wear will result. While it is true that in general a leaner mixture will increase combustion temperatures and therefore increase valve, seat, and guide wear and deterioration, it is also true that further leaning out of the mixture to 16:1 or leaner will decrease combustion temperatures considerably. Other reasons for excessive combustion temperatures include retarded valve timing, incorrect ignition timing, detonation, and preignition.

Valve timing becomes retarded as a result of valve train wear and excessive valve lash adjustment. Retarded ignition timing will cause increased combustion chamber temperatures. Excessively advanced ignition timing can cause detonation and preignition, both of which cause extremely high combustion temperatures and early piston and valve failure.

Carbon buildup in the combustion chamber can cause preignition and detonation. Excessive valve lash or excessive valve spring pressures can cause high-velocity valve seating and rapid valve and seat wear. Valves can stick in the guides as a result of bent valves, insufficient stem-to-guide clearance, incorrect cylinder head tightening, guide distortion, and varnish and carbon deposits.

CRACKED, BURNED, AND BROKEN VALVES

High engine speeds and loads, poor valve seating due to carbon deposits, warpage, misalignment or insufficient lash, preignition, and detonation are all factors that cause excessive valve temperature and cracked, burned, or broken valves. High temperatures, combined with combustion by-products (acids and moisture) and high-velocity gases, can sufficiently erode the valve stem to result in valve breakage.

Carbon deposits on the valves are a result of fuel residues and engine oil being burned. Engine oil can enter the combustion chamber past the valve stems if valve stem and guide wear are excessive. Engine oil can also get past worn cylinders and piston rings. Carbon deposits in the combustion chamber can cause preignition and detonation, which increase combustion and valve temperatures.

Poorly seated valves allow hot-burning gases to escape past the valve with a cutting-torch effect and can cause valves to be burned and cracked. Severely cracked valves can have pieces break away and cause piston and cylinder damage. A valve head that has separated from the stem also causes severe piston, cylinder, and cylinder head damage. Stem erosion and acid etching can cause this kind of valve breakage.

Excessive oil consumption can be caused by excessive valve stem-to-guide clearance or hardened and cracked valve stem seals. Excessive guide and valve stem wear will allow too much heat to get to the seals and cause them to harden and crack.

VALVE FAILURE ANALYSIS

For detailed treatment and illustrations of valve failures and their causes, refer to Chapter 5.

INSPECTING AND MEASURING VALVES

Valves should be inspected and measured for wear or damage as follows. See Figure 17-19 for valve inspection areas.

1. *Margin.* Minimum margin width after grinding should be $\frac{1}{32}$ in. A valve with too little margin has reduced heat capacity and will warp and burn.

2. *Stem wear.* Stem wear is measured with a micrometer in the area of contact with the valve guide. Maximum wear usually occurs at each end of this area. Maximum valve stem wear should be no more than one-half

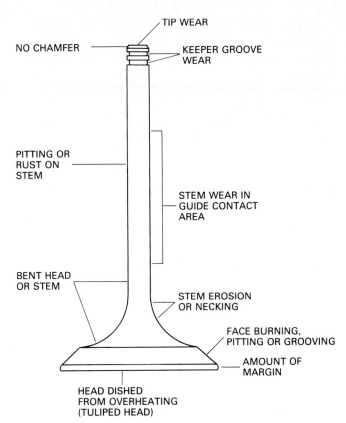

FIGURE 17-19 Valve inspection points. (Courtesy of FT Enterprises).

the maximum guide clearance specification given in the service manual.

3. *Bent valve.* A bent valve has a head that is tilted off square from the valve stem. Bend usually occurs in the area between the head and the guide contact area of the stem. Valve bend is most easily checked with the valve chucked in the valve refacer. Rotating the valve will reveal head wobble or runout if the valve is bent.

4. *Keeper groove wear.* Keeper grooves that are worn have the potential for valve breakage. Valves with worn keeper grooves should be replaced and new keepers used.

Valves that pass these checks can be reconditioned by dressing and chamfering the tip and grinding the valve face on valve grinding equipment shown in Figures 17-20 to 17-23.

VALVE GRINDING

The valve grinding procedure is generally similar for most types of valve grinding equipment. One of the major differences is in how the valve stem is held in the grinding chuck. Sioux machines use a coned shaft

FIGURE 17-20 Sioux valve refacer. (Courtesy of Sioux Tools, Inc.)

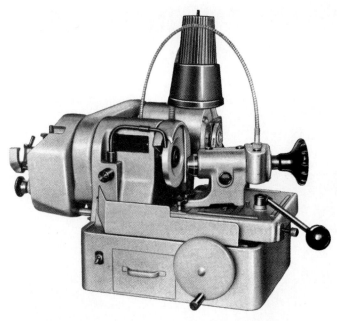

FIGURE 17-21 Van Norman valve refacer. (Courtesy of Winona Van Norman Machine Company.)

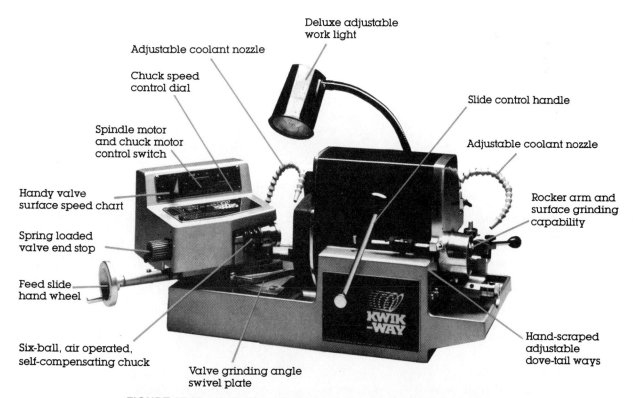

Deluxe adjustable work light

Adjustable coolant nozzle

Chuck speed control dial

Spindle motor and chuck motor control switch

Handy valve surface speed chart

Spring loaded valve end stop

Feed slide hand wheel

Six-ball, air operated, self-compensating chuck

Valve grinding angle swivel plate

Slide control handle

Adjustable coolant nozzle

Rocker arm and surface grinding capability

Hand-scraped adjustable dove-tail ways

FIGURE 17-22 Kwik-Way valve refacer. (Courtesy of Kwik-Way Manufacturing Company.)

FIGURE 17-23 Tobin Arp valve refacer. (Courtesy of Sunnen Products Company.)

FIGURE 17-25 Kwik-Way six-ball valve chuck. Hollow shaft accommodates different-length valve stems. (Courtesy of Kwik-Way Manufacturing Company.)

(Figure 17-25). Most valve grinding equipment features a quick-release chuck, which allows the valve to be quickly removed and installed without having to reset the chuck each time. Some valve grinders are equipped with two grinding wheels, while others have only one.

The following valve grinding procedure is typical for Sioux grinding equipment. The procedural steps required are the same, however, regardless of the type of equipment used. Always use eye protection.

Dressing the Grinding Wheel for Tip Grinding

1. Position the chuck carriage to the extreme left.
2. Adjust the diamond-tipped dresser in the at-

and three roller chuck to clamp the valve (Figure 17-24). The valve stem tip must be refaced and chamfered before chucking the valve to ensure proper valve alignment. Dressing and chamfering the tip should be done regardless of the type of grinding equipment used. Kwik-Way uses a six-ball chuck to clamp the valve

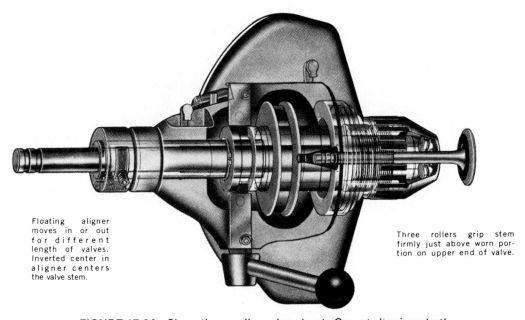

Floating aligner moves in or out for different length of valves. Inverted center in aligner centers the valve stem.

Three rollers grip stem firmly just above worn portion on upper end of valve.

FIGURE 17-24 Sioux three-roller valve chuck. Coned aligning shaft is adjustable for different length valve stems. (Courtesy of Sioux Tools, Inc.)

tachment on the right side of the machine to overhang the post about $\frac{3}{8}$ in.

3. With the machine not running, adjust the diamond dresser to just barely touch the side of the right-hand grinding wheel.

4. Start the machine, apply coolant, and move the diamond dresser slowly across the side face of the wheel. With the diamond dresser off the wheel, adjust it slightly (no more than 0.0005 in.) for a deeper cut. Make several passes across the stone. A rapid pass leaves a fairly rough surface on the wheel. A slow pass leaves a smoother surface and provides a smoother ground finish on the valve. The chuck motor should be switched off while dressing grinding wheels to reduce any vibrations.

Caution Regarding Refacing the Valve Tip

Grinding the valve stem tip reduces the installed valve stem height. Valve stem height specifications must be maintained to ensure proper valve operation. Detailed measuring procedures are described later. When stem height specifications are not available measure the installed stem height of each valve prior to grinding the valves and seats and record these measurements. Keep all the valves in order in relation to their position in the head. After grinding the valves and seats measure the stem height of each valve again. The difference in the before and after measurements is the amount that must be ground from the valve stem tip to restore the installed valve stem height. If the heads have been resurfaced an adjustment must be made to the stem height at a ratio of about 1.5 to 1. For example if .006 in. has been removed from the head the valve stem should be shortened by 0.009 in.

Excessive valve stem height may be caused by severe valve seat recession. In this case a new seat must be installed to reduce valve stem height.

Never remove too much metal from the valve tip since this may remove all the wear resistant heat treated tip material. A valve tip that is lower than the spring retainer can cause the rocker arm to strike the retainer allowing the valve keepers to be dislodged. The uncontrolled valve can be struck by the piston causing serious engine damage. Usually no more than about 0.030 in. should be removed from a previously unground stem tip.

Grinding the Valve Tip

1. With the machine shut off, clamp the valve stem securely in the holding fixture with the tip almost touching the grinding wheel.

2. Turn on the grinder motor. Slowly advance

the valve toward the wheel with the micrometer, and feed until the valve tip just touches the grinding wheel. Note the reading on the micrometer feed at that point.

3. Advance the micrometer feed 0.001 in. and move the valve across the grinder several times. Repeat until the valve tip is trued. Re-

FIGURE 17-26 Dressing the valve stem tip. One hand (top) is on the valve holding clamp and the other is on the micrometer feed. Note the diamond stone dresser just below the valve. (Courtesy of Sioux Tools, Inc.)

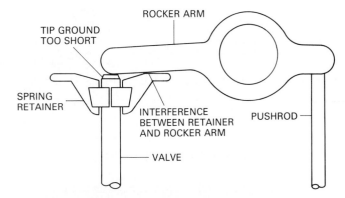

FIGURE 17-27 A valve stem tip ground too short may cause rocker arm to spring retainer interference. (Courtesy of FT Enterprises).

move as little metal as possible. When completed, note the reading on the micrometer feed once more. The difference in the two readings is the amount removed from the valve stem tip. Removing metal from the valve tip affects valve stem height. Removing too much may cause interference between the rocker arm and spring retainer (Figures 17-26 and 17-27).

Chamfering the Valve Tip

1. Clamp the V-type bracket securely in the holding fixture.
2. Turn on the grinder motor.
3. Place the valve stem in the V bracket and hold it in place by hand with the tip bottomed against the stop (Figure 17-28).

FIGURE 17-29 Dressing attachment in place, ready to dress the valve refacing stone. (Courtesy of Sioux Tools, Inc.)

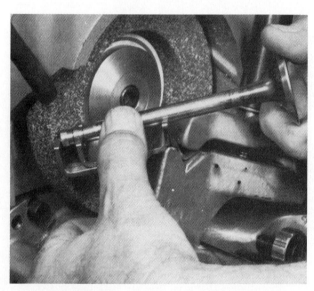

FIGURE 17-28 Chamfering the valve stem tip. The valve is being held down in the V bracket with one hand while the other hand is used to turn the valve while holding it against the stem stop. (Courtesy of Sioux Tools, Inc.)

4. Advance the feed until the valve just contacts the grinding wheel. Rotate the valve in the V bracket, keeping it bottomed against the stop. Repeat the procedure until the tip has a chamfer no more than $\frac{1}{32}$ in. wide.

Dressing the Valve Refacing Stone

1. Move the chuck carriage to the extreme left.
2. Adjust the diamond-tipped dressing screw in the dressing attachment to about $\frac{3}{8}$ in. overhang on the tip side of the post.

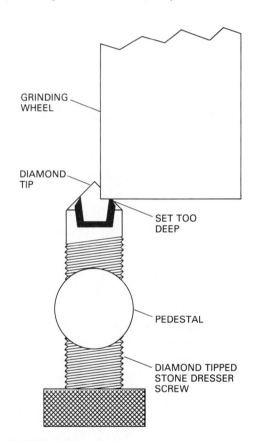

FIGURE 17-30 If set too deep, the grinding wheel will knock the diamond tip out of the stone dresser. (Courtesy of FT Enterprises).

3. Mount the attachment on the carriage table with the diamond dresser at a slight angle to the wheel face and about $\frac{1}{8}$ in. to the left of the wheel. Tighten the mounting nut securely (Figures 17-29 and 17-30).

4. Use the rubber shield to prevent grit from entering the chuck.

5. Adjust the grinder head to place the diamond tip in a position almost touching the wheel face.

6. Start the grinder motor and apply coolant.

7. Pass the diamond across the wheel face with cuts of no more than 0.0005 in. deep. The feed screw is graduated in 0.001-in. increments.

8. A fast pass produces a rougher stone face for fast grinding and grinding of hard-faced valves. (In some cases special grinding stones for hard-faced valves must be used.) A slower pass produces a smoother stone face for finish grinding of valves.

Setting the Grinding Angle

1. Loosen the chuck carriage angle holding clamp and pivot nut.

2. Move the carriage to index with the proper angle for the valve to be refaced. Take into consideration any interference angle required (44° for 45° valves and 29° for 30° valves). Follow the engine manufacturer's specifications for refacing angles.

3. Tighten the nut on the pivot and holding clamp to maintain the desired setting.

Chucking the Valve

1. Loosen the setscrew holding the depth setting shaft.

2. Adjust the chuck and insert the valve. Use the rubber shield on the valve to prevent grinding grit from entering the chuck.

3. With one hand on the valve and one on the shaft, slide them into position to provide clamping of the valve just below the fillet area on the unworn machined area of the stem (Figure 17-31). Tighten the lock screws on the depth-setting shaft collar. Once this depth adjustment has been made, no further depth adjustment is required for valves with the same stem length.

4. With the quick-release lever in the release position, insert the valve in the chuck. Adjust the chuck to clamp the valve, then back it off just enough to allow the valve to slide in and out freely. This setting can be left unchanged and the valve clamped and released just by using the quick-release lever. When chucking a valve, be sure to keep it firmly

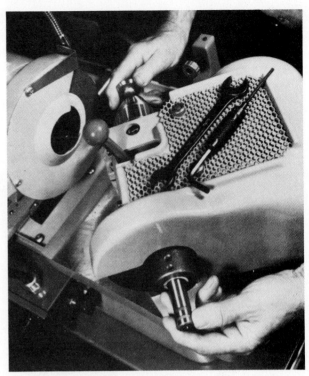

FIGURE 17-31 Adjusting the depth-setting shaft on a Sioux valve refacer. (Courtesy of Sioux Tools, Inc.)

pushed in the chuck while setting the release lever to clamp the valve.

Adjusting Coolant Flow

Coolant is required for cooling the valve and to flush away any grinding grit. Coolant flow should be directed at the valve face as much as possible. Proper application of coolant will also aid in producing a smoother surface on the valve face.

Valve Grinding Procedure

1. Move the chuck carriage to the extreme left and the grinder head to the rear.

2. Chuck the valve as outlined earlier.

3. Turn on the grinder motor and chuck motor.

4. Move the chuck carriage to the right to position the valve face in front of the grinding wheel. Adjust the travel stop to prevent moving the valve off the right side of the wheel. This prevents accidental grinding of the fillet. A valve with a nicked fillet must be replaced.

5. Advance the grinding head toward the valve until it nearly touches.

6. Turn on the coolant and adjust the nozzle to direct coolant at the valve face.

7. While moving the valve face across the face

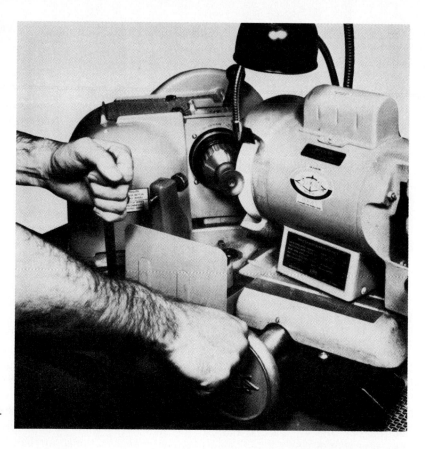

FIGURE 17-32 Grinding a valve.
(Courtesy of Sioux Tools, Inc.)

of the grinding wheel, advance the grinding wheel until a very light cut is made (Figure 17-32). Do not allow the valve face to move off the wheel face on either side. Doing so rounds off the edges of the stone and the valve face. Do not allow the edge of the grinding wheel to contact the valve fillet area. Even the slightest contact damages the valve and it must be replaced. This type of valve damage results in valve breakage if installed in an engine. The smaller the valve head diameter, the more critical the clearance in this area becomes. As valve grinding progresses, this clearance is reduced. Careful attention must be given to this clearance throughout the grinding procedure. During grinding, keep stroking the valve across the face of the grinding wheel with careful even movement across the entire width of the stone face each time. The grinding wheel must be dressed as necessary to provide an even grinding surface.

8. Advance the depth of cut very slowly, about 0.001 to 0.002 in. at a time while continuing the same stroking speed across the face of the stone. Advancing the depth of cut too much produces a rough finish. Do not remove any more metal than necessary to pro-duce a smooth finish over the entire face of the valve. Once grinding is completed, make a few more passes across the face of the stone until grinding is very light. This produces a better surface finish. Do not move the valve off the face of the stone after grinding is finished unless there is very little or no contact. It is safer to move the stone away from the valve instead of sliding the valve off the stone face. Once there is no more stone to valve contact, move the chuck carriage to the extreme left and remove the valve.

Make sure that the grinding wheel is backed off before chucking and grinding the next valve. If the grinding wheel is not backed off, severe gouging of the grinding wheel and valve will occur when the valve is moved into contact with the wheel.

Valve Inspection While Grinding

The valve can be inspected during grinding by backing the grinder away from the valve, then moving the chuck carriage to the extreme left. This automatically stops the valve from turning, to allow inspection. The valve must be ground until all pitting and blemishes are removed from the valve face, leaving a smoothly ground surface. After grinding, check that the valve has sufficient margin left to provide good heat capacity. There should be at least $\frac{1}{32}$ in. of margin left after grinding.

VALVE STEM HEIGHT

Grinding the valves and seats increases the height of the valve stem. Valve stem height is the distance from the spring seat to the valve tip with the valve seated (Figure 17-33). The increased valve stem height

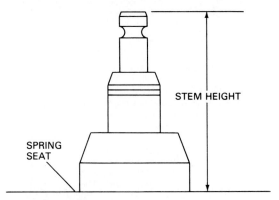

FIGURE 17-33 Valve stem height dimension. (Courtesy of Chrysler Canada Ltd.)

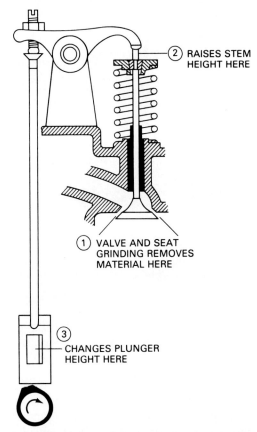

FIGURE 17-34 Effects of increased valve stem height on valve train. (Courtesy of FT Enterprises.)

changes the operating angle of the rocker arm. The valve end of the rocker arm is higher and the pushrod end is lower (Figure 17-34). The plunger in the hydraulic lifter therefore operates lower down in the lifter body. If the stem height is too high, the lifter plunger may bottom out and not allow the valve to close completely. To ensure that the lifter plunger operates in midrange on pushrod engines, the valve stem height must be checked and corrected if necessary. On OHC engines with hydraulic lash adjusters the effect is the same and valve stem height must be adjusted as well.

Although valve stem height specifications are not always provided in the service manual, this information is available from other sources. Manufacturers of valve stem height gauges can usually provide these specifications. One way to determine the correct stem height dimension is to measure the distance from the bottom of the spring retainer to the tip of the valve stem on a new valve. Add this figure to the valve spring installed height specification listed in the service manual. In some cases the proper valve stem height gauge establishes the range of stem height tolerance (Figures 17-35 and 17-36). A micrometer type of height gauge is shown in Figure 17-37.

If the valve stem height exceeds specifications, material must be ground from the stem tip. As a general rule, no more than 0.030 in. should be removed from the tip. If too much is ground off, all the hardened material will be removed and rapid wear will result. If the stem tip is too short, there may also be interference between the rocker arm and the spring retainer. During the stem height adjusting procedure, the position of

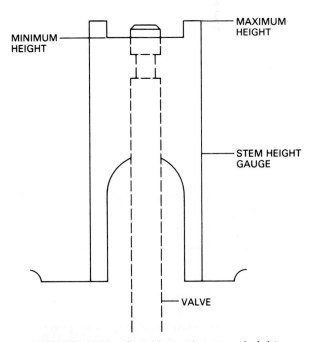

FIGURE 17-35 Checking valve stem height with stepped gauge. (Courtesy of Chrysler Canada Ltd.)

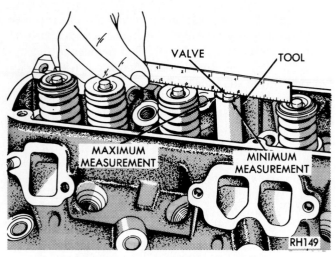

FIGURE 17-36 Stem height must fall between minimum and maximum levels on stepped gauge. (Courtesy of FT Enterprises.)

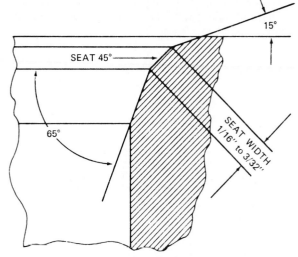

FIGURE 17-38 Valve seat reconditioning dimensions. (Courtesy of Chrysler Canada Ltd.)

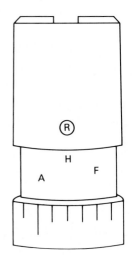

FIGURE 17-37 Micrometer-type stem height gauge. (Courtesy of FT Enterprises.)

the valves in the head must be maintained. Mixing the order of the valves changes the stem height dimension obtained for each valve.

VALVE SEAT RECONDITIONING

Before any seat reconditioning or grinding is done, the valve guides must be serviced. Since the guide is the base from which seats are ground, valve seat and guide concentricity is possible only if the guides are serviced first. See Chapter 15 for detailed procedures. Seat reconditioning includes refacing the seat and correcting seat width in a manner that will provide proper seat-to-valve face contact as well as correct seat width (Figure 17-38).

Special equipment is required to grind valve

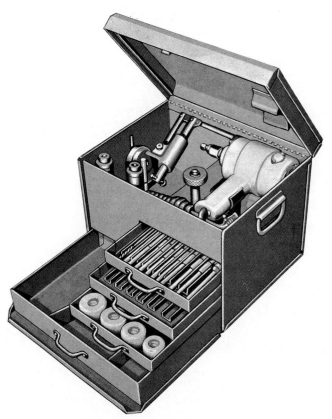

FIGURE 17-39 Valve seat grinding equipment includes driver, stone holders, stone dresser, and a selection of different-sized pilots and seat grinding stones. (Courtesy of Winona Van Norman Machine Company.)

seats. A high-speed driver (8000 to 12,000 rpm) is used to drive a grinding stone mounted on a pilot inserted tightly in the valve guide (Figures 17-39 and 17-40). Roughing stones, finishing stones, and special stones for induction-hardened or stellite seats are available. When seats only require a little grinding, only the fin-

FIGURE 17-40 Grinding valve seats using Sunnen equipment. (Courtesy of Sunnen Products Company.)

FIGURE 17-41 Expanding pilots. (Courtesy of Red River Community College.)

ishing stone is used. Seats that require more grinding may require the use of a roughing stone first and then a finishing stone to complete the job. Seat grinding stones are available in a number of diameters and cutting angles, to suit the various valve seat diameters and angles. Special diamond wheels are available that maintain their angle and do not need dressing. Three-angle seat milling equipment is used in another method of seat reconditioning. Valve seats (both integral and replaceable) that are worn down too far or are too badly damaged can be restored by installing new valve seat inserts. For details on valve seat repair and insert installation, see Chapter 15.

Valve Seat Grinding Pilots

Valve seat grinding pilots are required to support the seat grinder when grinding the valve seats. The pilot is inserted firmly into the guide and the seat grinder is then placed on the pilot. There are two basic types of pilots: tapered and expanding. The tapered pilot wedges tightly into the guide when properly installed. The expanding pilot is tapered at the top end and expands at the bottom (Figure 17-41). It is inserted tightly into the guide at the top and then expanded at the bottom by tightening the adjusting screw at the top. This locks the pilot into place. Valve seat pilots are available in a wide range of fractional inch, decimal inch, and metric sizes. Typical sizes are shown in Figures 17-42 and 17-43.

Seat Grinding Stones

Seat grinding stones or wheels are available in a variety of sizes and types (Figure 17-44). Different valve seat diameters require seat grinding stones of different diameters. Valve seat angles determine the stone angle required. The grit size required depends on the valve seat material to be ground.

Seat grinding stones have a threaded metal insert in the center for assembly to the stone holder. The hole diameter and thread type of the stone must match that of the stone holder. Common thread sizes are $\frac{1}{2}$ in. $\times$ 20, $\frac{9}{16}$ in. $\times$ 16 and $\frac{11}{16}$ in. $\times$ 16. Stone diameters range from 1 to 2 in. in $\frac{1}{16}$ in. steps. Stones over 2 in. in diameter are usually available in $\frac{1}{8}$ in. steps. The correct stone size to use is determined by the valve seat diameter and the clearance around the seat. The proper size stone should be $\frac{1}{16}$ to $\frac{1}{8}$ in. larger than the valve head diameter. The stone is usually either 30 or 45°, depending on the valve seat angle. For topping (overcutting) the valve seat a 15° stone should be used. For undercutting (throating) a 65° stone is required (Figure 17-45).

Seat grinding stones come in rough, medium, and fine or finish grit. Roughing stones are used to rough

Size	(in.)	Size	(in.)
$\frac{1}{4}$		0.380	
$\frac{1}{4}$	+0.001	0.385	
$\frac{1}{4}$	+0.002		
$\frac{1}{4}$	+0.003	$\frac{25}{64}$	
$\frac{1}{4}$	+0.004		
		0.395	
$\frac{9}{32}$		0.396	
$\frac{9}{32}$	+0.001	0.397	
$\frac{9}{32}$	+0.002	0.398	
$\frac{9}{32}$	+0.003	0.399	
$\frac{9}{32}$	+0.004		
		$\frac{13}{32}$	−0.003
$\frac{5}{16}$	−0.002	$\frac{13}{32}$	−0.002
$\frac{5}{16}$	−0.001	$\frac{13}{32}$	−0.001
$\frac{5}{16}$		$\frac{13}{32}$	
$\frac{5}{16}$	+0.001	$\frac{13}{32}$	+0.001
$\frac{5}{16}$	+0.002	$\frac{13}{32}$	+0.002
$\frac{5}{16}$	+0.003	$\frac{13}{32}$	+0.003
$\frac{5}{16}$	+0.004	$\frac{13}{32}$	+0.004
$\frac{21}{64}$		$\frac{7}{16}$	−0.003
		$\frac{7}{16}$	−0.002
$\frac{11}{32}$	−0.002	$\frac{7}{16}$	−0.001
$\frac{11}{32}$	−0.001	$\frac{7}{16}$	
$\frac{11}{32}$		$\frac{7}{16}$	+0.001
$\frac{11}{32}$	+0.001	$\frac{7}{16}$	+0.002
$\frac{11}{32}$	+0.002	$\frac{7}{16}$	+0.003
$\frac{11}{32}$	+0.003	$\frac{7}{16}$	+0.004
$\frac{11}{32}$	+0.004		
		$\frac{15}{32}$	
0.358			
		$\frac{1}{2}$	
$\frac{23}{64}$		$\frac{1}{2}$	+0.001
		$\frac{1}{2}$	+0.002
$\frac{3}{8}$	−0.003	$\frac{1}{2}$	+0.003
$\frac{3}{8}$	−0.002	$\frac{1}{2}$	+0.004
$\frac{3}{8}$	−0.001		
$\frac{3}{8}$		$\frac{17}{32}$	
$\frac{3}{8}$	+0.001	$\frac{17}{32}$	+0.001
$\frac{3}{8}$	+0.002	$\frac{17}{32}$	+0.002
$\frac{3}{8}$	+0.003	$\frac{17}{32}$	+0.003
$\frac{3}{8}$	+0.004	$\frac{17}{32}$	+0.004

FIGURE 17-42 Sioux inch-dimensioned pilot sizes. (Courtesy of Sioux Tools Inc.)

grind hard valve seats after which they are finish ground with a finishing stone. For stellite valve seats a stone with a special grit compound is required, after which they are finish ground. Finishing stones are used to grind cast-iron seats. Since cast iron is not very hard, a finishing stone is all that is normally required. Finishing stones are made of very fine grit and produce a good surface finish on the valve seat. Medium-grit stones are sometimes used as a compromise instead of using both a roughing and finishing stone. The success of this method depends on the type of seat material being ground. A finishing stone produces a smoother finish.

Diameter		Pilot No.	Diameter		Pilot No.
Inches	mm		Inches	mm	
.214	5.44	**VS-214**	.334	8.48	**VS-334**
.215	5.46	**VS-215**	.335	8.51	**VS-335**
.216	5.49	**VS-216**	.336	8.53	**VS-336**
.217	5.51	**VS-217**	.341	8.66	**VS-341**
.218	5.54	**VS-218**	.342	8.69	**VS-342**
.234	5.94	**VS-234**	.343	8.71	**VS-343**
.235	5.97	**VS-235**	.344	8.74	**VS-344**
.236	6.00	**VS-236**	.345	8.76	**VS-345**
.237	6.02	**VS-237**	.352	8.94	**VS-352**
.238	6.04	**VS-238**	.353	8.97	**VS-353**
.255	6.48	**VS-255**	.354	8.99	**VS-354**
.256	6.50	**VS-256**	.355	9.02	**VS-355**
.257	6.53	**VS-257**	.356	9.04	**VS-356**
.258	6.55	**VS-258**	.372	9.45	**VS-372**
.259	6.58	**VS-259**	.373	9.47	**VS-373**
.260	6.60	**VS-260**	.374	9.50	**VS-374**
.261	6.63	**VS-261**	.375	9.53	**VS-375**
.262	6.65	**VS-262**	.391	9.93	**VS-391**
.273	6.93	**VS-273**	.392	9.96	**VS-392**
.274	6.96	**VS-274**	.393	9.98	**VS-393**
.275	6.99	**VS-275**	.394	10.01	**VS-394**
.276	7.01	**VS-276**	.395	10.03	**VS-395**
.277	7.04	**VS-277**	.431	10.95	**VS-431**
.278	7.06	**VS-278**	.432	10.97	**VS-432**
.279	7.09	**VS-279**	.433	11.00	**VS-433**
.280	7.11	**VS-280**	.434	11.02	**VS-434**
.281	7.14	**VS-281**	.437	11.10	**VS-437**
.282	7.16	**VS-282**	.438	11.13	**VS-438**
.309	7.85	**VS-309**	.452	11.48	**VS-452**
.310	7.87	**VS-310**	.453	11.51	**VS-453**
.311	7.90	**VS-311**	.454	11.53	**VS-454**
.312	7.92	**VS-312**	.470	11.94	**VS-470**
.313	7.95	**VS-313**	.471	11.96	**VS-471**
.314	7.98	**VS-314**	.472	11.99	**VS-472**
.315	8.00	**VS-315**	.473	12.01	**VS-473**
.316	8.03	**VS-316**	.474	12.04	**VS-474**
.317	8.05	**VS-317**	.500	12.70	**VS-500**
.332	8.43	**VS-332**	.501	12.73	**VS-501**
.333	8.46	**VS-333**	.502	12.75	**VS-502**

VS-500, VS-501, and VS-502 pilots are 7½ in. in length. All other pilots are 6½ in. in length.

FIGURE 17-43 Sunnen pilot sizes. (Courtesy of Sunnen Products Company.)

Precision diamond grit grinding wheels are available which do not wear away like ordinary abrasives. These stones therefore do not require dressing (Figures 17-45 and 17-46). These grinding wheels are used with a special honing oil (Figure 17-40).

Stone Chatter

Stone chatter may be encountered when dressing a seat grinding stone or when grinding a valve seat. Stone chatter must be avoided to ensure accuracy in

FIGURE 17-44 Valve seat grinding stones for roughing and finishing. (Courtesy of Snap-On Tools of Canada Ltd.)

FIGURE 17-45 Installing a grinding wheel on a wheel holder. (Courtesy of Sunnen Products Company.)

seat grinding. If stone chatter is allowed, it can destroy a valve seat in seconds. Stone chatter may be caused by any of the following .

- Loose pilot
- Loosely mounted stone dressing fixture
- Dry pilot
- Dry drive connection
- Worn stone holder
- Worn drive connection
- Poorly dressed seat grinding stone
- Using the wrong stone for the seat being ground
- Side pressure on the seat grinder
- Incorrect stone pressure against the seat

To avoid chatter, the drive connection in the stone holder and the driver must be in good condition. The ID of the stone holder must not be worn excessively. The pilot must be tight in the guide. The stone dresser must be mounted securely. A drop of oil on the drive connection and on the pilot will reduce friction and sticking. The seat grinding stone must be dressed carefully and accurately. Select a hard seat grinding stone for hardened seats. Do not exert any side pressure on the seat grinder while grinding. Vary the downward pressure on the seat to establish the pressure required to avoid chatter.

FIGURE 17-46 Precision diamond grit grinding wheels. (Courtesy of Sunnen Products Company.)

Seat Grinding

To grind the seats, select the proper-diameter pilot and install it tightly in the valve guide (Figure 17-47). Select a seat grinding stone of the correct angle (same angle as the seat), usually 45 or 30° (Figure 17-48). The stone should be $\frac{1}{16}$ or $\frac{1}{8}$ in. larger in diameter than the valve head. Frequently dress the stone in the stone dressing fixture (Figure 17-49) during the seat grinding procedure, to ensure a good seat finish. Use eye protection, shield, and dust collector.

It is a good idea to wipe the seats clean by using a piece of fine emery cloth between the stone and the seat and giving it a good hard rub. This avoids contaminating the seat grinding stone with any oil or carbon residue that may be on the valve seat. Seat grinding stones should be handled in a manner that will keep

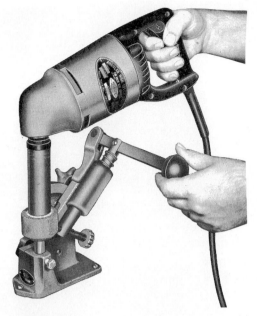

FIGURE 17-49 Stone dressing fixture with dust shield removed. (Courtesy of Sioux Tools, Inc.)

FIGURE 17-47 Installing a pilot in the valve guide. (Courtesy of Ford of Canada.)

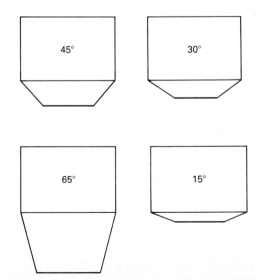

FIGURE 17-48 Common angles of seat grinding stones. (Courtesy of FT Enterprises).

FIGURE 17-50 Grinding a valve seat. Left hand under driver controls downward pressure of grinding wheel against seat. (Courtesy of Sioux Tools, Inc.)

them clean. Stones will soak up oil like a blotter. This causes them to become glazed and ineffective for seat grinding. Remove only as much material from the seat as is required to provide a good finish of sufficient width all the way around the seat. Avoid any side pressure during the grinding process.

Grind the seat with short bursts only, checking frequently to inspect progress. Pressure of the stone

FIGURE 17-51 Close-up of seat grinder and finished seat. (Courtesy of Jasper Engine and Transmission Exchange.)

against the seat must be precisely controlled to avoid chatter and to provide a good seat finish (Figures 17-50 and 17-51). Excessive pressure or chatter can destroy a valve seat very quickly. A drop of oil on the pilot and the star or hex drive can reduce the tendency to chatter. Avoid getting any oil on the seat or stone.

Note: If the valve seat is too wide after grinding, it must be narrowed to specifications, usually $\frac{1}{16}$ to $\frac{3}{32}$ in. (1.6 to 2.3 mm), with the exhaust seat being the wider for better heat transfer. Figures 17-52 and 17-53). This will require topping, throating, or both (Figure 17-54).

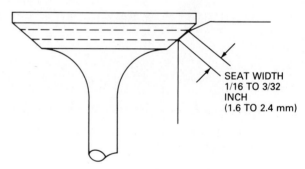

SEAT WIDTH
1/16 TO 3/32
INCH
(1.6 TO 2.4 mm)

FIGURE 17-53 Typical valve seat width dimensions. (Courtesy of FT Enterprises).

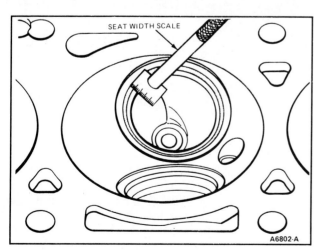

SEAT WIDTH SCALE

A6802-A

FIGURE 17-52 Measuring the width of a seat valve. (Courtesy of Ford of Canada.)

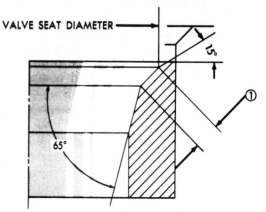

VALVE SEAT DIAMETER

15°

65°

①

FIGURE 17-54 Topping (overcutting) a valve seat is usually done with a 15° stone. Undercutting (throating) is done with a 65° stone. (Courtesy of Chrysler Canada Ltd.)

Three-Angle Seat Grinding

Three-angle seat grinding can be done with a seat milling tool that cuts all three angles at once or it can be done with seat grinding stones by grinding each angle separately. Figure 17-55 illustrates this procedure using seat grinding equipment. Figures 17-56 and 17-57 show milling a three-angle seat in a single operation. The objective in three-angle seat grinding is twofold: (1) The seat should be the correct width, and (2) it should contact the center of the valve face. To determine whether topping (overcutting) or throating (undercutting) is required to narrow the valve seat, a new or reconditioned valve must be used.

Overcutting is done in the same manner as seat grinding, except that a 15° stone is used for the pur-

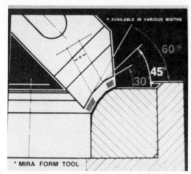

FIGURE 17-57 Close-up of three-angle seat milling. Cutters with different angles are available. (Courtesy of Peterson Machine Tool, Inc.)

pose. This narrows the seat from the combustion chamber side and lowers the point of contact on the valve face (farther from the margin). Undercutting is done similarly with a 65° stone, which narrows the seat from the port side.

VALVE SEAT-TO-FACE CONTACT AND CONCENTRICITY

To determine where the seat contacts the valve face, use a new or reconditioned valve. Mark the valve face with a series of pencil marks across the face of the valve all around the valve. Insert the valve in the guide, press down on the valve, and turn it one-quarter turn and back. Remove the valve and check the pencil marks. The pencil marks will be wiped out at the point where the seat contacts the valve. (Because of the interference angle, only the edge of the seat on the combustion chamber side will wipe out the pencil marks.) This should be about one-third of the way down on the face of the valve away from the margin to center the seat on the valve face (Figure 17-58). Turning the valve only one-quarter turn while in contact with the seat provides the means for checking whether or not the seat is concentric. If the pencil marks are wiped out at one point all the way around the valve, seat con-

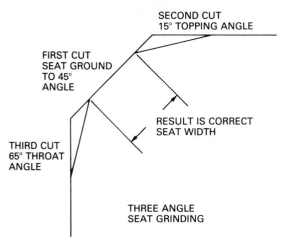

FIGURE 17-55 Typical seat narrowing and three-angle valve grinding procedure. (Courtesy of FT Enterprises).

FIGURE 17-56 A single-point carbide cutter used in a head reconditioning machine cuts all three seat angles in a single operation. (Courtesy of Peterson Machine Tool, Inc.)

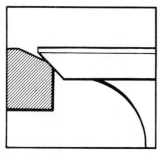

FIGURE 17-58 The valve seat should contact the middle of the valve face. (Courtesy of Sioux Tools, Inc.)

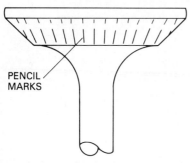

FIGURE 17-59 Valve face-to-seat contact is indicated at the point where pencil marks across valve face are wiped out. (Courtesy of FT Enterprises.)

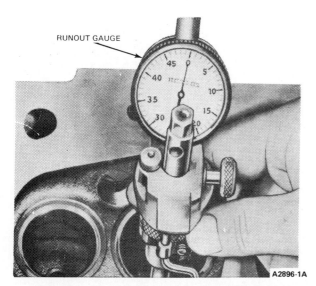

FIGURE 17-60 Dial gauge for checking valve seat concentricity. (Courtesy of Ford of Canada.)

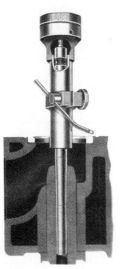

FIGURE 17-61 Cross section of concentricity gauge installed on pilot. (Courtesy of Sioux Tools, Inc.)

centricity is within limits (Figure 17-59). A dial-type seat concentricity gauge may be used to check seat runout (Figures 17-60 to 17-62). Prussian blue may also be used to check seat contact and concentricity (Figure 17-63). If the guide, seat, and valve have all been properly reconditioned, they will be concentric (centered in relation to each other). Concentricity is required to provide a good seal between the valve and seat.

FIGURE 17-62 Sunnen valve seat runout gauge. (Courtesy of Sunnen Products Company.)

FIGURE 17-63 Using Prussian blue to check valve seat-to-face contact. (Courtesy of TRW, Inc., Automotive Aftermarket Division.)

REVIEW QUESTIONS

1. What is the function of the intake and exhaust valves in the engine?

2. Valve head designs include:
   (a) dished, domed, crowned, and convex
   (b) convex, concave, curved, and crowned
   (c) domed, dished, flat, and recessed
   (d) flat, recessed, crowned, and convex

3. Intake and exhaust valve materials include:
   (a) cast iron, cast aluminum, steel, and chrome
   (b) cast steel, chrome, iron, and aluminum
   (c) alloy steel, chrome, nickel, and stellite
   (d) stellite, chrome, cast iron, and steel

4. The sodium in sodium filled valves:
   (a) liquefies at operating temperature
   (b) solidifies to strengthen the valve
   (c) bonds the valve metals through a fusion process
   (d) reduces carbon buildup

5. Why do some engines have four valves per cylinder? List four advantages.

6. Why is the intake valve usually larger than the exhaust valve in an engine with two valves per cylinder?

7. What is the reason behind the tapered valve stem design?

8. What are the objectives in the canted valve design?

9. What benefit is derived from swirl-finished valves?

10. Chrome surfacing of valve stems:
    (a) reduces stem wear
    (b) reduces stem warpage
    (c) reduces valve leakage
    (d) reduces blowby

11. Exhaust valve temperatures can reach:
    (a) 1400°F
    (b) the red-hot stage
    (c) both (a) and (b)
    (d) neither (a) nor (b)

12. What effect does the valve seat have on the temperature of a valve?

13. Common valve seat angles include:
    (a) 35° and 45°
    (b) 30° and 60°
    (c) 35° and 65°
    (d) 30° and 45°

14. Valve seat width is an important factor in:
    (a) heat transfer from the valve
    (b) the ability of the valve and seat to self-clean
    (c) both (a) and (b)
    (d) neither (a) nor (b)

15. The width of an exhaust valve seat should be from:
    (a) 0.80 to 0.100 in.
    (b) 0.08 to 0.010 in.
    (c) 0.0008 to 0.001 in.
    (d) none of the above

16. What is an interference angle between the valve and seat?

17. True or false: The interference angle should be negative.

18. What causes valve seat wear?

19. List five causes of excessive valve temperatures.

20. The valve margin should not be less than _____.

21. Valve stem wear should not exceed _____.

22. How deep a cut should be made when diamond dressing a valve refacing stone?

23. What effect does a rapid pass across the stone face (while dressing) have on the refacing stone?

24. Why should you not remove too much metal from the valve stem tip?

25. How wide should the chamfer be on a valve stem tip?

26. Where should coolant flow be directed when grinding a valve?

27. Why must the valve not be removed off the edge of the grinding stone while grinding?

28. What should be done with a valve that has been accidentally nicked in the fillet area during grinding?

29. How much should the depth of cut be advanced each time while grinding a valve?

30. What effect does grinding the valves and seats have on valve stem height?

31. Define the term "valve stem height."

32. How can excessive valve stem height be corrected?

33. What effect does increased valve stem height have on hydraulic valve lifter operation?

34. How can the correct valve stem height be established when specifications are not available?

35. The valve _____ must be serviced before grinding the valve seats.

36. Name the two basic types of valve seat pilots.

37. Seat grinding stones are available in what three grit types?

38. The threads in the seat grinding stone must match those of the stone _____.

39. The seat grinding stone turns at a speed of:
    (a) 2000 rpm
    (b) 200 rpm
    (c) 1000 rpm
    (d) 10,000 rpm

40. Precision diamond seat grinding stones do not require _____.

41. List six reasons why stone chatter can occur during seat grinding.

42. A valve seat that is too wide can be narrowed by using a _____° stone for topping and a _____° stone for throating.

43. On what part of the valve face should the valve seat make contact?

44. How can valve seat to face contact be determined?

45. Why is valve seat concentricity important?

46. How can valve seat concentricity be checked?

CHAPTER 18

Cylinder Head Assembly and Installation

OBJECTIVES

To develop the ability to assemble and install cylinder head to meet manufacturer's specifications, including the following operations:

1. Install core plugs.
2. Measure spring installed height.
3. Correct spring installed height.
4. Install valves, stem seals, springs, and retainers.
5. Leak test installed valves.
6. Install rocker arms and shafts.
7. Install overhead camshaft.
8. Install cylinder head gasket and cylinder head.

INTRODUCTION

After reconditioning the cylinder head, it must be assembled in a manner that will ensure trouble-free operation for a long time. Proper cylinder head assembly involves a series of steps and checks which require careful attention and accuracy. After assembly the cylinder head is installed on the engine block.

INSTALLING THE CORE PLUGS

Core plugs must be installed to ensure leak-free operation. An incorrectly installed core plug may leak and cause the engine to overheat. The following guidelines will help eliminate core plug problems.

1. Make sure that the core holes in the cylinder head are absolutely clean. Remove all old sealer, burrs, rust, and dirt.
2. Use a good hardening-type sealer on the core plug. Coat the sealing surface with sealer just before installing the plug.
3. Make sure to use the correct type and size of core plug. There are three basic types.
   a. *Disc-type (dished) plugs*. These are used in core holes that are counterbored (Figure 18-1). The crowned side of the plug faces out. The plug is installed with a driver that contacts the crowned center area of the plug. With the plug seated against the counterbore, flatten the plug enough to expand the plug tightly against the bore. Too little flattening can leave the plug loose. Flattening the plug too

FIGURE 18-1 Dished core plug installation. (Courtesy of FT Enterprises.)

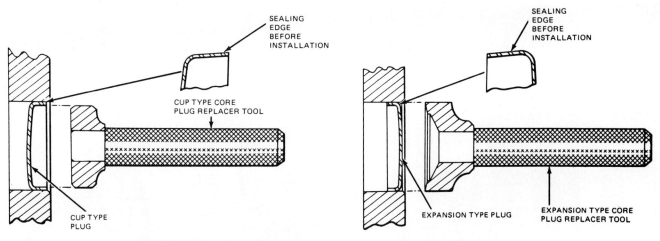

FIGURE 18-2 Installing cup-type core plugs. Note the different designs and installation methods. (Courtesy of Ford of Canada.)

much or dishing it inward will loosen the plug.

b. *Cup-type plug with open side out* (Figure 18-2). This plug has a tapered sealing edge. The plug is larger in diameter across the open edges than it is across the face. Use the proper-size driving tool to prevent cup distortion during installation. Keep the cup square during installation and install it to the specified depth, usually $\frac{1}{32}$ in. past the inner edge of the hole chamfer.

c. *Cup-type plug with open side in* (Figure 18-2). This plug has a reverse taper sealing edge. The open side of the cup is smaller in diameter than it is across the face. Use only the correct size of driver and install to the specified depth.

Core Plug Dimensions

Core plugs are available in sizes with as little as 0.010 in. difference in diameter. Accurate sizing of core plug holes and core plug selection is therefore critical to leak-free operation. Cup-type plugs are also selected on the basis of cup depth. Several cup depths are available in each size. Use only the type, diameter, and cup depth specified by the original-equipment manufacturer when installing core plugs (Figure 18-3).

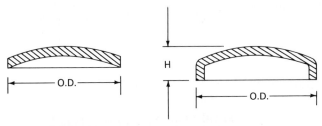

FIGURE 18-3 Critical core plug dimensions. (Courtesy of Ford of Canada.)

Threaded Plugs

Make sure that all threads are clean and in good condition. Use a good thread sealer on the threads of the plug and install to the specified torque.

INSTALLED SPRING HEIGHT

The height of the valve spring in its installed position is a determining factor in the closing pressure applied to the valve. Material removed from the valve seat and valve face cause the valve stem to extend farther above the spring seat. This results in increased installed spring height and reduced spring tension. Grinding the valves and seats can result in a loss of as much as 25% of spring tension if not corrected. Reduced spring tension can result in the following operating problems:

- Valve float
- Lifter pump-up
- Poor fuel economy
- Rough engine idle
- Missing at higher speeds
- Impact loading of valve train
- Rapid valve train wear
- Overheated valves

Measuring Installed Spring Height

Installed spring height can be measured with the valve and spring assembled or before installing the spring as follows using dividers, a small-scale rule, or calipers with a depth attachment.

1. Measuring assembled installed spring height. Measure from the base of the spring (surface of the spring seat) to the lower side of the

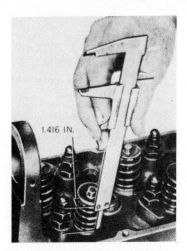

FIGURE 18-4 Measuring installed spring height after assembly. (Courtesy of Silver Seal Products Company, Inc.)

FIGURE 18-5 Measuring spring height before spring installation. (Courtesy of Silver Seal Products Company, Inc.)

spring retainer (top of the valve spring). Remember to measure only the actual spring height. Do not include oil shedders or spring seats in your measurement (Figure 18-4).

2. Measuring installed spring height prior to assembly. Install the valve in the guide. Install the spring retainer and keepers on the valve. Pull the valve against its seat by pulling on the spring retainer. With the valve and retainer held in this position, measure the distance between the spring seat and the lower face of the spring retainer. Compare with specifications (Figure 18-5).

Correcting Installed Spring Height

Valve spring shims are available in several thicknesses to correct installed spring height as follows (Figure 18-6):

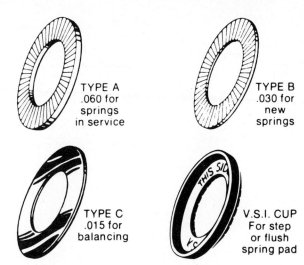

FIGURE 18-6 Valve spring shims. (Courtesy of Silver Seal Products Company, Inc.)

Type A: 0.060 in. thick for springs in service
Type B: 0.030 in. thick for new springs
Type C: 0.015 in. thick for balancing spring pressures
VSI cup: for stepped or flush spring seats

Shims with serrations should be installed with the serrated side toward the head (Figure 18-7). The serrations reduce heat transfer from the head to the spring. Cooler-running springs have less tension loss. Shimming should not exceed 0.060 in. Shims should never be used to increase the tension of weak springs since reducing installed spring height will cause spring coils to collide and cause breakage.

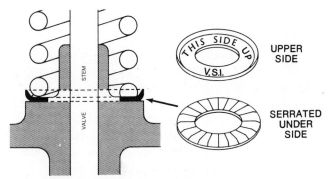

FIGURE 18-7 Proper shim installation. (Courtesy of Silver Seal Products Company, Inc.)

BALANCING VALVE SPRING PRESSURES

The objective in balancing valve spring pressures is to obtain equal static spring pressure for all valve springs, taking into consideration the difference in valve open spring heights due to variations in cam lobe, lifter, pushrod, rocker arm, and valve stem tip dimensions. For information on valve spring balancing procedures, refer to Chapter 4.

INSTALLING THE VALVES

Installing the valves includes assembling the valves, valve stem seals, springs, spring retainers, and valve locks or keepers. The sequence of assembly differs somewhat depending on the type of oil seal or oil shedder used. The following steps may be used as guidelines for assembling the valves and springs after all reconditioning and leak testing have been completed, as outlined in Chapters 15 to 17.

1. Prelubricate the valve stems and guides using an assembly lubricant that does not drain off, such as an engine oil supplement or special lube designed for the purpose. Motor oils are too light to remain on the parts for long and the danger of scoring the valve stems and guides because of dry start up is not worth the risk (Figure 18-8).

2. Install the valve spring seat (if so equipped). In some engines a valve rotator is used between the spring and spring seat. On this type of engine the rotator must be installed next to the head (Figure 18-9).

3. Install the valve stem seal and spring in the correct order, depending on design.
   a. Positive valve stem seals are installed

prior to installing the spring (Figures 18-10 and 18-11). This is also the case with rubber umbrella-type oil deflectors or oil shedders. After these seals are installed, the spring is placed in position (close spaced coils and large coils on conical springs toward the head), and then the spring retainer (Figure 18-12). The spring is then compressed just enough to allow the stem locks to be installed (Figure 18-13). (**Caution:** Overcompressing the spring can crush or damage the oil seal. After the locks are in place, slowly release the spring compressor, making sure

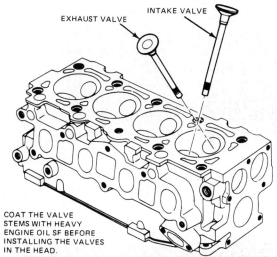

COAT THE VALVE STEMS WITH HEAVY ENGINE OIL SF BEFORE INSTALLING THE VALVES IN THE HEAD.

FIGURE 18-8 Prelubricate valve stems before installation. (Courtesy of Ford of Canada.)

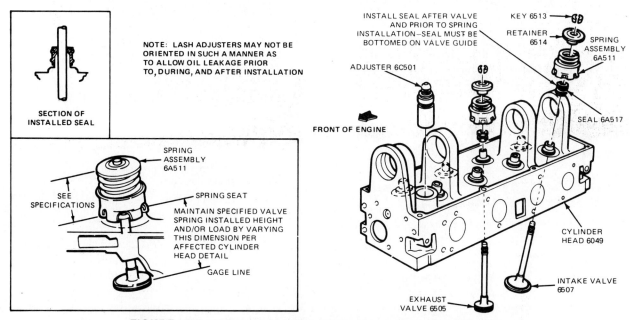

FIGURE 18-9 Valve, spring seat, stem seal, spring, and spring retainer installation on 2.3-liter Ford engine. (Courtesy of Ford of Canada.)

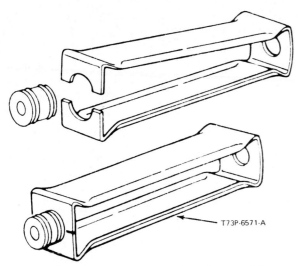

FIGURE 18-10 Positive valve stem seal installation tool. (Courtesy of Ford of Canada.)

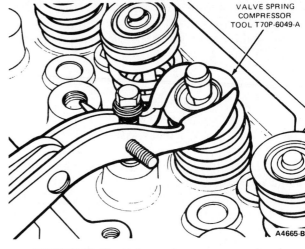

FIGURE 18-13 Valve spring compressed and keepers in place. (Courtesy of Ford of Canada.)

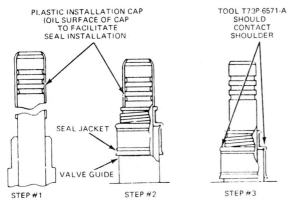

STEP #1 — WITH VALVES IN HEAD, PLACE PLASTIC INSTALLATION CAP OVER END OF VALVE STEM.
STEP #2 — START VALVE STEM SEAL CAREFULLY OVER CAP. PUSH SEAL DOWN UNTIL JACKET TOUCHES TOP OF GUIDE.
STEP #3 — REMOVE PLASTIC INSTALLATION CAP. USE INSTALLATION TOOL-T73P-6571-A OR SCREWDRIVERS TO BOTTOM SEAL ON VALVE GUIDE.

FIGURE 18-11 Installing a positive valve stem seal. (Courtesy of Ford of Canada.)

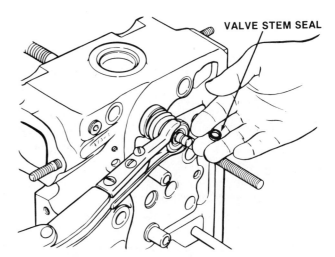

FIGURE 18-14 Installing an O-ring valve stem seal after the spring is compressed. (Courtesy of Ford of Canada.)

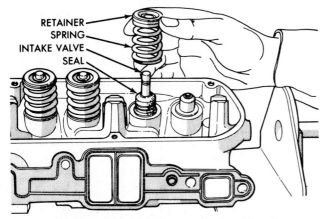

FIGURE 18-12 Installing the spring after the positive valve stem seal is in place. (Courtesy of Chrysler Canada Ltd.)

that the locks or keepers remain in place.)

b. General Motors O-ring seals are installed in the second keeper groove after the spring is compressed. Next install the keepers and carefully release the spring compressor. The O-ring seals between the spring retainer and valve stem. If the O-ring is installed before the spring and spring retainer, it will be pushed out of place down the valve stem when the spring is compressed (Figures 18-14 and 18-15).

c. On some Toyota engines the O-ring seal fits in a groove above the keepers. It is therefore installed after the keepers are in place.

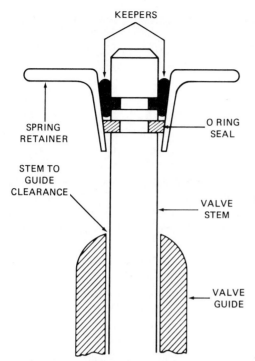

FIGURE 18-15 Position of GM O-ring valve stem seal. (Courtesy of FT Enterprises.)

Labels in figure: KEEPERS, SPRING RETAINER, O RING SEAL, STEM TO GUIDE CLEARANCE, VALVE STEM, VALVE GUIDE

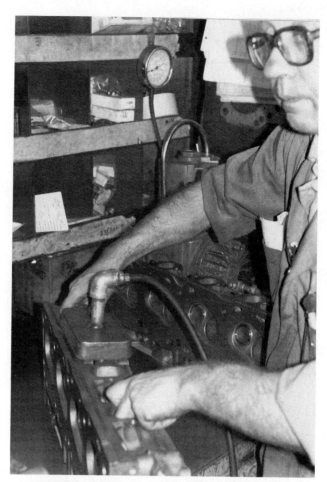

FIGURE 18-16 Vacuum leak testing the valves. (Courtesy of Western Engine Ltd.)

After the valves are installed, tap each valve stem squarely with a soft hammer to ensure proper seating of all components and that keepers do not pop out.

Leak Testing the Valves and Seats

After assembly the valves and seats should be leak tested with a vacuum tester designed for the purpose (Figure 18-16). The tester is equipped with a vacuum pump, vacuum gauge, control valve, and vacuum cup. To leak test the valve and seat, turn the vacuum pump on. Place the vacuum cups snugly over the port of the valve to be tested. Open the control valve to expel the air from the vacuum cup. Observe the vacuum gauge for a few seconds for loss of vacuum, which indicates leakage. On some heads the AIR passages may have to be closed by hand while the test is performed.

Caution: Aluminized valves will exhibit some leakage due to their somewhat rough surface finish.

INSTALLING THE CYLINDER HEAD

1. The cylinder head and block gasket surfaces should not be too smooth or too rough. If they are too smooth, there is too little friction between them and the gasket, making it easier for combustion pressures to cause a blown gasket—especially between cylin-

ders. A surface finish that is too rough results in poor sealing since the gasket cannot conform to the rough surface. A surface finish of 90 to 110 μin is preferred. (See Chapters 7 and 15 for resurfacing.) Sealing surfaces must be clean and free of any foreign material.

2. Check the gasket and locate the Top, Front, or This Side Up marking. (See Chapter 13 for gasket information.) If the gasket is not marked, it should be installed with the stamped ID number toward the head.

3. Check the gasket holes against the holes in the block and cylinder head. In some cases a water jacket hole may be covered by the gasket or the hole in the gasket may be much smaller than the water jacket hole. This aids in controlling and directing coolant flow in the block and head.

4. Be certain that all head bolt holes in the block are chamfered, clean, and do not contain any oil or coolant. Fluid or sealant in a blind bolt hole causes inaccurate torque readings and

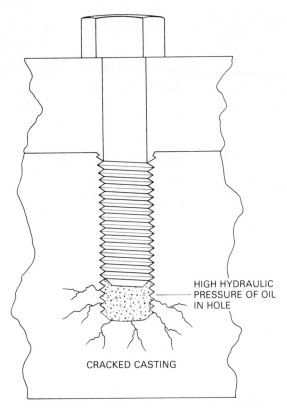

FIGURE 18-17 Hydraulic pressure in a blind hole can crack a cylinder block when the head bolts are tightened. (Courtesy of FT Enterprises.)

FIGURE 18-18 Studs keep head gasket in place and guide head into place. (Courtesy of Ford of Canada.)

can create enough hydraulic force to crack the block (Figure 18-17).

5. Thoroughly clean all the head bolt threads and install guide studs (Figure 18-18).

6. Most gaskets do not require that sealer be used. Embossed steel gaskets may require a spray sealer to be applied to both sides. Fol-

low the gasket manufacturer's recommendations.

7. Make certain that special head bolts or bolts of different length are installed in the proper location.

8. The head bolt holes on some engines extend into the water jacket. This includes most Chevrolet small blocks, some AMC six- and eight-cylinder engines, and some Chrysler V8 engines. The bolt holes in the block can be probed with a piece of wire or welding rod to determine whether they extend into the water jacket. The threads of head bolts on these engines must be coated with a good nonhardening sealer to prevent coolant leakage past the threads. Do not use sealer on bolts installed in blind holes. Lightly oil the threads only on bolts to be installed in blind holes. On some aluminum heads special washers are required to prevent galling the head when bolts are tightened. Make sure that these are used where required.

9. Turn all head bolts down until the bolt heads contact the head. Tighten the bolts in the recommended sequence and in several steps (one-half, one-third, then full torque or as specified in the service manual) to the torque

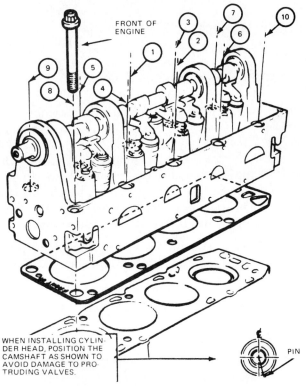

FIGURE 18-19 Cylinder head installation on Ford 2.3-liter engine. (Courtesy of Ford of Canada.)

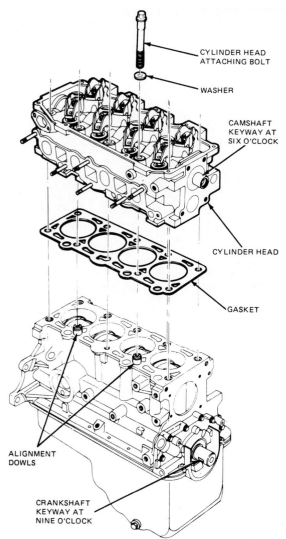

CYLINDER HEAD
ATTACHING BOLT

WASHER

CAMSHAFT
KEYWAY AT
SIX O'CLOCK

CYLINDER HEAD

GASKET

ALIGNMENT
DOWLS

CRANKSHAFT
KEYWAY AT
NINE O'CLOCK

FIGURE 18-20 Aluminum cylinder head installation on 1.6-liter Ford engine uses special washers with cylinder head bolts. (Courtesy of Ford of Canada.)

FIGURE 18-22 Tightening cylinder head bolts on a Chrysler 318-CID V-8. (Courtesy of Chrysler Canada Ltd.)

specified by the engine manufacturer (Figures 18-19 to 18-22). Head warpage and leaks are caused by improper tightening of head bolts.

INSTALLING THE OVERHEAD CAMSHAFT

Overhead camshaft installation varies depending on whether the bearings are of the full circle type or the split type with bearing caps. Thoroughly lubricate the cam lobes and bearing journals with the special lubri-

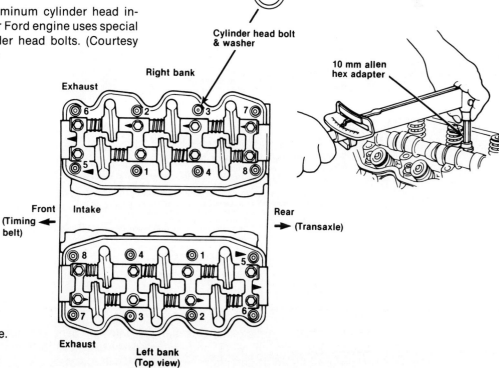

FIGURE 18-21 Cylinder head bolt tightening sequence on 3.0-liter V-8 Chrysler Mitsubishi engine. (Courtesy of Chrysler Canada Ltd.)

Right bank

Exhaust

Cylinder head bolt & washer

10 mm allen hex adapter

Front
(Timing belt)

Intake

Rear
(Transaxle)

Exhaust

Left bank
(Top view)

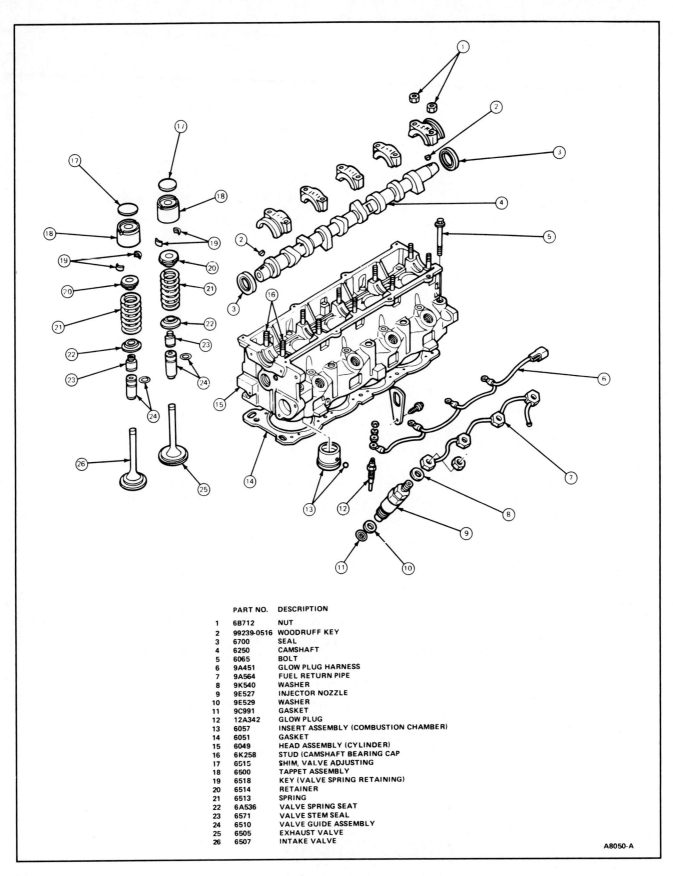

	PART NO.	DESCRIPTION
1	6B712	NUT
2	99239-0516	WOODRUFF KEY
3	6700	SEAL
4	6250	CAMSHAFT
5	6065	BOLT
6	9A451	GLOW PLUG HARNESS
7	9A564	FUEL RETURN PIPE
8	9K540	WASHER
9	9E527	INJECTOR NOZZLE
10	9E529	WASHER
11	9C991	GASKET
12	12A342	GLOW PLUG
13	6057	INSERT ASSEMBLY (COMBUSTION CHAMBER)
14	6051	GASKET
15	6049	HEAD ASSEMBLY (CYLINDER)
16	6K258	STUD (CAMSHAFT BEARING CAP
17	6515	SHIM, VALVE ADJUSTING
18	6500	TAPPET ASSEMBLY
19	6518	KEY (VALVE SPRING RETAINING)
20	6514	RETAINER
21	6513	SPRING
22	6A536	VALVE SPRING SEAT
23	6571	VALVE STEM SEAL
24	6510	VALVE GUIDE ASSEMBLY
25	6505	EXHAUST VALVE
26	6507	INTAKE VALVE

A8050-A

FIGURE 18-23 Ford 2.0-liter diesel engine with split cam bearings, bucket type cam followers, and shimmed valve lash adjustment. (Courtesy of Ford of Canada.)

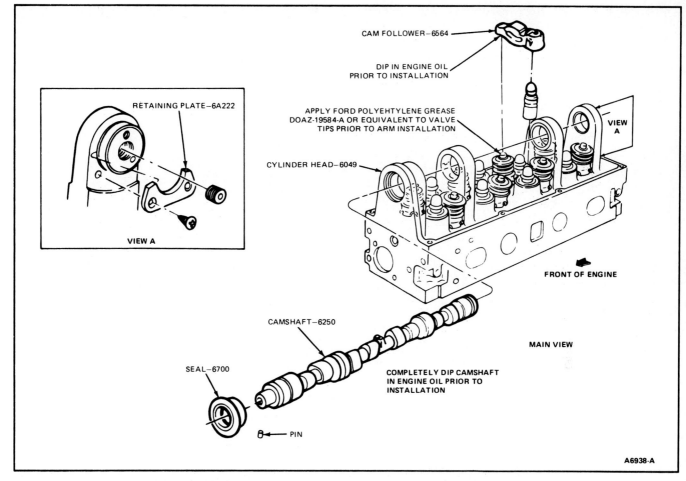

FIGURE 18-24 Installing the overhead camshaft in a Ford 2.3-liter engine. (Courtesy of Ford of Canada.)

cant provided with the new camshaft kit. Proper lubrication is critical to good camshaft break-in. Poor lubrication can cause serious scoring of the camshaft and lifters during engine startup. Rapid wear will take place if scoring has occurred.

On the type with bearing caps, first install the bearing inserts in the head and bearing caps. Make sure that each insert is in its proper location and that oil holes are aligned. Lubricate the bearings thoroughly. Place the camshaft into position on the head. Install the bearing caps, making sure that each cap is in its proper location and is not reversed. Install the cap bolts and start them all by hand. Tighten the bolts in steps in the order specified in the manual to the specified torque. Make sure that the camshaft can be turned by hand with moderate force. If any binding occurs, loosen one cap and try to turn the camshaft again. If it still binds, loosen the remaining caps one at a time until the camshaft can be turned. Correct the cause of the binding, install and tighten the bearing caps, and check to ensure that the binding has been eliminated (Figure 18-23).

On overhead camshafts with full circle cam towers, insert the camshaft carefully, making sure that it is well supported and that the sharp edges of the cam lobe and bearing journals do not scrape the bearing surfaces (Figure 18-24). A long bolt screwed into the drive end of the camshaft is helpful in supporting the camshaft during installation. The camshaft must be controlled during installation, whether the head is in the upright position or if it is stood on end. Turn the camshaft slowly while it is inserted. If binding is encountered, do not force the camshaft. Remove the camshaft, determine the cause of binding, and make the required correction before installing the camshaft.

Camshaft fit is acceptable if the camshaft can be turned by hand with moderate force. If binding is present, it will show up as a shiny area in the affected bearing. Very careful scraping of the area with a bearing scraper will correct the problem. An alternative method is to line hone the bearings lightly. All metal scrapings and honing abrasives must, of course, be removed prior to final assembly. See Chapter 11 for more information on camshafts.

INSTALLING THE PUSHRODS AND ROCKER ARMS

Insert the pushrods, making sure that the right end is up and that the bottom ends are seated in the lifters. On engines with rocker arm shafts, make sure that the proper end of the shaft faces forward and the correct side of the shaft faces up. Refer to the appropriate service manual for this information. Thoroughly lubricate all parts before assembly. Assemble the shafts, rocker arms, springs, and washers (if any), in the proper sequence. Make sure that differences in rocker arm design and location are taken into account during assembly. If incorrectly installed, rocker arms will be misaligned.

Loosen any valve lash adjusters before installing the rocker shaft assembly. Place the assembly on the head and start all the bolts by hand (Figure 18-25). Make sure that all the pushrods engage the lifters and rocker arms properly. Tighten the mounting bolts slowly and evenly a little at a time to allow hydraulic lifters to bleed down and to prevent the rocker shaft from bending. Tighten all the bolts to specified torque.

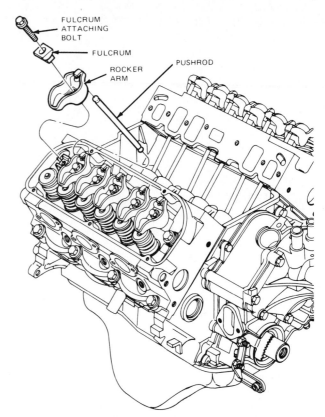

FIGURE 18-26 Installing the pushrods and pedestal-mounted rocker arms on a Ford 3.8-liter V-6 engine. (Courtesy of Ford of Canada.)

FIGURE 18-25 Installing the pushrods and rocker arm assembly on a Ford V-6 engine. (Courtesy of Ford of Canada.)

Place stud-mounted rocker arms on the studs, observing any differences in design and location. Place the pivot balls on the studs and start the stud nuts. Turn the stud nuts down while making sure that the pushrods engage the rocker arms. Turn the nuts down just enough (on the adjustable type) to eliminate any free play and perform the lash adjustment as outlined in Chapter 19. On engines with positive-stop rocker studs, tighten the nuts down to specified torque.

On engines with pedestal-mounted rocker arms, assemble the pedestal and rocker arms and install them on the head. Start the bolts by hand and tighten them to specified torque (Figure 18-26). For valve adjustment procedures on both mechanical and hydraulic lifter valve trains, refer to Chapter 19.

REVIEW QUESTIONS

1. What type of sealer should be used on core plugs?

2. What three types of core plugs are used in automotive engines?

3. Core plug size differences may be as little as:
    (a) 0.100 in.
    (b) 0.010 in.
    (c) 0.001 in.
    (d) 0.0001 in.

4. What effect does excessive installed spring height have on spring pressure?

5. List five engine operating problems that can occur as a result of excessive installed spring height.

6. Name two ways to measure installed spring height.

7. Valve spring shims for correcting installed spring height are available in the following sizes:

(a) 0.060, 0.030, 0.015 in.
(b) 0.015, 0.030, 0.045 in.
(c) 0.60, 0.30, 0.15 in.
(d) 0.15, 0.30, 0.45 in.

8. True or false: Valve springs that are too weak can be used if shimmed to increase spring pressure.

9. What is meant by ''balancing'' valve spring pressures?

10. Valve shimming should not exceed _____ in.

11. The valve should be _____ just before installation.

12. Why should valve springs not be overcompressed during assembly?

13. On General Motors heads with O-ring valve stem seals, the order of assembly is as follows:
(a) valve, seal, spring, retainer, locks
(b) valve, spring, retainer, seal, locks
(c) valve, spring, seal, retainer, locks
(d) valve, retainer, spring, seal, locks

14. After all valves and springs are assembled, it is a good idea to strike each valve stem squarely with a soft hammer. Why?

15. What method is used to leak test valves after assembly?

16. Why do aluminized valves exhibit some leakage during the leak test?

17. Cylinder head and block gasket surfaces should have a microinch finish from:
(a) 50 to 80

(b) 30 to 50
(c) 100 to 150
(d) 90 to 110

18. True or false: If a head gasket is not marked to identify the top side, it should be installed with the stamped number toward the block.

19. True or false: All the head gasket holes must match all the holes in the head as to both size and location.

20. Why should guide studs be used when installing a cylinder head?

21. True or false: On some engines the threaded head bolt holes extend into the water jacket.

22. Why should cylinder head bolts be tightened in a specific order?

23. True or false: Pushrods may be installed with either end up.

24. Before installing a rocker arm shaft assembly:
(a) loosen the lash adjusters if so equipped
(b) make sure that right and left rocker arms are correctly located
(c) place all the pushrods in position
(d) all of the above

25. Why should rocker arm bolts be tightened slowly and evenly?

26. True or false: When assembling positive-stop stud-mounted rocker arms, tighten the nut to the torque specified.

CHAPTER 19

Final Assembly, Installation, and Break-in

OBJECTIVES

To develop the ability to:

1. Assemble and install the engine and prepare it for startup.
2. Perform proper break-in procedures to ensure good piston ring seating and long engine life.

FINAL ASSEMBLY

Final assembly includes installation of such items as the motor mounts, starter, water pump, manifolds, distributor, oil filter, fuel pump, carburetor or fuel injectors, alternator, AIR and EGR systems, fan, shroud, radiator, A/C compressor, power steering pump, clutch and transmission, battery, exhaust, fuel, electrical and vacuum line connections, and the hood. Some of these items may be assembled before the engine is installed, while others can only be installed after the engine is in place. With the wide range of models and the differences in engine mounting (transverse or longitudinal), the procedures will vary with specific vehicles. This section provides general guidelines for final assembly. For specific models, refer to the appropriate service manual.

NEW PARTS VERSUS USED PARTS

Critical to good engine performance, fuel economy, emission levels, and reliability is the decision as to whether to replace certain engine parts on a rebuilt engine. While it may not be necessary to replace all the items referred to in this section, only those used items that are in good condition and can be expected to last a long time should be used. The following parts should be inspected and tested to ensure that their performance and service life are adequate or they should be replaced. Using old parts whose performance and reliability are in doubt can lead to failure and comebacks.

- Radiator hoses, heater hoses, and clamps
- Radiator cap
- Thermostat
- Fan belts
- Heat riser valve (and vacuum actuator if so equipped)
- Spark plugs
- Spark plug wires
- Distributor cap and rotor
- PCV valve
- Oil filter
- Fuel filter
- Air filter

- Clutch and pressure plate
- Engine mounts
- Transmission mount
- Water pump
- Fan
- Fan clutch
- Oil pump
- Radiator
- Carburetor
- Starter
- Alternator
- Fuel injectors

SERVICE AND REPLACEMENT GUIDELINES

Oil Pump

The engine's life blood is its lubricating oil. The oil pump must be able to deliver oil in sufficient quantity and under adequate pressure to the operating parts of the engine. If the pump is worn or damaged, it cannot do its job. An oil pump that does not measure up to the manufacturer's specifications should be replaced. In fact, the price of a new oil pump is cheap insurance against rebuilt engine failure due to inadequate lubrication. Many rebuilders will provide an engine warranty only if the oil pump is replaced. See Chapter 14 for detailed information on engine oil pumps and pump installation.

Water Pump

The water pump should be replaced when an engine is rebuilt. The ability of the engine to perform and last is only as good as its ability to maintain the proper operating temperature. The following guidelines should be followed when installing a water pump.

1. Make sure that the hoses and hose clamps to be used are in good condition. Replace any questionable parts.
2. Check the fan and fan clutch to ensure that they are in good condition. If there is any indication that the fan is bent or cracked, replace it. Never straighten or weld a fan. Replace a fan clutch that shows any sign of silicone leakage.
3. If the water pump is belt driven, inspect the pulleys and drive belt. Replace if worn or damaged.
4. Make certain that pump mounting surfaces are clean and smooth. Check the gasket to make sure that it is the correct one. If the

water pump has a back plate, check the attaching bolts to make sure that they are tight.
5. Clean all the attaching bolt threads and determine where the longer and shorter bolts go.
6. Use a good sealer on both sides of the water pump gasket and place it in position on the pump.
7. Install the pump and lightly tighten all the bolts, then torque them alternately in steps to specified torque.
8. Install the pulley and fan (if applicable) and tighten the bolts to specifications.
9. Install the drive belt (if applicable) and tighten the adjustment to specified tension using a belt tension gauge.

Cooling System

The radiator should be cleaned, descaled, and tested to provide good coolant flow and ensure against leakage. The radiator cap should be inspected for damage and the seal should be in good condition. The cap must be pressure tested to ensure that it holds rated pressure. Replace a faulty radiator cap with a new one that meets the specified pressure rating. Use a new thermostat rated at the recommended opening temperatures. Use a 50% solution of a good grade of ethylene glycol and water. Make sure that the antifreeze contains silicates to protect aluminum engine parts from corrosion. To prevent erosion of engine cooling system components, advise the owner to do the following.

1. Drain and flush the cooling system at least once every two years and install a new solution of 50% ethylene glycol antifreeze.
2. Maintain the coolant at the proper level and correct any leaks immediately.
3. Make sure that the engine to body (or frame) ground strap is securely attached and maintains a good electrical ground to reduce the formation of electrolysis in the engine coolant.

Starter

The starter should be disassembled, cleaned, and tested or replaced. If necessary, turn the armature and replace the drive pinion. Replace the brushes and bushings or bearings to ensure reliability.

Alternator

The alternator should be checked for any damaged parts, a worn pulley, or rough bearings. Rough bearings and worn brushes should be replaced.

Carburetor

It is good practice to disassemble, clean, and adjust the carburetor using a new carburetor kit. Follow the equipment manufacturer's specifications for adjustments. The following guidelines may be used when installing the carburetor.

1. Make sure that the throttle body base and manifold mounting surfaces are clean and smooth.
2. Install a new base gasket on the manifold, making sure it is the correct one. Install a new steel shim or spacer (if so equipped). Install the EGR valve assembly (if so equipped).
3. Mount the carburetor on the manifold. Start all the nuts or cap screws by hand, leaving them loose.
4. Connect the fuel and vacuum lines. Avoid any fuel line pressure against the carburetor and use a new fuel filter.
5. Tighten the base nuts or cap screws alternately and evenly in steps to prevent cracking the throttle body base.
6. Connect the choke, throttle, and downshift linkages (as applicable) and make sure that they do not bind through their complete travel range.
7. Set the idle mixture screws at about 1 to $1\frac{1}{2}$ turns open from their lightly seated position if not yet sealed with tamperproof devices.
8. Set the idle speed screw, idle control solenoid, or idle speed control motor to the manufacturer's specifications. This will ensure that the engine will not race or stall when it is started.

Throttle Body and Fuel Injectors

Installation of the throttle body on port injection systems is similar to carburetor installation described above. For throttle body injection equipment installation and adjustment, refer to the manufacturer's service manual. Installation of fuel injectors and fuel rails on multipoint injection systems must be done according to service manual specifications. Fuel injectors should be cleaned and tested or replaced.

Clutch and Flywheel

1. Inspect flywheel for damaged ring gear teeth, heat spots, cracks, scoring, and wear. Replace or resurface the flywheel if it is badly damaged. If the flywheel is serviceable, clean with fine emery cloth to remove the glazed finish. If ring gear teeth are badly damaged, replace the ring gear as outlined in the service manual.
2. Inspect the clutch pressure plate friction surface for heat spots, cracks, scoring, and wear. Inspect springs for damage due to overheating. Springs show burned paint or discoloration if overheated due to slippage. Inspect the cover for damage or warpage. Check the drive lug holes for wear. Replace the pressure plate if damaged. The usual procedure is to replace release bearing, pressure plate, and clutch disc as a matched assembly if one or the other requires replacement.
3. Inspect the clutch disc friction material for overheating, cracks, glazing, or oil contamination. Check to make sure that attaching rivets are tight and that enough lining (about two-thirds of the original thickness) remains. Check the wave springs and metal disc; they should not be cracked or warped. Check the torsional springs and drive pins. Springs should not be loose or broken. Drive pins and washers should be tight. Replace the disc if required.
4. Replace the pilot bushing (if required) and lubricate as recommended.
5. Check and correct the clutch housing alignment (bell housing) if required.
6. Use a clutch aligning arbor or suitable dummy shaft to assemble the clutch disc and pressure plate to the flywheel. (Make sure that the proper side of the disc faces the flywheel.) Tighten the special bolts progressively to the torque specified.
7. Inspect the release bearing for roughness or looseness. Replace if required.
8. Install the release bearing (lubricate the interior of the sleeve if recommended).
9. Install the transaxle or transmission, cross member, and drive line or half-shafts as applicable.
10. Adjust the pedal height, clutch linkage, and transmission linkage as specified in the manufacturer's manual.
11. Check the transmission lubricant level and correct if required.

TEST STAND RUN-IN

Engine rebuilding and remanufacturing shops usually "motor" the rebuilt engine assembly on a test stand with an electric motor drive (Figure 19-1). The engine

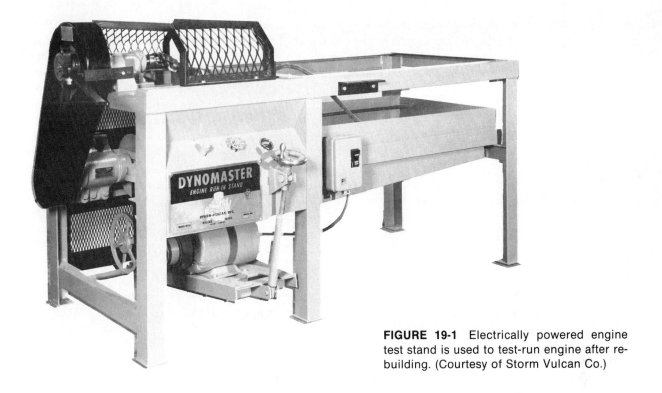

FIGURE 19-1 Electrically powered engine test stand is used to test-run engine after rebuilding. (Courtesy of Storm Vulcan Co.)

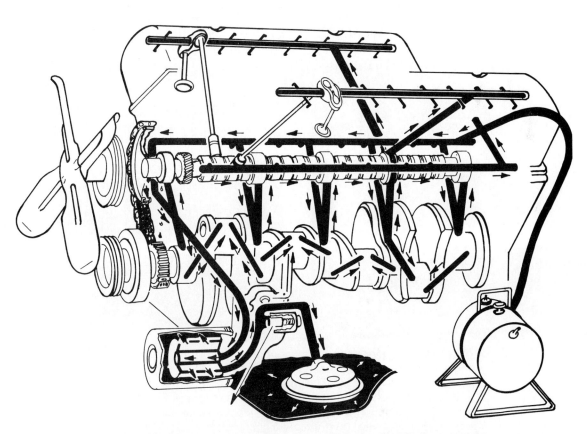

FIGURE 19-2 The importance of priming the lubrication system cannot be overemphasized. A pressure tank with a hose and control valve connected to an engine oil gallery are used. (Courtesy of Hastings Manufacturing Company.)

is first prelubricated (Figure 19-2) to prevent dry friction when the engine is being test run. During the test run the operator observes the oil pressure and watches for any abnormalities. Minor valve adjustments may be made at this time as well. After run-in the engine is removed from the test stand and prepared for shipment.

ADJUSTING THE VALVES

The objective in adjusting the solid lifter valve train is to provide sufficient valve lash (clearance) to allow for any expansion of parts due to heat and still ensure that valves will be fully seated when closed. At the same time, there must not be excessive lash, which would retard valve timing and cause rapid wear of valve train parts. The objective in adjusting the hydraulic lifter valve train is to center the plunger in the hydraulic lifter. This provides zero lash in operation and allows the lifters to adjust automatically to compensate for wear or expansion and contraction of valve train parts due to temperature changes.

A static valve adjustment is required after an engine overhaul to ensure proper engine starting and prevent damage to the valves from pistons hitting the valves. Normally, if the static valves are adjusted accurately, no further valve adjustment is required unless the cylinder heads are tightened. Valve adjustment is done as a routine service procedure when there is no

engine overhaul involved. This is done with the engine stopped or running at a slow idle speed. In this case provision must be made to prevent oil from squirting and spraying over other engine parts. Also, when adjusting hydraulic lifters under these conditions, the adjusting screw or nut must be turned down slowly to allow for lifter leakdown and to prevent the valve from being held open when it should be closed. Care must be taken not to overadjust and cause the lifters to bottom out and damage valve train parts. Adjustment should only be made within the limits prescribed in the manufacturer's shop manual.

ADJUSTING SOLID LIFTER VALVES (TAPPET CLEARANCE)

This may be done with the engine cold or hot. Clearance specifications are usually 0.002 in. (0.05 mm) greater for cold engines than for hot engines. Follow the recommendation for hot or cold engine adjustment and the clearance specifications in the appropriate service manual.

Select the feeler gauge thickness for the clearance specified. Clearance is usually greater for exhaust valves than intake valves. Turn the crankshaft until the number 1 piston is near the top of the compression stroke. You can feel air coming out of the spark plug hole as the piston comes up on this stroke. Do not turn too far. The intake and exhaust valves are both closed

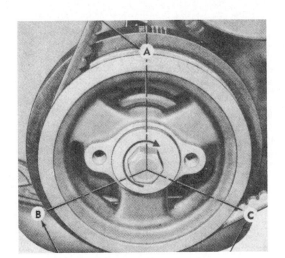 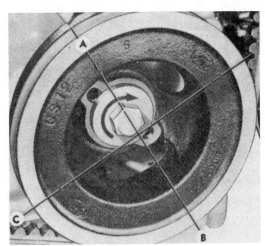

FIGURE 19-3 A static valve adjustment must be performed prior to starting an engine that has been overhauled. To adjust valves, proceed as follows. (1) Put piston number 1 in firing position (TDC at top of compression stroke). (2) Adjust both valves for cylinder number 1 to specifications. (3) Turn crankshaft clockwise (as above) 120° for six-cylinder engines, 90° for eight-cylinder engines, and 180° for four-cylinder engines. This places the next cylinder in the firing order sequence in position for valve adjustment. (4) Adjust both valves for this cylinder. (5) Repeat steps 3 and 4 until all valves for all cylinders have been adjusted. (Courtesy of Ford of Canada.)

with the piston in this position (lifters on base circle of cam lobes). Both valves can now be adjusted.

Slide the correct feeler gauge between the valve stem tip and the rocker arm. If it will not go in, back off the rocker arm adjusting screw until it does. If there is a locknut on the screw, it must be loosened first. Tighten the adjusting screw until you feel a slight drag as you slide the feeler gauge back and forth. Then tighten the locknut, making sure that the adjustment does not change. Adjust both valves in the same manner. Follow the same procedure on the remaining cylinders.

When you adjust the number 1 cylinder first, you can proceed to do the other cylinders in the order of the engine's firing order. This allows you to turn the crankshaft only half a turn to place the next piston in the right position for adjustment and so on through the firing order on a four-cylinder engine. On a six-cylinder engine it would be one-third of a turn, and a quarter-turn on an eight-cylinder engine (Figures 19-3 and 19-4).

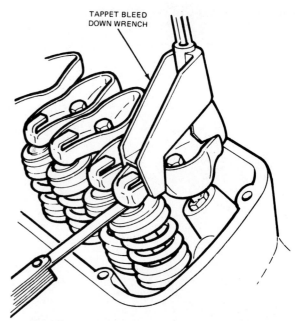

FIGURE 19-5 Checking tappet clearance while holding lifter in bled-down bottomed position. If clearance is incorrect, pushrod length must be changed. Typical for some 302- and 351-CID Ford V-8 engines. (Courtesy of Ford of Canada.)

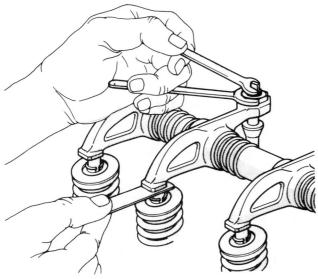

FIGURE 19-4 Adjusting valve lash on solid lifter valve train. (Courtesy of Chrysler Canada Ltd.)

NONADJUSTABLE HYDRAULIC LIFTERS

Some nonadjustable valve trains use selective-length pushrods to center the lifter plunger after valves and seats are ground. Special tools are used to determine the pushrod length required. Refer to the appropriate service manual for procedures (Figure 19-5). On other nonadjustable stud- or pedestal-mounted rocker arms, the nuts or bolts are simply tightened to the specified torque. Adjustment is achieved during the valve grinding procedure by correcting the valve stem length.

ADJUSTING HYDRAULIC LIFTERS (ADJUSTABLE TYPE)

Hydraulic lifters may be adjusted with the engine running or with the engine off. Many technicians prefer to adjust lifters with the engine running. To adjust hydraulic lifters with the engine running, first install oil spray stoppers or otherwise shield the oil spray from the rocker arms before starting the engine (Figure 19-6). With the engine running at operating temperature, tighten any noisy rockers just enough to stop all rocker clatter. Working from front to rear, adjust each rocker as follows.

1. Loosen the adjustment until it clatters, then slowly tighten the adjuster just enough until it quiets down. This is the zero lash position. This position can also be established by turning the pushrod between the thumb and forefinger while slowly tightening the adjuster. Turning effort will increase at precisely the zero lash point (Figure 19-7).

2. From the zero lash point, tighten the adjuster the additional specified amount. This is most often one additional turn of the adjuster. (Check in the service manual.) This will center the lifter plunger midway between its upper and lower travel limits in the lifter

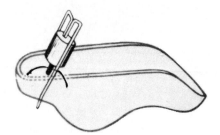

FIGURE 19-6 Rocker arm oil spray stopper. (Courtesy of Mac Tools, Inc.)

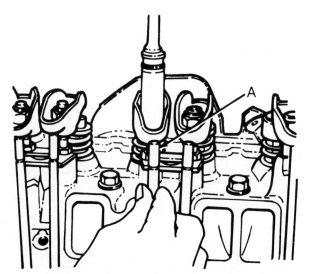

FIGURE 19-7 Establishing the zero lash point by turning the pushrod (A) while adjusting the rocker stud nut. (Courtesy of Chrysler Canada Ltd.)

body. Be sure to tighten the adjuster slowly to prevent the engine from missing or stalling and to prevent the valve from being held open and struck by the piston.

ADJUSTING THE OVERHEAD CAM VALVES

Solid lifter valve adjustment may be by an adjusting screw in the rocker arm or by selective thickness adjusting shims. Clearance is measured with a feeler gauge. The old shim is removed and measured with an outside micrometer to calculate the required thickness of the new shim and the new shim installed. The procedure for adjusting shim removal and installation varies. Refer to the appropriate service manual for procedure. One example of adjusting shim location is shown in Figures 19-8 and 19-9. A third method uses a screw adjustment which is part of the lifter.

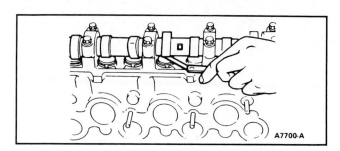

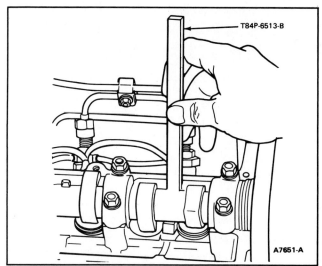

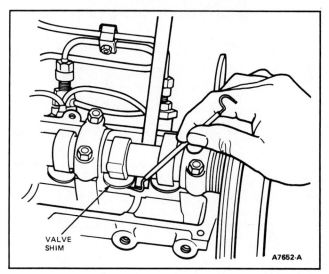

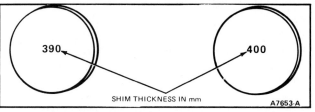

FIGURE 19-8 Adjusting shimmed lash adjustment on OHC engine. Shim adjustment is used on several engine makes and models. Top, checking clearance; second from top, installing cam follower retaining tool; third from top, replacing adjusting shim; bottom, different shim sizes. (Courtesy of Ford of Canada.)

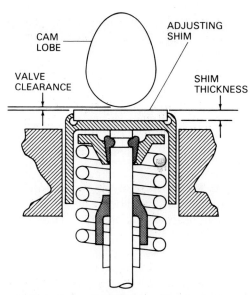

FIGURE 19-9 Close-up of adjusting shim location on OHC engine. (Courtesy of FT Enterprises.)

INSTALLING THE VALVE COVER

Leaking valve covers are one of the more common complaints of oil leakage. Valve covers are easily distorted, and if not straightened or replaced, may leak shortly after installation. Overtightening of screws or bolts is the most common problem. Valve covers are made of stamped steel, cast aluminum, or plastic. Valve covers use either silicone sealer or gaskets. Gaskets may be cork, rubberized cork, or synthetic rubber.

Make sure that the valve cover is clean and that there is no old sealer or gasket material left on the sealing surfaces. Check that the cover is not warped by placing it sealing surface down on a flat surface. The cover should make contact with the flat surface all the way around. Stamped steel covers can be straightened with a hammer. A pair of pliers can also be used. Be careful not to kink the metal. Warped plastic covers must be replaced. Cast aluminum covers can sometimes be trued by machining if not too severely warped.

On covers with gasket retaining clips or notches, position the gasket on the cover so that it will stay in position. On others apply a thin coat of adhesive on the contact surface around the cover. Allow it to become tacky, then stick the gasket in place on the cover. Do not use silicone sealer instead of a gasket on covers designed for use with a gasket. The gasket thickness provides clearance between the cover and the rocker arms in many cases. If silicone sealer is to be used, shimming between the cover and head at the screw holes is required in such instances.

On covers designed for use with silicone or RTV sealer, make sure that the sealing surfaces are clean and dry. Apply a continuous bead of sealer about $\frac{3}{32}$ in. (2.38 mm) to the sealing surface of the cover. Follow the instructions on the sealer package regarding time allowed to set. Install the valve covers and start the screws or bolts by hand. Tighten the screws in the specified sequence and to the specified torque. Overtightening can crack the gasket or bend the cover, causing leakage. Covers with center-mount bolts are easily spread wider at the base by overtightening, allowing oil leakage. See Chapter 13 for more information on gaskets.

ENGINE INSTALLATION

The following guidelines should be used to install the engine.

1. Clean the engine compartment of the vehicle in which the engine is to be installed. Steam cleaning is an effective method.
2. Attach the lifter brackets to the engine and suspend it from the arm of the portable crane.
3. Position the crane with the engine over the engine compartment.
4. Lower the engine carefully into the engine compartment with the engine at an angle which allows it to be positioned properly.
5. Carefully guide the engine into place as it is being lowered, at the same time leveling the engine to its normal position.
6. If the engine and transmission are being installed as an assembly, carefully raise the rear of the assembly to allow cross-member installation.
7. If the transmission was left in the vehicle, avoid damage to the clutch disc (standard transmission) or flex plate (automatic transmission) during alignment procedure.
8. Connect the bell housing to the engine and tighten all attaching bolts to specifications (automatic transmission).
9. Connect the transmission to the bell housing (standard transmission) and tighten all bolts to specifications.
10. Align and connect the torque converter; tighten the converter drive bolts to specifications.
11. Align the motor mounts and tighten bolts to specifications.
12. Remove the engine lift attachment.
13. Attach the exhaust pipes with new gaskets.

14. Install the carburetor or fuel injectors and fuel rail.

15. Install the distributor to initial timing specifications to ensure startup.

16. Make all electrical, fuel, and emission control connections; attach all linkages and install new oil filter.

17. Install the radiator and hoses. Tighten the clamps securely.

18. Fill the cooling system (including the reserve tank) with the correct antifreeze solution.

19. Install the specified amount of recommended engine oil, including an allowance for the oil filter.

20. Prime the engine lubrication system with the prelubricator.

21. Install the fully charged battery; make sure that all connections are clean and tight.

22. Make sure that all connections are properly made and are tight before starting the engine.

ENGINE STARTUP

1. Start the engine and observe engine oil pressure to be sure that the system is operating normally.

2. Set the engine idle speed to approximately 1000 rpm.

3. Observe the oil pressure, coolant temperature indicator, and charge indicator to make sure that all systems are operating properly.

4. While the engine is running, carefully check for any oil or coolant leaks (correct if necessary); the engine should be run until normal operating temperature is reached.

5. If specified in the shop manual, retorque the cylinder heads after the engine has reached operating temperature. (If heads are retorqued, valve adjustment may have to be corrected.)

6. Install the hood in accordance with scribe marks made when the hood was removed.

7. Correct the coolant level if required.

ENGINE BREAK-IN

To assure proper seating of the rings, proper engine break-in procedures should be followed. The following procedure can be used to achieve good break-in. Road test the vehicle as follows:

1. Select a section of the road where the procedures can safely be followed.

2. While driving at 30 mph (48 kmh), accelerate rapidly to 50 mph (80 khm); then immediately decelerate to 30 mph (48 kmh).

3. Repeat this procedure at least 10 times.

The break-in procedure may also be performed on a chassis dynamometer when available. This procedure loads the rings on acceleration (high ring-to-cylinder wall pressure) and assures adequate lubrication of rings and cylinders on deceleration. This assures good seating of the rings. If proper ring seating is not achieved, an oil consumption problem may result. After the break-in procedure, the vehicle should be returned to the shop and a thorough inspection made to ensure that there are no oil or coolant leaks and that the fluid levels are correct.

The vehicle should now be prepared for customer acceptance by making sure that it is clean both inside and out. The customer should be informed that it is good practice to change the engine oil and filter after the first 500 miles (800 kilometers) of driving. This removes any minor impurities and bits of metal that may have remained as a result of the overhaul. The vehicle can be driven normally after the break-in procedure.

REVIEW QUESTIONS

1. List at least 20 engine-related items that should be checked or replaced before engine installation.

2. The life blood of an engine is its _____.

3. True or false: A bent or cracked steel fan should be straightened and welded.

4. A _____ or _____ water pump should be installed on a rebuilt engine.

5. Why are the demands on the cooling system greater after a rebuilt engine is installed?

6. What should be done with the radiator before it is installed?

7. The radiator cap should be _____ tested.

8. The carburetor should be cleaned and a new _____ installed.

9. Why is it important to tighten the carburetor base mounting bolts evenly?

10. List five flywheel inspection points.

11. What is the purpose of "motoring" a rebuilt engine on a test stand?

12. What is the effect of insufficient valve lash on engine operation on an engine with mechanical lifters?

13. Why is a static valve adjustment required before starting a rebuilt engine?

14. Where should the piston be in the cylinder, and on what stroke should it be in order to perform a static valve adjustment?

15. What two methods are used to correct valve train length on engines where no valve train adjustment is provided?

16. Describe the method used to adjust the valve train on an engine with adjustable rocker stud nuts and hydraulic valve lifters.

17. What is the objective in adjusting a hydraulic lifter valve train?

18. What three methods of valve train adjustment are used on OHC engines?

19. Why should engine lift brackets be mounted so that they are tight against the engine?

20. Why should the lubrication system be primed before engine startup?

21. Engine break-in procedures are designed to:
    (a) properly seat the bearings
    (b) baby the engine so that it will last longer
    (c) achieve good ring seating
    (d) all of the above
    (e) none of the above

22. Why should the oil and filter be changed after 500 miles of driving a rebuilt engine?

Appendix

Decimal equivalents and tap drill sizes

Drill size	Decimal	Tap size	Drill size	Decimal	Tap size	Drill size	Decimal	Tap size
$\frac{1}{64}$	.0156		21			T	.3580	
$\frac{1}{32}$	.0312		20	.1610		$\frac{23}{64}$	.3594	
60	.0400		19	.1660		U	.3680	$\frac{7}{16}$-14
59	.0410		18	.1695		$\frac{3}{8}$	.3750	
58	.0420		$\frac{11}{64}$	.1719		V	.3770	
57	.0430		17	.1730		W	.3860	
56	.0465		16	.1770	12-24	$\frac{25}{64}$	.3906	$\frac{7}{16}$-20
$\frac{3}{64}$ 55	.0469	0-80	15	.1800		X	.3970	
54	.0520		14	.1820	12-28	Y	.4040	
	.0550	1-56	13	.1850	12-32	$\frac{13}{32}$	.4062	
53	.0595	1-64, 72	$\frac{3}{16}$	.1875		Z	.4130	
$\frac{1}{16}$	.0625		12	.1890		$\frac{27}{64}$	.4219	$\frac{1}{2}$-13
52	.0635		11	.1910		$\frac{7}{16}$	.4375	
51	.0670		10	.1935		$\frac{29}{64}$	.4531	$\frac{1}{2}$-20
50	.0700	2-56, 64	9	.1960		$\frac{15}{32}$	.4687	
49	.0730		8	.1990		$\frac{31}{64}$	.4844	$\frac{9}{16}$-12
48	.0760		7	.2010	$\frac{1}{4}$-20	$\frac{1}{2}$	.5000	
$\frac{5}{64}$	.0781		$\frac{13}{64}$	.2031		$\frac{33}{64}$	.5156	$\frac{9}{16}$-18
47	.0785	3-48	6	.2040		$\frac{17}{32}$	.5312	$\frac{5}{8}$-11
46	.0810		5	.2055		$\frac{35}{64}$	.5469	
45	.0820	3-56, 4-	4	.2090		$\frac{9}{16}$	.5625	
44	.0860	32	3	.2130	$\frac{1}{4}$-28	$\frac{37}{64}$	.5781	$\frac{5}{8}$-18
43	.0890	4-36	$\frac{7}{32}$	.2187		$\frac{19}{32}$	.5937	$\frac{11}{16}$-11
42	.0935	4-40	2	.2210		$\frac{39}{64}$	.6094	
$\frac{3}{32}$	.0937	4-48	1	.2280		$\frac{5}{8}$	.6250	$\frac{11}{16}$-16
41	.0960		A	.2340		$\frac{41}{64}$	.6406	
40	.0980		$\frac{15}{64}$	.2344		$\frac{21}{32}$	.6562	$\frac{3}{4}$-10
39	.0995		B	.2380		$\frac{43}{64}$	.6719	
38	.1015		C	.2420		$\frac{11}{16}$	.6875	$\frac{3}{4}$-16
37	.1040	5-40	D	.2460		$\frac{45}{64}$	.7031	
		5-44	E, $\frac{1}{4}$	.2500		$\frac{23}{32}$	.7187	
36	.1065	6-32	F	.2570	$\frac{5}{16}$-18	$\frac{47}{64}$	.7344	
$\frac{7}{64}$	.1093		G	.2610		$\frac{3}{4}$	.7500	
35	.1100		$\frac{17}{64}$	.2656		$\frac{49}{64}$	.7656	$\frac{7}{8}$-9
34	.1110	6-36	H	.2660		$\frac{25}{32}$	.7812	
33	.1130	6-40	I	.2720	$\frac{5}{16}$-24	$\frac{51}{64}$	.7969	
32	.1160		J	.2770		$\frac{13}{16}$	.8125	$\frac{7}{8}$-14
31	.1200		K	.2810		$\frac{53}{64}$	.8281	
$\frac{1}{8}$	.1250		$\frac{9}{32}$	.2812		$\frac{27}{32}$	.8437	
30	.1285		L	.2900		$\frac{55}{64}$	.8594	
29	.1360	8-32, 36	M	.2950		$\frac{7}{8}$	.8750	1-8
28	.1405	8-40	$\frac{19}{64}$	.2968		$\frac{57}{64}$	.8906	
$\frac{9}{64}$	.1406		N	.3020		$\frac{29}{32}$	.9062	
27	.1440		$\frac{5}{16}$	.3125	$\frac{3}{8}$-16	$\frac{59}{64}$	.9219	
26	.1470		O	.3160		$\frac{15}{16}$	.9375	1-12, 14
25	.1495	10-24	P	.3230		$\frac{61}{64}$	.9531	
	.1520		$\frac{21}{64}$	3281		$\frac{31}{32}$	.9687	
24	.1540		Q	.3320	$\frac{3}{8}$-24	$\frac{63}{64}$	.9844	
23	.1562		R	.3390		1	1.000	
$\frac{5}{32}$	.1570	10-30	$\frac{11}{32}$	.3437				
22	.1590	10-32	S	.3480				

Pipe thread sizes

Thread	Drill	Thread	Drill
$\frac{1}{8}$-27	R	$1\frac{1}{2}$-$11\frac{1}{2}$	$1\frac{47}{64}$
$\frac{1}{4}$-18	$\frac{7}{16}$	2-$11\frac{1}{2}$	$2\frac{7}{32}$
$\frac{3}{8}$-18	$\frac{37}{64}$	$2\frac{1}{2}$-8	$2\frac{5}{8}$
$\frac{1}{2}$-14	$\frac{23}{32}$	3-8	$3\frac{1}{4}$
$\frac{3}{4}$-14	$\frac{59}{64}$	$3\frac{1}{2}$-8	$3\frac{3}{4}$
1-$11\frac{1}{2}$	$1\frac{5}{32}$	4-8	$4\frac{1}{4}$
$1\frac{1}{4}$-$11\frac{1}{2}$	$1\frac{1}{2}$		

Source: Frank J. Thiessen and Davis Dales, *Diesel Fundamentals*, 2nd ed. (Englewood Cliffs, NJ; Prentice-Hall, Inc., 1986), p. 680.

Drill sizes

Letter sizes	Drill diam. inches	Wire gauge sizes	Drill diam. inches	Wire gauge sizes	Drill diam. inches	Wire gauge sizes	Drill diam. inches
Z	0.413	1	0.2280	28	0.1405	55	0.0520
Y	0.404	2	0.2210	29	0.1360	56	0.0465
X	0.397	3	0.2130	30	0.1285	57	0.0430
W	0.386	4	0.2090	31	0.1200	58	0.0420
V	0.377	5	0.2055	32	0.1160	59	0.0410
U	0.368	6	0.2040	33	0.1130	60	0.0400
T	0.358	7	0.2010	34	0.1110	61	0.0390
S	0.348	8	0.1990	35	0.1100	62	0.0380
R	0.339	9	0.1960	36	0.1065	63	0.0370
Q	0.332	10	0.1935	37	0.1040	64	0.0360
P	0.323	11	0.1910	38	0.1015	65	0.0350
O	0.316	12	0.1890	39	0.0995	66	0.0330
N	0.302	13	0.1850	40	0.0980	67	0.0320
M	0.295	14	0.1820	41	0.0960	68	0.0310
L	0.290	15	0.1800	42	0.0935	69	0.0292
K	0.281	16	0.1770	43	0.0890	70	0.0280
J	0.277	17	0.1730	44	0.0860	71	0.0260
I	0.272	18	0.1695	45	0.0820	72	0.0250
H	0.266	19	0.1660	46	0.0810	73	0.0240
G	0.261	20	0.1610	47	0.0785	74	0.0225
F	0.257	21	0.1590	48	0.0760	75	0.0210
E	0.250	22	0.1570	49	0.0730	76	0.0200
D	0.246	23	0.1540	50	0.0700	77	0.0180
C	0.242	24	0.1520	51	0.0670	78	0.0160
B	0.238	25	0.1495	52	0.0635	79	0.0145
A	0.234	26	0.1470	53	0.0595	80	0.0135
		27	0.1440	54	0.0550		

Source: Ford of Canada.

Torque conversion

Newton metres (N-m)	Pound-feet (lb-ft)	Pound-feet (lb-ft)	Newton metres (N-m)
1	0.7376	1	1.356
2	1.5	2	2.7
3	2.2	3	4.0
4	3.0	4	5.4
5	3.7	5	6.8
6	4.4	6	8.1
7	5.2	7	9.5
8	5.9	8	10.8
9	6.6	9	12.2
10	7.4	10	13.6
15	11.1	15	20.3
20	14.8	20	27.1
25	18.4	25	33.9
30	22.1	30	40.7
35	25.8	35	47.5
40	29.5	40	54.2
50	36.9	45	61.0
60	44.3	50	67.8
70	51.6	55	74.6
80	59.0	60	81.4
90	66.4	65	88.1
100	73.8	70	94.9
110	81.1	75	101.7
120	88.5	80	108.5
130	95.9	90	122.0
140	103.3	100	135.6
150	110.6	110	149.1
160	118.0	120	162.7
170	125.4	130	176.3
180	132.8	140	189.8
190	140.1	150	203.4
200	147.5	160	216.9
225	166.0	170	230.5
250	184.4	180	244.0

Source: Ford of Canada.

English–metric equivalents

Fractions	Decimal Inch	Metric mm	Fractions	Decimal Inch	Metric mm
1/64	.015625	.397	33/64	.515625	13.097
1/32	.03125	.794	17/32	.53125	13.494
3/64	.046875	1.191	35/64	.546875	13.891
1/16	.0625	1.588	9/16	.5625	14.288
5/64	.078125	1.984	37/64	.578125	14.684
3/32	.09375	2.381	19/32	.59375	15.081
7/64	.109375	2.778	39/64	.609375	15.478
1/8	.125	3.175	5/8	.625	15.875
9/64	.140625	3.572	41/64	.640625	16.272
5/32	.15625	3.969	21/32	.65625	16.669
11/64	.171875	4.366	43/64	.671875	17.066
3/16	.1875	4.763	11/16	.6875	17.463
13/64	.203125	5.159	45/64	.703125	17.859
7/32	.21875	5.556	23/32	.71875	18.256
15/64	.234375	5.953	47/64	.734375	18.653
1/4	.250	6.35	3/4	.750	19.05
17/64	.265625	6.747	49/64	.765625	19.447
9/32	.28125	7.144	25/32	.78125	19.844
19/64	.296875	7.54	51/64	.796875	20.241
5/16	.3125	7.938	13/16	.8125	20.638
21/64	.328125	8.334	53/64	.828125	21.034
11/32	.34375	8.731	27/32	.84375	21.431
23/64	.359375	9.128	55/64	.859375	21.828
3/8	.375	9.525	7/8	.875	22.225
25/64	.390625	9.922	57/64	.890625	22.622
13/32	.40625	10.319	29/32	.90625	23.019
27/64	.421875	10.716	59/64	.921875	23.416
7/16	.4375	11.113	15/16	.9375	23.813
29/64	.453125	11.509	61/64	.953125	24.209
15/32	.46875	11.906	31/32	.96875	24.606
31/64	.484375	12.303	63/64	.984375	25.003
1/2	.500	12.7	1	1.00	25.4

Source: Ford of Canada.

English–metric conversion

Description	Multiply	By	For Metric Equivalent
ACCELERATION	foot/sec^2	0.304 8	metre/sec^2 (m/s^2)
	inch/sec^2	0.025 4	metre/sec^2
TORQUE	pound-inch	0.112 98	newton-metres (N-m)
	pound-foot	1.355 8	newton-metres
POWER	horsepower	0.746	kilowatts (kw)
PRESSURE or STRESS	inches of water	0.2488	kilopascals (kPa)
	pounds/sq. in.	6.895	kilopascals (kPa)
ENERGY or WORK	BTU	1 055.	joules (J)
	foot-pound	1.355 8	joules (J)
	kilowatt-hour	3 600 000. or 3.6 × 10^6	joules (J = one W's)
LIGHT	foot candle	10.76	lumens/metre^2 (lm/m^2)
FUEL PERFORMANCE	miles/gal	0.425 1	kilometres/litre (km/l)
	gal/mile	2.352 7	litres/kilometre (l/km)
VELOCITY	miles/hour	1.609 3	kilometres/hr. (km/h)
LENGTH	inch	25.4	millimetres (mm)
	foot	0.304 8	metres (m)
	yard	0.914 4	meres (m)
	mile	1.609	kilometres (km)
AREA	inch^2	645.2	millimetres^2 (mm^2)
		6.45	centimetres^2 (cm^2)
	foot^2	0.092 9	metres^2 (m^2)
	yard^2	0.836 1	metres^2
VOLUME	inch^3	16 387.	mm^3
	inch^3	16.387	cm^3
	quart	0.016 4	litres (1)
	quart	0.946 4	litres
	gallon	3.785 4	litres
	yard^3	0.764 6	metres^3 (m^3)
MASS	pound	0.453 6	kilograms (kg)
	ton	907.18	kilograms (kg)
	ton	0.90718	tonne
FORCE	kilogram	9.807	newtons (N)
	ounce	0.278 0	newtons
	pound	4.448	newtons
TEMPERATURE	degree Farenheit	0.556 ($^\circ$F − 32)	degree Celsius ($^\circ$C)

Source: Ford of Canada.

Index

Accessory installation, 316
Acid dipping, 190
AERA, 7
Aerobic sealant, 214
Airless shot blasting, 21
Align boring, 47, 104
Align honing, 104
Aluminized valves, 280, 309
Anaerobic sealant, 215
Angle milling, 301

Balance bosses, 144, 161
Balance shafts, 125, 126
Balancing:
    Combustion chamber volumes, 50
    engine, 56
    valve springs, 55, 306
Base circle, 181, 188
Bead blasting, 20, 21
Bearing:
    camshaft, 206
    characteristics, 205
    clearances, 202, 207
    conformability, 205
    corrosion resistance, 206
    crankshaft, 202
    crown wall thickness, 207
    crush, 204
    design, 204
    embedability, 205
    failure analysis, 65
    fatigue resistance, 206
    function, 202
    grooving, 204
    installation, 206
    locating lug, 203
    lubrication, 202
    materials, 205, 206
    oil holes, 204
    parting relief, 204
    sizes, 207
    spin, 102
    spread, 204
    surface action, 206
    thrust, 122, 179, 207
Belt inspection, 82
Belt installation, 199
Belt tension, 200
Bending stress, 65
Block:
    cleaning, 16, 116
    core plugs, 117, 304
    deburring, 116
    deck clearance, 28, 99
    deck height, 28, 99
    decking, 99
    painting, 118
    reconditioning, 98

Blueprinting, 43
Boring bar, 110
Break-in procedure, 318, 324
Broaching, 150
Bronze valve guide replacement, 255
Brushes, cleaning, 12

Cam action, 181
Camshaft:
  base circle runout, 188
  bearing installation, 208, 209
  bearings, 187, 206
  clearance ramp, 181
  closing ramp, 181
  degreeing in a, 51
  design, 179, 181, 182
  drive methods, 193
  end thrust, 179
  failure analysis, 79
  function, 177
  high performance, 50
  inspection, 187
  installation, 190, 311
  lobe design, 180
  lobe lift, 181, 187
  location, 177
  materials, 178
  measurement, 187
  opening ramp, 181
  regrinding, 189
  runout, 188
  straightening, 189
  surface treatment, 190
  thrust plate, 179, 180
  timing, 182
  wear, 187
Camshaft bearing:
  align honing, 313
  alignment, 247
  installation, 209
  lubrication, 228
  wear, 187
Canted valves, 281
Cap and rod grinder, 148
Casting numbers, 5
Ceramic sealing, 93
Chain tensioner, 199
Chamfering:
  cylinder bores, 114
  cylinder head bolt holes, 115
  valve stem tips, 290
Chemicals, cleaning, 18
Chrome plating, 132
Chrome rings, 168
Cleaning engine parts, 16
Clutch pilot bushing, 140
Cold soak cleaning, 20
Combustion chamber:
  balancing, 50
  volume, 50
Compression ratio, 49
Connecting rod:
  alignment, 145
  balance, 144
  bearings, 202
  bend, 145

bleed holes, 145
bolt covers, 174
bolt nuts, 174
bolt replacement, 174
bore stretch, 145
bushing replacement, 150
cap and rod grinding, 147
design, 143
failure analysis, 73
function, 143
heater, 157
honing, 149
installation, 154
materials, 143
measurement, 145
offset, 144
oil holes, 145
reconditioning, 147
side clearance, 135, 174
twist, 145
Core credits, 10
Core plugs, 117, 304
Crack(s):
  causes, 86
  detection, 87
  repair, 88
Crankshaft:
  balance, 122
  bearings, 202
  chrome plating, 132
  counterweights, 122
  crack detection, 87
  cross drilling, 132
  design, 120
  diesel engine, 123
  failure analysis, 64
  fillet radius, 134
  finishing, 136
  grinding, 132
  grinding direction, 133
  grinding limits, 132
  hardening, 130, 132
  inspection, 126
  journal rebuilding, 130
  materials, 120
  measuring, 127
  metal spraying, 131
  micro-inch finish, 136, 137
  nitriding, 130
  polishing, 136
  radius grinding, 132, 134
  runout, 128
  sizes, 127
  straightening, 129
  submerged arc welding, 131
  thrust surfaces, 128
  undersizes, 130
Crosshatch pattern, 106
Cutting tool lubricants, 95
Cylinder:
  bore chamfering, 114
  bore deglazing, 114
  bore gauge, 26
  bore sizes, 26
  boring, 108
  honing, 111
  liner, 110

measuring, 26
out of round, 27
reconditioning, 106
ridge, 25
sleeving, 110
taper, 27
wall crosshatch, 106
wall finish, 106
wall thickness, 25, 98
waviness, 27
wear, 25
Cylinder head:
    assembly, 304
    bolt tightening, 310
    gasket, 217, 309
    installation, 309
    port alignment, 245
    porting, 48
    reconditioning, 239
    resurfacing, 239
    straightening, 248

Deburring cylinder block, 116
Deck:
    clearance, 28, 99
    height, 28, 99
    resurfacing, 99
    plate, 108
Deglazing cylinders, 114
Degreeing in a cam, 51
Descaling, 20
Detonation damage, 73
Dial bore gauge, 26, 32
Displacement, 49
Drive belt, camshaft, 195
Dye penetrant, 88

Electrolysis, 317
Engine:
    balancing, 56
    blueprinting, 43
    break-in, 324
    identification, 3
    installation, 323
    kits, 8
    prelubrication, 318
    test stand, 318
English and metric dimensions, 24
Epoxy repair, 88
Exhaust manifold, 41
Exhaust valve:
    cleaning, 14
    design, 279
    failure analysis, 75
    inspection, 286
    materials, 280
    measurement, 286
    temperatures, 283

Failure analysis, 58
Flexible hone, 114
Flex plate, 124
Flywheel:
    design, 124
    inspection, 138

materials, 124
overheating, 138
resurfacing, 140
ring gear replacement, 140
runout, 138
Freeze plugs, 117, 304

Gasket:
    cylinder head, 217
    handling, 214
    installation, 218
    materials, 212
    manifold, 221
    oil pan, 219
    packages, 214
    properties, 212
    sealants, 214
    valve cover, 218
Glass bead cleaning, 20, 21

Hand cleaning tools, 12
Harmonic balancer:
    design, 124
    inspection, 138
    repair sleeve, 138
    wear, 138
Head bolt:
    hole chamfering, 115
    hole cleaning, 115
    tightening, 310
Head gasket, 217
Head resurfacing, 239
Head straightening, 248
Heat stress, 77
Heat tabs, 118
High pressure spray cleaning, 23
Honing:
    cam bearing bores, 313
    connecting rods, 149
    cylinders, 111
    lifter bores, 116
    main bearing bores, 104
    plate, 108
    valve guides, 255
Hot tank cleaning, 16, 19, 20
Hydraulic lifter:
    design, 184
    failure analysis, 79
    function, 183
    leakdown, 79
    noise, 79
    operation, 183
    rotation, 184
    testing, 37
    wear, 79

Insert:
    ring groove, 162, 165
    valve guide, 250, 253, 255
    valve seat, 261, 262, 263
Installed spring height, 305
Installed valve stem height, 293
Intake manifold resurfacing, 245
Interference angle, 284

Keepers, valve, 276, 280
Knurling:
 pistons, 166
 valve guides, 251

Lash adjusters, 184
Lifter:
 bore honing, 116
 design, 184
 function, 183
 leakdown, 79
 noise, 79
 operation, 183
 oversize, 184
 roller, 187
 rotation, 184
 testing, 37
 wear, 79
Liner, cylinder, 110
Lobe lift, 181, 187
Locating lug, 203
Lubrication system:
 components, 227
 function, 227
 service, 233
 types, 228

Main bearing:
 caps, 102, 103
 characteristics, 205
 clearance, 202
 design, 204
Main bearing bore:
 align boring, 104
 align honing, 104
 alignment, 102
 measuring, 103
 out of round, 102
 stretch, 102
Manifold resurfacing, 245
Manuals, service, 6
Measuring:
 block deck clearance, 28
 block deck height, 28
 block deck surface, 40
 camshaft, 35
 combustion chamber volume, 50
 connecting rods, 32
 crankshaft, 30
 cylinder head surface, 40
 cylinder wall thickness, 25
 cylinder wear, 25, 26, 27
 installed valve spring height, 305
 installed valve stem height, 293
 main bearing bores, 27
 manifolds, 41
 pistons, 34
 ring grooves, 35
 valve guides, 38
 valve margin, 286
 valve seat width, 284
 valve springs, 37
Metric conversion, 24, 329, 331, 332
Moly piston rings, 168

Necking, valve stem, 76
Nitriding, 130

Oil control ring, 167, 168, 171, 172
Oil cooler, 233
Oil filter, 233
Oil pan:
 flange straightening, 219, 220
 gasket, 219
 installation, 220
 materials, 219
 sealant, 220
Oil pump:
 clearance, 235
 design, 230
 drives, 230, 233, 234
 failure analysis, 233
 operation, 230
 pickup tube, 234
 pressure regulation, 230
 relief valve, 230, 234
Oil seal:
 design, 223
 installation, 224
Oil slinger, 121
Oil throwoff, 61, 62

Painting the block, 118
Parts:
 failure analysis, 58
 ordering, 9
 pricing, 10
PERA, 7
Pilot bushing, 140
Pinning cracks, 90
Piston:
 balance pads, 161
 bosses, 160
 cam action, 160
 cast, 163
 clearance, 163
 compensated, 164
 expansion control, 160, 161
 forged, 163
 function, 159
 head design, 159
 heat dam, 160
 installation, 172, 173, 174
 knurling, 166
 noncompensated, 164
 oil drain back holes, 163
 pin offset, 162
 pin lubrication, 162
 regrooving, 165
 ring grooves, 159
 ring lands, 159
 selection, 165
 sizes, 164
 skirt finish, 163
 temperatures, 163
 thrust, 162
Piston pin:
 assembly, 154
 clearances, 154
 design, 151

failure analysis, 70, 74
fitting, 154
function, 151
installation, 154, 155
materials, 151
mounting methods, 152
offset, 153
Piston ring:
    action, 167
    back clearance, 171
    chrome facing, 168
    clearance, 170
    coatings, 168
    compressor, 173
    design, 167
    expander spacer, 168, 169
    expander tool, 173
    failure analysis, 58
    gaps, 169
    groove cleaner, 16
    groove machining, 165
    groove wear, 165
    installation, 172
    materials, 168
    moly facing, 168
    selection, 171
    side clearance, 170
    sizes, 172
    torsional twist, 167
    types, 167
Piston to valve clearance, 247
Port alignment, 245
Porting, 48
Positive valve stem seals, 259, 307
Power cleaning tools, 13
Preignition damage, 60, 61
Prelubricator, 319, 320
Pressure testing, 88
Publications, 6
Pushrod:
    adjustable, 267
    design, 266
    function, 266
    guide plate, 266
    inspection, 267
Pyrolytic oven cleaning, 23

Rear main oil seal:
    design, 224
    installation, 224, 225
Replaceable valve guides, 248
Replaceable valve seats, 261
Rigid honing, 111
Ring gear replacement, 140
RMS finish, 107, 108
Rocker arm:
    adjustable, 272
    function, 267
    inspection, 272
    lubrication, 271
    materials, 267
    mounting methods, 268
    pivots, 268
    ratio, 267
    refacing, 272
    studs, 256, 273

Rocker arm shaft, 268
Rocker arm stud, 256, 273
Rod heater, 157
Rotary files, 116
RTV sealant, 214

Scale removal, 20
Seal:
    rear main, 224, 225
    valve stem, 258
Sealants, 214
Service manuals, 6
Short block, 8
Shot peening, 121, 132
Sleeving:
    cylinder, 110
    harmonic balancer, 138
Sodium valves, 280
Spark plug hole repair, 94
Starter ring gear replacement, 140
Steam cleaning, 17
Stellite valves, 280
Straightedge, 28, 40, 99
Submerged arc welding, 131
Surfacing:
    block deck, 99
    cylinder head, 239
    manifolds, 245

Thermal cleaning, 23
Thread:
    inserts, 93, 94
    pull, 115
    repair, 93, 94
Timing belt:
    adjustment, 199
    design, 195
    inspection, 82
    installation, 199
    tension, 199
Timing chain:
    design, 194
    failure analysis, 80
    inspection, 80
    installation, 196
    types, 194
Timing gears:
    design, 193
    failure analysis, 80
    inspection, 80
    installation, 196

Vacuum testing valves, 309
Valve:
    adjustment, 320
    aluminized, 281
    canted, 281
    chrome, 281
    cleaning, 14
    concentricity, 301
    design, 279, 281
    face angles, 284
    face to seat contact, 301
    face to seat seal test, 309

Valve (*cont.*)
  failure analysis, 75
  float, 273
  function, 279
  grinding, 286
  hardfacing, 282
  inspection, 286
  installation, 307
  interference angle, 284
  lift, 181
  margin, 286
  materials, 280
  measuring, 286
  nomenclature, 282
  overlap, 182
  rotation, 276
  sodium, 283
  stem chamfering, 290
  stem height, 293
  stem seals, 258
  stem taper, 281
  stem tip grinding, 289
  stem to guide clearance, 38, 248
  swirl finished, 281
  temperatures, 283
  timing, 182
Valve cover gasket, 323
Valve guide:
  bronze walling, 255
  chamfering, 250
  cleaning, 14
  drivers, 248
  honing, 255
  installation, 248
  installed height, 248
  knurling, 251
  measuring, 38, 248
  reaming, 251
  removal, 248
  sleeving, 253
Valve lifter:
  anti-pump-up, 185
  design, 184
  failure analysis, 79
  function, 183
  hydraulic, 183
  pump-up, 273
  solid, 183
  testing, 37
  wear, 79
Valve rotators, 276

Valve seat:
  angles, 284
  concentricity, 301
  grinding, 294, 299
  grinding stones, 296
  inserts, 261
  narrowing, 301
  pilots, 296
  repair, 261
  runout, 301
  three angle, 301
  wear, 285
  width, 284
Valve spring:
  balancing, 55, 306
  cleaning, 16
  color coding, 274
  failure analysis, 82
  function, 273
  inspection, 82, 274
  installation, 307
  installed height, 305
  materials, 273
  measuring, 37
  pressure, 37
  retainers, 275
  seat, 274
  shims, 306
  testing, 274
  vibration control, 273
Valve stem:
  failure analysis, 75
  inspection, 286
  taper, 281
  tip chamfering, 290
  tip grinding, 289
Valve stem seal:
  design, 258
  replacement, 260
Valve timing, 182
Valve train:
  components, 265
  noise, 79
  requirements, 265
Vehicle identification number (VIN), 1
Vibration damper:
  design, 124
  inspection, 138
  repair sleeve, 138
  wear, 138

Water pump, 317